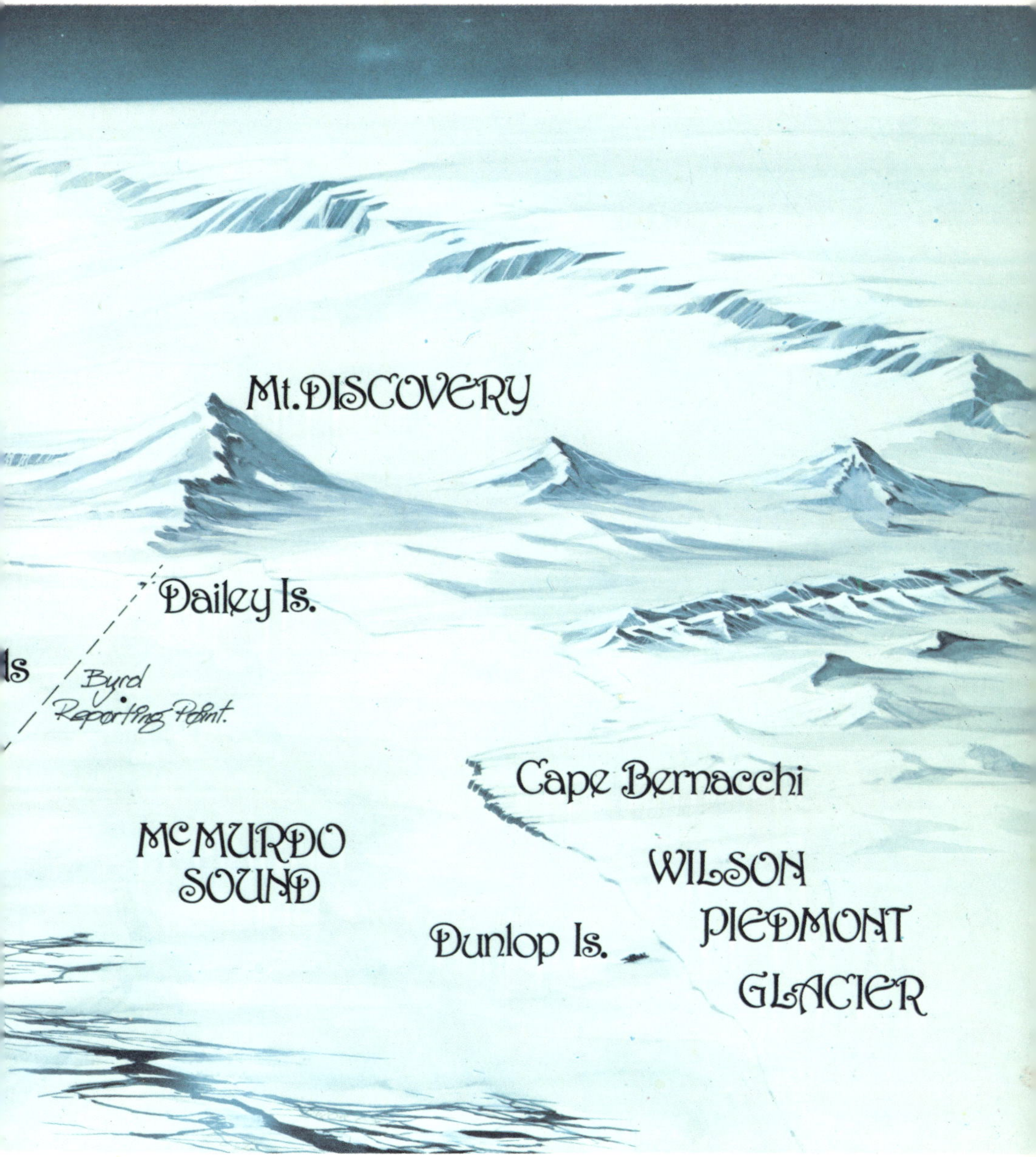
Mt. DISCOVERY
Dailey Is.
Byrd Reporting Point.
Cape Bernacchi
McMURDO SOUND
WILSON PIEDMONT GLACIER
Dunlop Is.

IMPACT EREBUS

The crash of Flight 901 was the loneliest of the world's worst air disasters. On 28 November 1979, the Air New Zealand DC10 with 257 people aboard took off from Auckland International Airport and flew 2000 miles southwards to the Antarctic, to plunge into the slopes of Mt Erebus, a 12,000-foot volcano. Nine hours later, a US Navy aircraft from McMurdo Station sighted the wreckage — a brown smear on the ice. Nobody survived.

Yet for all its isolation it was one of the best documented catastrophes. The aircraft's electronic sensors were working and decipherable. Almost every passenger on the sightseeing trip carried cameras and shot film up to the last second. This was painstakingly salvaged and developed. And Antarctic weather scientists were monitoring local weather patterns, and receiving sophisticated film from satellites. But still the cause eluded investigators. Why should a skilled crew, with an Antarctic explorer on the flight deck, fly straight into a mountain wall in clear weather? The crew was blamed.

It took nearly two years for the fog to lift from the mystery. Gordon Vette, a fellow pilot of Captain Jim Collins, the man in command, could not accept the 'pilot error' verdict, and began his own study and investigation. A Royal Commission headed by a forthright High Court Judge, dug deep into the planning and execution of the flight.

The result was a story which is eerie in its implications for airmen. Even with the most modern instruments available, nature can still spring traps beyond prediction and even the best run airline could become the victim of a computer error.

Dedicated to
the Improvement of
Flight Safety

IMPACT EREBUS

GORDON VETTE
with John Macdonald

SHERIDAN HOUSE

First published 1983 by Hodder & Stoughton Ltd.,
Auckland, New Zealand

First published in The United States of America 1984 by
Sheridan House Inc.,
145 Palisade Street, Dobbs Ferry,
New York, 10522

ISBN 0 911378 49 9

Printed in Hong Kong.

Contents

Again and again, pilots are found to show little sympathy for their colleagues who are hurt or die; some simplifying explanation is immediately hit upon to reassure the others that the same fate will not be theirs.

Peter Garrison, *Pilot Error*

Foreword

by

the Hon. P.T. Mahon Q.C.

The author of this book, Captain Alwyn Gordon Vette, is by any standards a remarkable individual. In telling the story of the Antarctic aircraft disaster of 28 November 1979, when a DC10 operated by Air New Zealand collided with Mt Erebus involving the loss of 257 lives, he displays a diversity of talents.

In the first place, Gordon Vette is an international commercial jet pilot of long experience, with aviation skills which for years have been respected by his pilot colleagues in Air New Zealand. In the second place, as his book shows, he has that combination of deductive ability and perseverance which every competent investigator must either be born with or acquire. And finally, as every reader of his book will soon perceive, he is a writer of incisive clarity.

The prime purpose of the book is to explore and expose the basic elements of human vision as applied to the specialised skills of piloting a big high-speed jet aircraft, and to demonstrate, as illustrated by this awesome Antarctic disaster, how an ocular illusion created by weather can deceive not only the pilot, but every member of an aircrew.

When the news of this disaster was released to the world, and it became known that a DC10 airliner had flown in broad daylight into a mountainside, there was an outbreak of alarm at the McDonnell-Douglas Corporation in California because they thought, as did Air New Zealand, that an engine must have fallen off the plane, just as an engine had fallen off a DC10 at Chicago only six months before, an event which had led to a directive from the U.S. Federal Aviation authorities for the grounding of all DC10 aircraft.

Experts from McDonnell-Douglas therefore arrived at the far-away crash site with remarkable speed. But within days of the disaster the electronic flight recorder (the 'black box') was transcribed in Washington and to the intense relief of McDonnell-Douglas, it was found that the aircraft had been functioning perfectly when it crashed.

So what had caused the disaster? There seemed to be only one explanation. The aircraft must have been flying in cloud. Such was the reasoning, and it was not unnatural, of the New Zealand Office of Air Accidents Investigation. But the investigators then received a severe rebuff. Almost every passenger on the DC10 had been carrying a camera. Many prints were developed of photographs taken in the last few seconds of flight. And the astounding fact was revealed that in those last few seconds, the aircraft had been flying in clear air.

This revelation was received with surprise by the New Zealand Chief Inspector of Air Accidents and his staff, and by his employers the Civil Aviation Division, and with surprise and dismay by Air New Zealand itself. For it so happened that the Flight Operations Division of Air New Zealand had made a grievous mistake in programming the Antarctic flight and it had been hoped that proof of pilot error might obscure this technical mistake.

For years past, all modern jet airliners have been navigated not by human navigators but by a computer-based electronic navigation system built into the aircraft itself. The flight route is pre-selected and the geographical data fed into the aircraft's computer system, and the aircraft will then navigate itself, without possibility of error, along the pre-determined track. But whereas the computer track for previous flights had led aircraft on a course taking them some 27 miles to the west of Mt Erebus, along the broad flat expanse of McMurdo Sound, which is 40 miles wide, and whereas the crew of the fatal flight had been supplied at their briefing with the standard computer flight plan which repeated this flight track, the navigation staff of the Company, on the eve of the flight, had altered the computer track 27 miles to the east. So Flight TE 901 was now electronically programmed on a collision course with Mt Erebus, and the Company staff had been well aware from the McMurdo weather forecast that the DC10 would probably be approaching McMurdo at an altitude of about 2000 feet. But by an inexplicable blunder, the change in the flight track was never notified to the flight crew.

The management of Air New Zealand, who had become aware of the computer mistake within a day or so of the disaster, had naturally intended to rely on the 'flying in cloud' theory as amounting to pilot error which could be announced as the sole cause of the disaster, because the computer mistake had not been publicly revealed, but even when it became publicly known, as it was sure to be, it could validly be claimed not to be an effective cause.

One of those who had accepted the 'flying in cloud' theory had been the Chief Inspector of Air Accidents, but he also had to revise his opinion when the passengers' photographs were produced. Nevertheless, he and his staff were noticeably reluctant to abandon the 'visibility' concept. So they propounded the alternative theory that the DC10 had been flying 'towards' an area of impaired visibility, and in his statutory report the Chief Inspector stated that the 'probable cause' of the disaster was the decision of the pilot to fly onwards 'towards' (apparently) an area which he could see was obscured by cloud.

It only requires a moment's reflection to see that the Chief Inspector's final opinion was no different from the original 'flying in cloud' theory. To fly 'towards' an area of known cloud, and to collide with a mountain behind that cloud, is surely the same, in the air accident context, as flying in cloud throughout. From the aspect of accident causation, there can be no possible distinction. So in truth, the Chief Inspector never abandoned his 'flying in cloud' concept, and his final opinion was enthusiastically though privately endorsed by the management of Air New Zealand.

Only a matter of days after public distribution of the Chief Inspector's opinion, I was appointed by the New Zealand Government to investigate the cause of the disaster, and it was popularly expected that I would merely endorse the Chief Inspector's indictment of the aircrew, and indeed I thought myself that this would be the probable sequel.

But Gordon Vette and those of his pilot colleagues who were not part of the Air New Zealand management were profoundly disturbed at the Chief Inspector's report. Neither the Chief Inspector nor any of his staff had any expertise in the piloting or navigation of sophisticated jet airliners, and it became apparent as my inquiry proceeded that the Chief Inspector's report was constructed partly of technical data supplied to him from America, and partly by the dangerous process of relying upon verbal hearsay. Nevertheless, his report referred in its text to other possible factors which may have contributed to the disaster, and his inquiries had certainly been indefatigable, a fact which I came to recognise myself. And the technical data in the report, supplied by American experts, was of invaluable assistance. But the fact of the matter was that the Chief Inspector's opinion as to 'probable cause' was really untenable, and this is what disturbed Gordon Vette and his colleagues.

It so happened that the pilot in command of the fatal flight was known throughout Flight Operations Division as one of the most meticulous and conscientious pilots in the company. His experience with jet airliners was vast. His co-pilot was equally conscientious, and was also highly experienced and highly skilled, as were the two flight engineers. There was not the slightest chance that this aircrew, flying in unfamiliar territory at 2000 feet, would 'fly in cloud' or would fly 'towards' a fast-approaching area of bad visibility. So what had happened? Gordon Vette embarked on a wide-ranging private investigation, determined to find out.

There are two golden rules for an investigator, and they become well-known to people like trial lawyers and police officers who spend most of their time inquiring into the disputed facts of accidents and crimes. The first is, eliminate the impossible. If something could not have occurred, however attractive its possibility may be, then forget it. The second rule is, and it applies especially to judges along with lawyers and police officers, do not enter upon any inquiry with a pre-conceived view. Find out all the facts first, and then construct your conclusion. In my time as a trial lawyer I investigated many accidents and occurrences of which a great number, if I may say so, were far more complicated than the Mt Erebus inquiry, and I came to respect these two fundamental rules more and more as the years went by. And they were followed by Gordon Vette as a combination of his instinct and intelligence.

It was quite clearly not possible that the aircrew of the fatal flight were flying either in or towards cloud. Yet the flight-deck tape recorder conclusively proved that no one on the flight deck ever saw, even at the last second, the white slopes of the mountain in front of the plane. Therefore, something had occurred to deceive the aircrew as to what they were seeing in front of them. In Gordon Vette's reasoning, the answer could only be the occurrence of an ocular illusion. In his report the Chief Inspector had briefly quoted from a publication describing the polar 'whiteout' phenomenon, and had indicated a possibility, inspired I think by what someone told him in Antarctica, that a blending of pale horizon and white snow may have been a contributing factor. Gordon Vette came to suspect that it was the one and only final key to the disaster. Even assuming the computer mistake, an error regarded as nearly incredible by all overseas experts whom I consulted, an ocular illusion of some kind must have been the decisive last cause.

So Gordon Vette put on one side the Chief Inspector's honest but misconceived opinion. This was a prophetic act, because during my enquiry the Chief Inspector's views gradually came to be discarded by everyone except Air New Zealand and his own employers. I may also say that after all the evidence was concluded, counsel for Air New Zealand, in their lengthy final submissions, did not attempt to support the Chief Inspector's opinion as to causation. Thus it was that Gordon Vette started out on the quest which is described so vividly in his book and which was to make him so unpopular with the Chief Inspector and with the Company management.

You will read of his studies into the mechanics of sight, into the psychological settings which control and influence the reception of visual signals by the mind, and you will see a dramatic example of the accepted medical thesis that the gift of sight is always a combined function of the eyes and the brain. And of course the main thrust of the book is to demonstrate that the professional pilot may find himself, on rare occasions one may hope, in situations where his eyes will play him false, and how important it is to extend the current official criteria of operational visibility in the air so as to recognise this factor which the Mt Erebus disaster so tragically demonstrated.

But the book is by no means a technical book. The ordinary reader will find that a part of it, indeed the major part, consists of a description of this flight from New Zealand to Antarctica. The author knew the aircrew and their personalities, he knew the cabin staff. He had piloted an Antarctic flight himself. Over the last half hour of the flight, he has had the aid of the tape-recorded conversation of the aircrew, together with a record of the aircraft's movements as recorded by the 'black box'. So when he tells the story of the flight he does so in the present tense, and with his flying expertise and his gift of written expression he places the reader right there in the plane. You are a witness of all that happens, a listener to all that is said.

So apart from the account of the author's researches and discoveries in relation to visibility in the air, a fascinating recital in itself, the reader will find himself on the very flight deck of this DC10 which went to Antarctica and never returned. It is an enthralling account and must take its place, in due time, high in the annals of aviation literature.

P. T. Mahon
Christchurch,
February 1983

Author's Note

I hope the real purpose of my book will not be lost. It was positively not my intention to simply clear the aircrew of blame for this accident. It is my sincere belief that to do that, if they had been to blame would really have done the Airline Pilots' Association and my colleagues, a great disservice.

On the other hand it is most important that the aircrew should not be used as scapegoats for the underlying cause of an accident which is being concealed. I believe that the rest of the professionals in the aviation industry will benefit from the exposure of the true causes of the accident.

I believe that I have discovered the causes of this accident which most certainly could have trapped any one of the rest of us pilots.

I hope that any who may initially believe that I have vehemently defended the crew of this airliner at all costs, will take into account the fact that though I was friendly with Collins and the rest of the crew, the fact is that I had a much closer friendship with others within the airline whose views I had to strongly oppose. This, to my very great personal regret, has put a considerable strain on friendships of long standing, some extending for almost 35 years.

When I came to realise how difficult it would be to achieve improvement in certain of those inadequate areas, I decided that I ought myself to endeavour to assist in bringing about improvement in those areas. I felt that if I could participate in expanding public knowledge and awareness of these fields, then that knowledge and awareness might provide a further incentive toward taking the action which is necessary.

Some fields which are in need of improvement, are the following. Much more effort should be devoted to achieving rapid progress in obtaining a forward facing ground proximity device. If such a device had been on the DC10 it should have been of such a design as to give the crew an excellent chance of avoiding the crash which occurred. As it was, the standard device operated in accordance with its specifications, but was quite useless, and may as well not have been installed, as it operated vertically only.

A timely review of the pilot training syllabus is overdue, with particular reference to the field of visual perceptual psychology, etc. The structure of the Civil Aviation Division ought to be redesigned in order to give it the means which it needs to provide the public with the protection they deserve.

I was certain at the outset that although the Chief Inspector of Aircraft accidents had a great deal of experience in investigating light aircraft accidents, his total lack of flying experience on large civil airliners, and in investigating any accidents to them, would put him at a serious disadvantage, even with the best intentions, when he tried to understand the psychology of the crew loop concept, or to analyse correctly the contents of a voice recorder.

Even worse, I felt that this lack of experience, if he recognised it as a limitation, would put him at a disadvantage alongside the top Air New Zealand technical advisors, and they may be able to exert undue influence even though he would be well aware that they had an interest in the outcome. If these considerations were valid, I felt certain that even though Chippindale would do his usual meticulous job of sifting through the technical details, he would inevitably be forced to return the almost standard finding of 'pilot error'.

I am somewhat sad that the price I have had to pay for my attempts to find the truth of the Erebus disaster has been much greater than I expected. I had hoped that we might all be able to admit that in ignorance, we made a terrible mistake, and get on with rebuilding and learning from our mistakes.

The post-Erebus actions of some of my most highly esteemed colleagues was brought about by the simple fact that they could not grasp the scientifically proven fact that the crew of Flight 901 would be totally unaware that their eyes could misfunction under the conditions they encountered.

I have written this book with a varied readership in mind.

The narrative text which is contained in Chapters 1 to 8, tells the story for the general reader. The illustrated section from pages 32 to 80 contains illustrations together with explanatory captions, which clarify the text. That section may be read as an independent entity before one reads the narrative, in order to gain a quick understanding of what were, in my opinion, the causes of the crash.

The epilogue follows the narrative text, and here I deal at length with whether the crew were justified in descending beneath the height at which the airline said they should have been flying. I conclude the epilogue by considering the dilemma faced by pilots who fly visually in Antarctica.

Though there is no reason why the general reader should not read the remainder of the book, and doing so, will greatly increase his understanding of the crash, my reason for including the appendices and notes was to enable the specialised reader in the aviation and legal fields to study analyses of the crash in greater depth than was possible to include in the text.

The appendices consist of a memorandum of evidence obtained by the Royal Commissioner, Justice Mahon, and of briefs of evidence given by witnesses who appeared before the Commission. This evidence supports propositions which I advance elsewhere in the text. I concede in advance that I selected those witnesses who do support my propositions, and that witnesses appeared on behalf of the airline who strongly attacked almost everything I have said. It would have been impractical for me to have attempted to reproduce in its entirety the evidence tendered to the Commission. Those who wish to read alternative interpretations of the transcript of what was said by the crew before the crash, are referred to the Chippindale Report, which is the Aircraft Accident Report No. 79-139 of the Office of Air Accidents Investigation, and to the evidence of Captain Ian Gemmell tendered to the Royal Commission.

The case which I presented to the commission, and which I have set out in greater detail in this book, rested on the work and research of many others. I offer my thanks to them in my acknowledgements section.

The notes at the end of the book were independently researched and written by a senior lecturer at the University of Auckland, who was asked to study every piece of information available on the Erebus disaster, together with this manuscript, and to provide the uncommitted public with an unbiased overview. The notes elaborate certain points in the text chosen by the lecturer.

Gordon Vette
Taupo, 1983

Acknowledgements

My wife Eleanor, for the past three years of total disruption to our social life, for reading scripts and submissions, and for her loyal support of our team when all around us believed we were wrong.
My daughter Catherine for loyal support; my sons Mark, M.Sc Hons, and Karl, B.Sc, RAAF Pilot on supersonic F111s, for their loyal support, positive guidance through the varied study material, and for being sounding boards as I developed my hypothesis.
The legal team headed by Alister Macalister, who has attended almost every transport enquiry since the War.
Members of the library staffs of Auckland University and University of California, Los Angeles, for assistance.
Members of the Auckland University photography laboratory for their support and assistance.
Captain Peter Bressy, contributor to the book *Pilot Error,* for reading submissions to Commission, for comments and support and for allowing the use of extracts from his book in my manuscript.
Dr Ted Calvert, of the Royal Aircraft Establishment, Farnborough, for comments and support of submissions to Royal Commission.
Mr Bill Cole, Assistant Lecturer, Department of Psychology, Auckland University, for his audio-visual presentation of visual perceptual phenomena in polar conditions, and for his supportive testimony before Royal Commission, and for assistance and editing of Pilot Profile section.
Captains Arthur Cooper and Tony Foley, and First Officer Peter Rhodes for their splendid team work in putting together the submissions to the Royal Commission that formed such a substantial part of the final findings. I thank them for the opportunity of working with them. Their high spirits and professionalism helped us through the early stages where we were very much the underdogs. Their inputs and contribution to the manuscript are very much appreciated.
Professor Mike Corballis, Head of Department of Psychology, Auckland University, for reading and making corrections to the manuscript and making submissions to the Royal Commission, also for general support and encouragement.
Deputy Chief of Air Staff, Air Commodore David Crooks, and Auckland Base Commander, Group Captain Pete Adamson, for help and advice.

Jocelyn Davies for typing services.
Paul Davison and Roger MacLaren, a brilliant team, whose endless efforts and skill ensured our case was given the very best presentation. I am extremely grateful for their comments and support. I thank Paul's wife, Anne Marie, for tolerating the endless hours spent in their home, researching and preparing the case.
Professor Ross Day, Head of Department of Psychology, Monash University, Melbourne, for examining my hypothesis prior to submission to the Royal Commission, for contributing to those submissions and to the manuscript, for giving supportive testimony to the Royal Commission, for general encouragement and assistance, and for vastly improving my education in the field of visual perceptual phenomena.
Molly and Ron Donovan, for secretarial, typing and proof reading of submissions and manuscript, and for support and help as we agonised through this long period of time.
Captain E.P.M. Dowd, President of the Airline Pilots' Association, delegate to the International Federation of Airline Pilots, and member of the Aviation Safety Board of New Zealand, for assistance and advice in preparation of the manuscript.
Mr Arthur Ellis of Auckland University Pathology laboratory, for assistance in my submissions to the Royal Commission.
Bob Farrell, Flight Engineers' Association, for submissions to the Commission, and for assistance.
Carmel Friedlander, a freelance journalist who covered the Erebus inquiry, gave me much needed encouragement.
Captains Brian Gault and Ross Gordon for expert and valuable comments on the manuscript.
Dr J.A. Gribben, Senior Lecturer, Department of Psychology, Auckland University, for reading submissions to Commission and comments.
Professor R.J. Irwin, Senior Lecturer, Department of Psychology, Auckland University, for reading and making corrections to my submissions to Royal Commission, and for help and advice.
The late Professor Michael Joseph for reading the manuscript and for his comments.
Dr Barry Kirkwood, Senior Lecturer, Department of Psychology, Auckland University, for encouragement and compiling a reading list of texts on visual perceptual psychology to study prior to making submissions to Royal Commission.
Dr J.C. Lane, Director of Aviation Medicine, Department of Transport, Commonwealth of Australia, for support of hypothesis.
I felt it would assist the serious student of the Erebus affair if I

appended to this book a section of detailed notes from an independent source. The contributor I obtained, Stuart Macfarlane, has no connection with any of the parties, nor with aviation in general, and so could take an unbiased view of the events. I have not altered or amended these notes in any way, and they represent the views of the contributor. Stuart had the legal training (being a Senior Lecturer in Law at Auckland University) to carry out an extensive analysis of the evidence. His editorial experience gained through editing a legal publication, assisted in the editing of *Impact Erebus*. His wife Alison (a former women's editor of the *Christchurch Star*) also assisted in the editing.

Flight Lieutenant Ian McLeod, Antarctic helicopter pilot, for assisting with my education on polar operations, and for reading and assisting with my submissions, and for his supportive submissions to the Commission. To his long-suffering wife, for putting up with us cluttering up her lounge for many evenings, wading through endless photographs and maps.

Mr John T. Moehring of Cincinnati, U.S.A. for use of some of his photographs taken during the recovery operations at Mt Erebus.

The late Dr Fred Platts for help and advice on impact details.

Ray Richards, publishing agent, for encouragement and support.

Sir Geoffrey Roberts, who shares my concern about the Civil Aviation Division within the Ministry of Transport. I have quoted from his book, and my only disagreement with him is that I believe his views on the Erebus crash are mistaken. I would be happy if *Impact Erebus* revealed sufficient detail for him to reconsider the case.

Members of the Royal New Zealand Air Force associated with Antarctic operations for supportive inputs to the Commission.

Kevin Ryan, for off-course support and encouragement.

Mr Geoff Saville, for art work assistance in submissions to Royal Commission.

Captain Robert Sinclair, First Officer Tom Enright, Mr Jim McCourtie, and cabin crew member Lana Simpson, for reading the manuscript, for their comments, and for support throughout.

Professor Mel. Thistle, Ontario University, Canada for reading the manuscript, advice and comments.

Dr Bob Thompson, Head of Antarctic Division of Department of Scientific and Industrial Research, for encouragement and assistance and supportive testimony before the Commission.

The sole responsibility for the views of those above, as finally expressed, is mine. I only hope I have not misunderstood the advice of any of those who have assisted me.

Glossary

AINS (Area Inertial Navigation System): describes the complete system provided by the unit based on inertial platforms. Over long distances and at cruising height, the inertial systems independently provide the aircraft with safe and accurate guidance. However, what small error may creep into the system, effecting its ability to sense its exact position, is corrected by ground-based radio navigation aids when the aircraft comes within range of these.
Air Traffic Clearance: authorisation for a pilot flying in controlled airspace to carry out a particular action — climb, descend, take-off, land, etc.
Altitude: true height of an aircraft as shown on the altimeter adjusted to local millibar pressure.
ATC (Air Traffic Control): the ground-based controllers monitor and direct pilots of aircraft leaving and approaching airports and while flying in controlled airspace. These directions have the authority of a command except in the odd case where they can only exercise 'advice'. The McMurdo ATC was considered such a case when dealing with over-flying civilian aircraft (though any pilot disregarding such advice would be considered extremely foolish.)
Control Area: defined airspace encompassing routes on which traffic density warrants strict control of aircraft by ATC.
Control Zone: defined airspace encompassing the terminal area of an airport possessing instrument descent facilities.
'Clean Aircraft': an aircraft flying in the normal cruising configuration, with high lift devices and undercarriages retracted.
Co-ordinates: the latitude and longitude of a place or point on the map estimated to the minute. For example, McMurdo airfield was originally listed as 164.58 degrees east, and 77.53 degrees south. Waypoints are sets of latitude and longitude co-ordinates, whether inserted into the navigation track as figures or, for convenience, as code-words.
CVR (Cockpit Voice Recorder): a tape recorder device which records everything said by the technical crew during flight, and automatically erases the tape 30 minutes later. The device, with its record of the previous 30 minutes, is a valued aid in accident investigation.
DME (Distance Measuring Equipment): an electronic navigation

aid which provides a direct indication to the pilot of the distance to the DME station to which his aircraft is tuned. (In the Antarctic flights, the ground DME facility was incorporated in the military TACAN.)

DFDR (Flight Data Recorder): like the Cockpit Voice Recorder, a valued aid in accident investigation. The DFDR (often known as the 'Black Box') records on metallic tape basic operating information including height, speed, flight control movements, headings, engine performance, etc.

'Seven-Eights', etc: an expression of cloud amount. Eight-eights represents a completely clouded sky; four-eights a half-clouded sky.

Endurance: the length of time an aircraft's fuel load will allow it to remain airborne.

Flight Level (FL): an expression of height based on a standard altimeter setting of 1013.2 millibars. It is expressed in hundreds of feet; thus 12,000 ft would be FL 120. Leaving or approaching airports, a more exact altitude is used by relating the altimeter to the local millibar reading.

Flight Plan: A plan prepared before departure of a flight which records computations of wind effects, endurance, etc, in relation to speed and heading. The modern flight plan progressively lists the track to be followed, the waypoints, and technical information vital to the crew including its all-up weight at departure, and the progressive fuel burn.

Greenwich Mean Time: the time used in this book is GMT, which was chosen to facilitate comparison with the transcript of the CVR printed in the Chippindale Report, which uses GMT. The Mahon Report time is McMurdo time which is New Zealand standard time, 12 hours ahead of GMT. On the day of the crash New Zealand itself was on New Zealand summer time, which is 13 hours ahead of GMT. The time of the crash was 49 minutes 50 seconds past midnight (0049.50) GMT, 12.49.50 pm McMurdo time, 1.49.50 pm New Zealand summer time.

Grid Navigation: a navigation system used in polar and sub-polar regions where magnetic compass readings become inaccurate and longitudinal lines close in markedly. The system is based on a pure grid centered on the Greenwich zero longitude. Pilots monitoring their track switch to grid beyond 60 degrees latitude and use grid positons when communicating with ground control. However, the grid system has no effect on a route track programmed into an inertial system.

HF (High Frequency): as distinct from Very High Frequency (see VHF), HF radio makes possible communication over long distances, and is used by aircrew during long distance flights.

HSI (Horizontal Situation Indicator): an instrument mounted on the pilot's flight instrument panel capable, by selection, of displaying position information in the ILS, VOR or NAV modes. When selected in NAV mode, it displays distance to the next way-point in the top left-hand corner, ground speed in the top right-hand corner, and a plan view of the aircraft position to left, right or on track. It is designed to provide instant information to the pilot of his position on a map along his track to the next waypoint.

IFR (Instrument Flight Rules): standard rules governing flights if weather conditions preclude the use of visual navigation. However, commercial jets invariably fly IFR except on departure from or approaching airports and in unusual situations — such as special scenic flights.

IMC (Instrument Meteorological Conditions): conditions where visibility is less than specified for visual flying, and where it is mandatory to fly by instruments.

ILS (Instrument Landing System): an electronic approach aid which enables a pilot to carry out an approach for landing when weather conditions preclude visual 'contact' with the ground.

Inertial System: the long range navigation system universally used by modern jet airliners. The inertial platform, based on extremely sophisticated gyros and inertial weights, has the ability to sense and measure virtually every movement of the aircraft, and with its associated computer, plot the machine's position and track. Linked to the autopilot, the system flies the aircraft along a designated track. Most airliners, including the DC10, contain three inertial platforms, each monitoring each other, and capable of isolating one if it malfunctions. (See AINS.)

MSA (Minimum Safe Altitude): every commercial route and long distance oceanic flight has an MSA, below which the pilots are forbidden to descend unless specifically cleared by ground controllers. The MSA takes into account high terrain lying along a route track, and adds a margin of safety to the high points.

NDB (Non-Directional Beacon): a medium frequency radio transmitter capable of reception from any direction by the Automatic Direction Finder (ADF) in aircraft. Pilots can plot a heading from an NDB or can steer a heading towards it.

NOTAM (Notice to Airmen): messages relating to changes in the serviceability of aerodromes, radio and navigation facilities.

Radar: the ground-based radar at McMurdo had two modes, the first capable of detecting approaching aircraft from a considerable distance and plotting its bearing; the second, of higher resolution, for plotting aircraft in the vicinity of the station (up to about 40 miles). Weather radar, carried by the DC10, was primarily designed to detect turbulent cloud and rain storms ahead. Weather radar can be 'tilted' to (under some conditions) show up terrain, but reliance on it as a navigation aid is forbidden.

TACAN (UHF Tactical Air Navigation Aid): a military navigation and communications unit, usually mobile. Limited use can usually be made of TACAN units by civilian aircraft. The McMurdo TACAN contained Distance Measuring Equipment (DME) which could be interrogated by civilian aircraft.

Transponder: radio device carried by aircraft, designed to be triggered off by certain radar wavelengths. When this happens, the transponder sends a powerful signal which shows up clearly on a ground radar screen, and usually includes additional information such as the height of the aircraft.

UHF (Ultra High Frequency): frequency band of 300 to 3000 MHZ.

VFR (Visual Flight Rules): standard rules specifying procedures under which visual navigation and/or traffic separation are permissible.

VHF (Very High Frequency): this frequency band (30 to 300 MHZ) is largely free from interference and static, but is limited to 'line of sight' range. VHF radio communication is used within a limited vicinity of an airport or radio station.

VMC (Visual Meteorological Conditions): flying conditions which provide a specified range of forward visibility making it possible and permissible for the pilots to fly using a visual means in avoiding obstacles such as cloud, terrain, and other aircraft.

Visual Approach: An approach by an IFR flight when either part or all of an instrument approach procedure is not completed and the approach is executed in visual reference to terrain.

Introduction

ON 28 NOVEMBER 1979, an Air New Zealand DC10 airliner crashed into the northern flanks of Mt Erebus while on a scenic flight over Antarctica. All 257 people aboard died instantly.

Nine hours later a searching US Navy aircraft sighted the wreckage: 'a brown smear' dwarfed in the immensity of the mountain.

It would have been an appalling disaster for any country, but in New Zealand with a population of just three million, the horror was doubly felt. Almost everybody knew or had some association with somebody among the victims. Most were nationals, but there were also 24 Japanese aboard, 10 Americans, two Britons and an Australian.

There followed the gargantuan job of accident investigation, led by New Zealand's Chief Inspector of Air Accidents, Ron Chippindale.

Chippindale reported to the public on 30 May 1980, that Captain Jim Collins had erred by continuing the flight at low level toward an area of poor surface and horizon definition when the crew was not certain of their position.

First Officer Greg Cassin was criticised for failing to adequately monitor the flight because he was engrossed in trying to establish radio contact with McMurdo. The airline received virtually no criticism, barring mild censure for lapses in the briefing it gave the crew.

Polar whiteout was touched on in one sentence — as a possible cause of deception in the final moments of the aircraft's life.

The finding was hardly a surprise to the public. Leaks and hints to the press had led them to suspect blame would be apportioned this way, especially since any possibility of technical failure had been ruled out.

If the crash and the subsequent technical investigation form the first two segments of the Erebus story, the legal inquiry forms a powerful conclusion.

The work of the Royal Commission, set up to inquire into the Mt Erebus crash, amounted to a brilliant achievement for the legal profession.

Under New Zealand law, the catastrophe could well have ended with the accident inspector's report, as it had in previous serious air crashes. The matter would have been taken no further, though the build-up in litigation over dependants' claims would have gone on for years.

Instead, the Government decided to set up the commission. If nothing more, it would help assuage public uneasiness (which was growing) and the rising anger felt by the Airline Pilots' investigating team and next-of-kin of crew members. It might also more clearly define areas of blame, on which the lawyers could work. Above all, the magnitude of the disaster called for some measure beyond the norm.

The inquiry got under way in late June, 1980. Mr Justice Mahon was assisted by two barristers, David Baragwanath and Gary Harrison. They drew on the very able and proven technical and organisational expertise of a retired Air Marshal of the RAF, Sir Rochford Hughes.

The terms of reference stretched to ten points, but the main questions it set out to answer were: What caused the crash, and who — if anybody — was to blame?

All parties were represented: the airline, the Civil Aviation authority, Mr Chippindale's office (see note 1 on page 326, headed *Was Inspector represented before Royal Commission?*), the New Zealand Airline Pilots' Association, the Flight Engineers' Association, McDonnell-Douglas (the makers of the DC10) and a consortium representing the estates of some 100 passengers. The families of Captain Collins and First Officer Cassin were individually represented by barristers Paul Davison and Roger MacLaren.

David Baragwanath, Gary Harrison and Paul Davison applied the major brunt of the questioning and cross-examination. They had done their homework soundly; their grasp of navigational techniques and airline procedures was etched deeply enough for them to push past the incomplete and inconclusive answers which were, with some frequency, supplied by expert witnesses.

The Royal Commission sat for 75 days, seeking four extensions beyond the October 1980 deadline. Justice Mahon and David Baragwanath travelled to USA, Canada and Britain to obtain further evidence. A trip was made to Antarctica in November 1980. Finally, on 1 May 1981, the Commission's report was tabled in Parliament, and burst like a bomb over the airline and the community.

Up until the Royal Commission hearing was well advanced, the airline management was confident the report would largely reflect the Chippindale findings, no residue of blame going to the company and the Civil Aviation Division, but with the main cause being attributed to poor airmanship.

Instead, the Mahon Report put the blame squarely on the airline: 'The single effective cause of the crash was the act of personnel in the Flight Operations Division in altering the latitudinal and longitudinal co-ordinates of the destination waypoint without the knowledge of the aircrew, and in omitting to notify the aircrew, either before departure or in flight, of the fact that an alteration had been made.'

Although this was the single effective cause, there were two contributing factors for which the Civil Aviation Division was held responsible.

Civil Aviation had failed to ensure that the pilot in command of unscheduled flights to the Antarctic was provided at his pre-despatch briefing with a topographical map on which the programmed flight path had been plotted.

Civil Aviation had also dispensed with the requirement that the pilot in command of an Antarctic flight had flown the route before.

Justice Mahon also ordered the airline to pay $150,000 as a contribution to the public cost of the inquiry. The company had failed to 'put all its cards on the table', he said. It had denied every allegation of fault and had counter-attacked by ascribing total culpability to the crew, against whom it made the untenable allegation that there were no less than 13 separate varieties of pilot error.

'I have little doubt there are still one or two (cards) hidden in the pack. In such circumstances the airline should make a contribution towards the cost of the inquiry.' In the financial context of airlines it was miniscule. Symbolically, it was a sharp rap.

The report also carried a short chapter titled "The Stance Adopted by the Airline Before the Commission of Inquiry". It ended with two paragraphs now firmly inscribed on the New Zealand consciousness:

'No judicial officer ever wishes to be compelled to say that he has listened to evidence which is false. He always prefers to say, as I hope the hundreds of judgments which I have written will illustrate, that he cannot accept the relevant explanation, or that he prefers a contrary version set out in the evidence.

'But in this case the palpably false sections of evidence which I heard could not be the result of mistake, or faulty recollection. They originated, I am compelled to say, in a pre-determined plan of deception. They were clearly part of an attempt to conceal a series of disastrous administrative blunders and so, in regard to particular items of evidence to which I have referred, I am forced reluctantly to say that I had to listen to an orchestrated litany of lies.'

I know that Justice Mahon is writing a book, due for publication late in 1983, on the Royal Commission and its aftermath. Thus, much of the legal proceedings surrounding the Erebus disaster are touched upon but briefly in this book. I am certain that Justice Mahon will deal with this aspect with the same comprehensive wisdom he displayed at the Commission hearings and in compiling his widely admired report.

News of the disappearance of Flight 901 reached me as I was about to fly a DC10 out of Honolulu for Los Angeles. I was stunned, almost beyond belief. I had flown with every crew member over the entire period each had been employed by Air New Zealand, and had watched the progress of Jim Collins for nearly 25 years, from the time of his induction as a student pilot in the RNZAF at Wigram when I was one of his instructors. Some of the crew I had flown with only days earlier.

I believed that the crew of Flight 901 was as skilled, experienced and dedicated as any airline could muster. They were flying a machine whose navigation equipment was more advanced than that fitted to any of the other wide-bodied airliner types in service.

The only answers that could validly suggest themselves were that the machine had suffered some sort of structural failure, or had been sabotaged. The possibility of structural failure was one I could accept only with reluctance. True, the DC10 had come to assume a somewhat sinister reputation with the public and the press since it had been involved in three of the world's four worst air disasters.

But in six years of flying the DC10, I had developed a strong admiration for the machine — its sure and responsive handling qualities, its reliability and the long-legged ease with which it handled the vast Pacific routes.

Soon after I returned home to Auckland the possibilities of structural failure or sabotage were disproven. Investigators were probing the wreckage and two invaluable pieces of equipment had been recovered, the flight data recorder and cockpit voice recorder. Early read-outs of these showed that the aircraft was functioning in a perfectly normal way when it hit the mountain.

This only made the mystery more inexplicable. Already rumours were beginning to circulate of what is loosely called 'pilot error', or of some sort of flight deck mismanagement. All indications were that this would eventually be cited as the official reason for the crash.

It was at this point that I made my own commitment to an independent study of the crash. It was an undertaking which would last 18 months, carried out between flying assignments and hedged with frustrations.

I had noticed that when an accident investigator was predisposed towards devaluing a crew's previously unblemished record, he tended to come up with a quick result which implied that the crew had acted in an uncharacteristic manner, and that this resulted in the crash.

This was a false — and illogical — approach, I believed. Would it not be more accurate to accept as axiomatic (until the last moment) the high quality of the crew's airmanship, their care and professionalism, and then start looking at all other possible causes for the accident? In other words, look for a situation which could mislead even an experienced crew.

This should not be construed as meaning that the integrity of the crew must be defended at all costs by members of the Pilots' Association. To finally arrive at the true cause of the accident is the only proper way to serve the Pilots' Association and the public alike.

As the media stories, often based on 'leaks', accumulated after the crash, I realised that the inspector of air accidents and the airline executive seemed to have no difficulty in finding fault with the professional conduct of Captain Collins, and his crew.

By adopting the opposite approach I found I was being led to the inescapable conclusion that the crew had been trapped in a unique set of circumstances. They were in a situation where the usual confident application of their normal flying techniques would have been totally inappropriate.

In many years of instructing both military and civilian pilots, I had noticed that once airmen reached a certain level of experience, they tended to develop a very set technique for dealing with situations.

This could be seen in ground simulator training when a crew was progressively loaded with one emergency on top of another until, without warning, a point of total saturation was reached. At this point, the crew's sampling rate was operating at maximum, their discrimination of priorities was operating only at the most critical end of the scale.

But even on reaching this point of severe overload, the experienced crew would continue to act fairly closely within a pre-determined response pattern and in accordance with previous experience of similar situations. Rarely did they act irrationally.

My research followed a tenuous path at first; going back over the notes of one of the Antarctic flights I had commanded, talking with others who had flown the service, and piecing together what information I could glean from the investigative team.

Guided by my sons (both students of psychology) — Mark, a Master of Science with Honours from Auckland University, and Karl, a Bachelor of Science now flying supersonic F111's with the Royal Australian Air Force — I began reading. I approached psychology academics who, in turn, recommended more advanced texts. Eventually my quest led me to experts overseas — at Monash University, Melbourne, and to scientists in the United States.

Over several months I studied approximately 20 texts on visual perceptual psychology and related subjects. This increased comprehension assisted me in my communications with international experts in the field.

What I learnt supported my hypothesis — that the Erebus crew had become the innocent victims, trapped by a unique web of circumstances.

As the scientific support for the hypothesis increased, I found myself in complete disagreement with the Chief Inspector's interim report. As a result I decided to publish a manuscript which had been drafted and lodged with my solicitors.

This was a fictionalised account of Flight 901, based upon my own Antarctic flight experience, my close association with Jim Collins, and his crew, and information 'captured' in the cockpit voice and data recorders. This 'in cockpit' impression of Flight 901 forms the heart of *Impact Erebus.*

However, I decided that since the Chippindale report would be followed by the setting up of a Royal Commission to inquire into the disaster, it would achieve more if the research which had gone into the manuscript was submitted as evidence to the inquiry. If supported by the Royal Commissioner's findings, my work could then be published.

I was well aware that I had to take every possible step before making my submissions to ensure that the evidence was soundly and scientifically based.

My research concentrated upon two specialised fields of knowledge. The first was the condition known as a 'mental set', the condition which tends to make a person take a predictable course while undertaking a certain job. This forms the basis of the chapter in this book called 'Profile'. I felt that the crew of Flight 901 had performed exactly as expected given their training, the pre-flight briefing and the information available to them visually and displayed on the cockpit instruments.

The second was the phenomenon of 'visual perception and deception', those tricks the eye can play on people, causing misconceptions about the ability of sight.

Polar 'whiteout' is a condition many uninformed people still relegate to the realm of suspect theory. They avoid associating it with anything so definite as a crash because they feel it is still too esoteric, not yet a 'hard' science.

But a number of scientists and psychologists have closely studied the condition, and simulated its effects under laboratory conditions. It was these people I sought out, together with experienced ice flyers, including US Navy pilots, whose survival depended on their knowledge of the peculiarities of this remote environment.

I was able to place before the inquiry visual models and evidence which I had devised and checked with experts. I learned it was possible to produce a three-dimensional scale model of the McMurdo and Ross Island area with cloud layers and correct lighting that would exactly recreate the conditions of 'sector whiteout'. If this model was built to the correct dimensions, it could be superimposed on the visual platform used for our DC8 simulator at Mangere and an almost exact in-flight simulation reproduced. My efforts were constrained by cost and time and I had to resort to cheaper two-dimensional visual models instead.

I was well aware that the public had been soaked for a long time in the belief that Jim Collins had carried his crew 'yelling and kicking' down through cloud until they smashed into the mountain.

If my theory was accepted, a complete reversal of this impression would result. But I would be going against old friends and colleagues of many years' standing, as well as the Chief Inspector, the Director General of Civil Aviation, and probably the airline's management and board of directors, to say nothing of the political considerations of disagreeing with the ultimate Head of Civil Aviation and chief shareholder of Air New Zealand — the Prime Minister. I had no doubt where I would stand if these submissions failed.

I took a lengthy brief to the inquiry (the evidence appears as Appendix A in this book), and spent three days in the witness box, much of it under close cross-examination.

My first submission was that a first-class crew had been misled over the route the aircraft was programmed to automatically follow. This was the result of a history of blunders and mismanagement in the airline's operations branch, and in no way could the crew have been expected to detect the misinformation. Collins and his crew had acted professionally and with caution during the course of the flight.

I then propounded a theory, backed with evidence, that the area they flew into bore an uncanny resemblance to the area they believed they were approaching, given the peculiar weather conditions of the time. This was labelled the 'Vette theory' by the press, but came to be regarded as undoubted fact.

Most important, I laid bare a fundamental misconception, held by Civil Aviation and the airline, about the nature and danger of whiteout in polar areas. Whiteout manifests itself in numerous ways. But its most insidious form is under flying conditions which, in more familiar climates, would be described as good and even excellent — flying beneath a light overcast when the atmosphere is so clear one can see for many miles.

Fortunately the Royal Commissioner accepted the hypothesis, and indeed tested it against other authorities. He was able to visit McMurdo and the region of the crash site, and by chance witnessed identical conditions to those propounded.

In their final moments the crew of Flight 901 were looking out on a scene which had every appearance of being a vast, snow-covered clear expanse of flat sea ice stretching 40 to 60 miles ahead, under a cloud ceiling of 2000 to 3000 feet. The terrain in front of them was uniformly white, except for two or three thin strips of black rock on the snowy ice floor which were indistinguishable from sea water or which may have been concealed by drifting ice fog. But in reality they were looking at an invisible coastline, with the flanks of a 12,000-foot mountain climbing steeply beyond it.

If that sounds like a writer's dramatisation, then forget the scientific evidence and consider the visit of the senior High Court judge and his similarly logical and sceptical counsel assisting, David Baragwanath. They retraced the path of the airliner first in a Hercules and then in a US Navy helicopter a year later almost to the hour — and emerged convinced and a little shaken by the eerie experience which bore uncanny resemblance to the visual models I used at the commission to demonstrate a hypothesis vigorously supported by leading scientific opinions and practical polar pilots alike.

The condemnation of Jim Collins and his crew was inevitable given the regulations set by the Civil Aviation Division and observed by Air New Zealand. There was no allowance made in these for the peculiar circumstances encountered by Flight 901.

According to the regulations the cockpit crew were flying in either visual metrological conditions (VMC) or instrument metrological conditions (IMC). The former called for a required distance of atmospheric transparency — clear visibility. Once this had eroded, when reliance upon sight could not provide adequate obstacle avoidance, IMC regulations would apply.

These rules are ingrained into all professional pilots who must balance safety and economics. They are totally appropriate to almost all flight conditions where there is adequate visual texture and the eye's ability to detect any deterioration can be assured.

My investigations revealed that there is a third category of flight conditions which appears to indicate VMC responses but in fact requires IMC regulations. In certain circumstances, not uncommon in polar regions, a number of visual illusions are caused by the eye misfunctioning due to lack of visual texture. The pilot is effectively in IMC with no way of knowing it. Such 'sector whiteout' causing loss of visual protection can occur in clear weather under a light overcast.

Had the operations specification for Antarctic flights contained the simple warning that visual separation from white surfaces should not be attempted, particularly below an overcast sky, the Erebus tragedy would not have occurred.

It is my opinion that we (pilots and public) should have expected protection from the effects of this phenomenon in the preparation of the operations specification by Civil Aviation. They would argue that they are under-staffed and underfunded (I agree) to research the route adequately with the International Civil Aviation Organisation. Had they done so, I believe they would have realised the folly of permitting visual flying under an overcast sky in the Antarctic.

In his book *To Fly a Desk,* my benefactor and ex-chairman of Air New Zealand, Sir Geoffrey Roberts, whom I greatly admire writes of the deterioration of Civil Aviation from a full government department to a division within the Ministry of Transport. He believes this caused a loss of status and a decline in its standards. Sir Geoffrey claims that the spirit of ICAO rules was 'ignored shamefully' in regulating the Antarctic flights. I can only concur.

We should now be looking at what needs to be done to ensure that the vital lessons that could be learned from Erebus are not lost in political dog fights.

To this end I have set up a flight safety fund with my solicitors, and have arranged to apply profits from this book to the fund. I intend to use this to finance a Masters student to research the visual perceptual field as it applies to the professional pilot — that unique interfacing between environment and machine. I hope to make recommendations to ICAO on including this subject within the pilot training syllabus, and I also intend to research a forward facing ground proximity warning device for civil airliners.

Pictorial Account of Crash

The following 49 pages set out in visual terms the causes lying behind the crash of flight TE 901. Reference may be made from the narrative text to this section in order to illustrate the points made in the text. This pictorial supplement also provides an independent explanation of the crash, and it may be studied as a whole separately from the remainder of the book.

Why crew did not perceive Mt Erebus before impact?

The possible reasons the crew did not perceive Erebus as a mountain which lay on the track and at the altitude which they were proposing to fly are that firstly they were flying in cloud, or secondly they were flying in clear air but perceived cloud on their proposed track and that cloud in fact obscured their view of the mountain ahead, but they consciously flew into the cloud, or thirdly they were flying in clear air and perceived ahead McMurdo Sound in clear air, but that perception was false, being a visual illusion caused by sector whiteout.

It is my belief, upheld by Justice Mahon in his report, that the crew were deceived by sector whiteout which rendered Mt Erebus invisible.

Fig. 1 Mt Erebus

Mt Erebus is visible in this photograph, yet when flight 901 flew towards it, in daylight, with 40 mile visibility, it was invisible to all on the flight deck. Again, it was first visible but then became invisible to Justice Mahon and the occupants of a helicopter as they retraced the flight path of flight 901. How is it that we can see ('perceive') the mountain but they could not?

The ability to perceive a land surface and to judge its form and distance from it depends on the extent to which its surface features, such as hollows, buildings etc., can be perceived. Contrast is a difference in brightness and/or colour. Without contrasts, the eye can distinguish no form at all. Contrasts create the lines, shapes and patterns that the eye recognises as things. Contrast arises through differences in colour, in reflectivity, and by shadows and shadings.

We can see Erebus because of the difference in colour between the white of the mountain and the blue of the sky. When flight 901 and later Justice Mahon flew toward Erebus, although they were not flying in cloud, the overcast ahead obscured the blue sky so that there was no colour contrast between sky and terrain.

We can also perceive Erebus because of the shapes and patterns on its surface caused by the contrast produced by shadows and shadings resulting from the sun shining on the mountain at roughly a right angle to our line of sight. As flight 901 approached Erebus the sun was substantially from behind the aircraft, so that shadows would not have been visible.

The only other surface contrast was provided by the ice cliffs on the edge of Ross Island and the evidence suggests they were covered by sea fog, or lost in sector whiteout.

When cloud occurs over a snow surface, multiple reflections of light tend to cause the cloud undersurface and the snow to appear equally bright, and when no contrast, e.g. shadows, exists on the surface it becomes impossible to perceive the boundary between land and cloud.

The inability to perceive either land surface, cloud, or the boundary between them is whiteout. Full whiteout exists when the observer is surrounded by whiteout in all directions. Because it appears so normal and innocent, the most insidious form of whiteout is sector whiteout, in which the observer can perceive contrast (and hence is not in whiteout) in all directions except in one sector.

Flight 901, misled at briefing as to the nature of whiteout, unsuspectingly faced sector whiteout in the only sector which mattered, Mt Erebus directly ahead of them, on a computer flight plan route altered by the airline, without telling the aircrew, by some 26 miles, away from the safety of McMurdo Sound.

Fig. 2 Illustration of McMurdo area

The major features of the McMurdo area are illustrated in this reconstruction.

The flight plan distributed to the pilots at briefing ran down McMurdo Sound to a point just west of West Dailey Island. The Capes at the entrance to McMurdo Sound which the crew would expect to see were Cape Royds to the east and Cape Bernacchi to the west.

Unknown to the crew, the flight path had been re-programmed by the airline to approach McMurdo across Lewis Bay and over Mt Erebus. The aircraft passed to the east of Beaufort Island instead of to the west as they would have done had they been tracking down McMurdo Sound as they believed. Why were they not alerted to danger? Some witnesses, including me, believed that if they 'saw' the island at all then they 'perceived' it as a break in the ice, and not as an island. This theory may not sound plausible until one examines the island illustrated in fig. 23. Other witnesses, such as G.W. Shannon p.345, believed that the crew would have mistaken it for Dunlop Island which they would expect to see to the west.

Also illustrated to the west is the Wilson Piedmont Glacier, running north from Cape Bernacchi up Victoria Land to Cape Roberts. The reference during the flight south recorded on the CVR to 'Wilson' must surely have been a reference to this prominent feature, despite Ron Chippindale's opinion to the contrary.

Did crew believe they were in Lewis Bay or McMurdo Sound?

The aircraft actually flew up Lewis Bay between Cape Tennyson to the east and Cape Bird to the west and impacted on the lower slopes of Mt Erebus, which is 12,455 feet, at a height of 1500 feet.

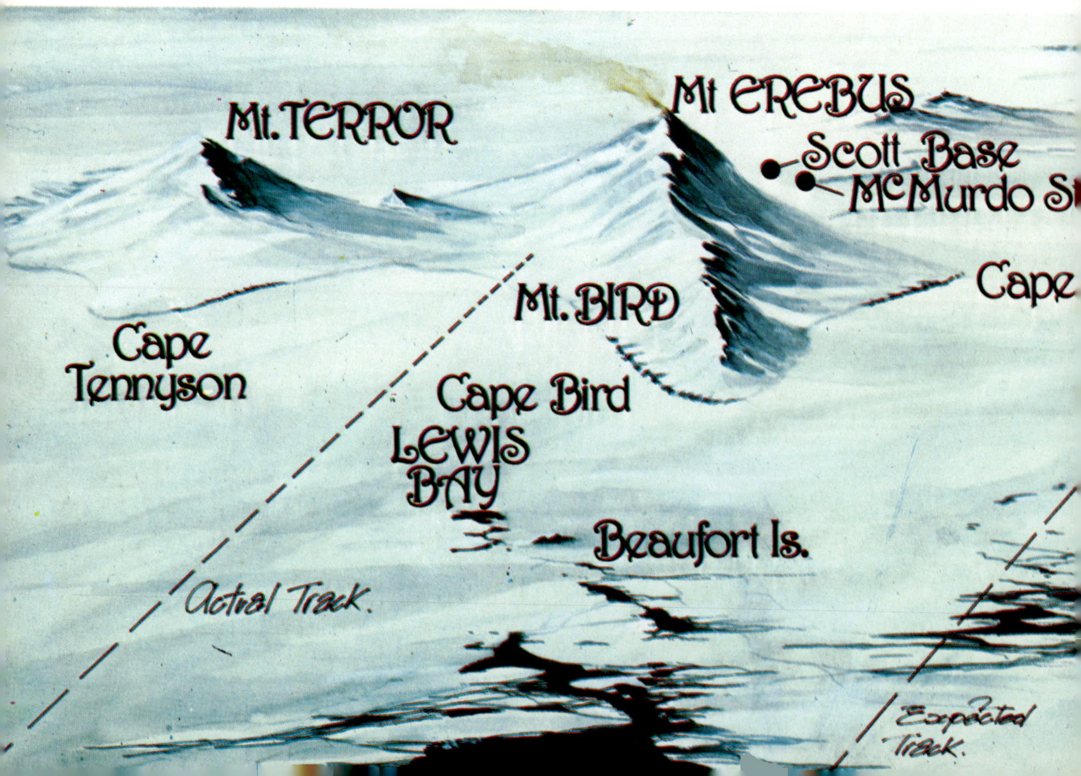

Captain Ian Gemmell testified that the crew might have known they were flying below 2000 feet up Lewis Bay directly towards Mt Erebus, but expected to be able to see Ross Island ahead and to be able to fly around it. In support he referred to Collins' statement 'we are 26 miles north we'll have to climb out of this' which, Gemmell testified, could be indicative of a sudden awareness or recollection of high ground on track ahead of the aircraft. Gemmell was confident that Peter Mulgrew's reference to 'some land ahead' referred to Beaufort Island, which Mulgrew could not have mistaken for a break in the ice because of his polar experience, but which he did not positively identify as Beaufort Island.

I do not believe that suggestion has any credibility at all. It implies an extraordinary degree of incompetence on the part of the crew, who took no evasive action at all until the ground proximity warning indicator sounded. It also implies a suicidal attitude on the part of Mulgrew. Owing to his excellent knowledge of the McMurdo area, had he seen Beaufort Island, he would have identified it and instantly located the aircraft as being in Lewis Bay and heading for Erebus.

I consider the evidence makes it clear that the crew believed they were flying up McMurdo Sound, between Cape Royds and Cape Bernacchi. What caused them to hold this belief?

The crew attended a briefing with three pilots from other flights. Those three formed the impression from the briefing that the flight path ran down McMurdo Sound. Setting aside that reason, Collins and Cassin had three reasons for believing their flight path was in McMurdo Sound: they were given maps showing such a route, they were probably given the track and distance diagram known as exhibit 164 showing a McMurdo Sound route, and finally they were shown a computer flight plan, used for the previous 14 months, which showed that route. I now illustrate these three last matters.

Fig. 3 Computer Flight Plans (page 37)

The left column is the flight plan distributed to the crew at briefing. The co-ordinates of the McMurdo waypoint, when plotted on a topographical map, reveal that the path runs down McMurdo Sound, close to the Dailey Islands. Collins planned his flight on the basis of that flight plan. The right column is the flight plan given to the crew just before their flight departed and which the crew inserted into their navigation computer. Without notifying the crew, the airline altered the plan on the morning of the flight from that on the left to that on the right so that it lay about 27 miles east of the left hand plan given at briefing and so that it took an aircraft, flying on navigation track, on a course intercepted by Mt Erebus.

Why did Collins not see that the flight plan given to him for inserting into the navigation computer differed from the one given to him earlier at briefing? Firstly, because if the aircrew followed standard practice Collins would probably not have been involved in loading the computer. Secondly, if Collins had been involved, he would have been concentrating, in accordance with the procedure laid down, on checking that the digits were being accurately transcribed from the plan into the computer. In fact the crew did transcribe them accurately.

Thirdly, before the accident, a pilot would have considered it inconceivable that he could be shown a flight plan at briefing and then handed another altered one for insertion, without being told it had been altered. Even the prime airline witness, Captain Gemmell, admitted that news of the change in waypoint coupled with non-disclosure to the crew came as a bombshell. Collins would never have suspected that the two flight plans might have differed.

Fourthly, the checking procedure laid down by the airline made no provision for such an event, which in itself suggested that such an event was never contemplated. Fifthly, flight plans were normally loaded into the aircraft by insertion of a cassette, with no suggestion by the airline that pilots should question whether the cassette represented the route demonstrated by printouts at briefing. Sixthly, Collins had one chance to check whether the two flight plans differed in any respect, whereas those who criticised him had 14 months to do this, and if the critics' evidence is believed, then during that time they made 54 mistakes in checking the flight path. (See note 3 on page 329.)

Fig. 4 Computer flight plan showed McMurdo Sound track, when plotted onto Collins' atlas (page 38)

The first reason which made the crew believe that when on nav track their flight would lie down the centre of McMurdo Sound was that, as noted on the preceding page, at the RCU briefing which the crew attended, printouts from the navigation computer were distributed.

The four computer flight plans distributed to the crews at briefing on 9 November were copies of Captain Dalziell's flight plans used three days before briefing. The McMurdo waypoint on each of the flight plans was

```
            CENTRE LANDING GEAR IS EXTENDED FOR TAKE OFF
OPS FLASH
NZN NZAA-NZCH RT NO       /        CAPT DALZIELL      RADIO LOG
06/11/79-1900Z  TRK.T  W/V  G/S  DIST ZEET FUELRM STN
M82  TE 901/07  TRK.M DDVVV FL   ZATA ZETA RQFUEL GMT

NZAA  AUCKLAND                                .  FREQ P
3700.6S17446.9E                  S/H ....  101.4      S

NP    NEWPLMTH  193.6       400   123   21
3900.2S17410.9E 174.3       CLB  .... ....   XX.X

NS    NELSON    199.3 23037 448   146   22     .
4117.8S17308.0E 179.3       FL31 .... ....   91.3

RY    MT MARY   216.2 24037 444   208   28     .
4408.2S17016.8E 195.2       FL31 .... ....   86.5

NV    INVRCRGL  211.8 27037 457   163   21     .
4624.8S16819.1E 189.2       FL31 .... ....   83.1

AUKIS AKLND IS  198.4 29078 478   271   34     .
5042.0S16610.0E 173.4       FL29 .... ....   77.5

55S   55S       185.7 29098 497   259   32     .
5500.0S16527.2E 156.2       FL29 .... ....   72.5

60S   60S       185.7 31060 504   302   36     .
6000.0S16431.1E 150.2       FL33 .... ....   66.8

BLYIS BALENYIS  185.7 31053 504   407   48     .
6645.0S16300.0E 349.5       FL31 .... ....   59.6

CPHLT C HALLET  155.8 31063 532   367   41     .
7220.0S17013.0E 322.4       FL31 .... ....   53.6

MCMDO MCMURDO   188.9 34054 517   337   40     .
7753.0S16448.0E 357.4       FL35 .... ....   47.9

CPHLT C HALLET  008.9 34054 425   337   47     .
7220.0S17013.0E 177.4       FL33 .... ....   41.5

70S   70S       358.8 33060 420   139   20     .
7000.0S17003.6E 168.9       FL33 .... ....   38.8

65S   65S       358.8 31068 425   300   42     .
6500.0S16946.6E 168.7       FL33 .... ....   33.2 .
```

```
ZKNZP ON GATE 2. CLG DOWN FOR DEPARTURE.
OPS FLASH
NZP NZAA-NZCH RT NO       /        CAPT COLLINS       RADIO LOG
27/11/79-1900Z  TRK.T  W/V  G/S  DIST ZEET FUELRM STN
M82  TE 901/28  TRK.M DDVVV FL   ZATA ZETA RQFUEL GMT

NZAA  AUCKLAND                                .  FREQ P
3700.6S17446.9E                  S/H ....  100.9      S

NP    NEWPLMTH  193.6       425   123   20
3900.2S17410.9E 174.3       CLB  .... ....   XX.X

NS    NELSON    199.3 30027 486   146   21     .
4117.8S17308.0E 179.3       FL31 .... ....   91.0

RY    MT MARY   216.2 31027 481   208   26     .
4408.2S17016.8E 195.2       FL31 .... ....   86.5

NV    INVRCRGL  211.8 31029 485   163   20     .
4624.8S16819.1E 189.2       FL31 .... ....   83.3

AUKIS AKLND IS  198.4 32029 495   271   33     .
5042.0S16610.0E 173.4       FL29 .... ....   77.9

55S   55S       185.7 31033 498   259   31     .
5500.0S16527.2E 156.2       FL29 .... ....   72.9

60S   60S       185.7 30034 487   302   37     .
6000.0S16431.1E 150.2       FL33 .... ....   66.9

BLYIS BALENYIS  185.7 29026 481   407   51     .
6645.0S16300.0E 349.5       FL31 .... ....   59.3

CPHLT C HALLET  155.8 29021 490   367   45     .
7220.0S17013.0E 322.4       FL31 .... ....   52.8

MCMDO MCMURDO   188.5 24015 463   336   43     .
7752.7S16658.0E 357.0       FL35 .... ....   46.5

CPHLT C HALLET  008.5 24015 483   336   42     .
7220.0S17013.0E 177.0       FL33 .... ....   40.8

70S   70S       358.8 29024 465   139   18     .
7000.0S17003.6E 168.9       FL33 .... ....   38.4

65S   65S       358.8 29024 465   300   39     .
6500.0S16946.6E 168.7       FL33 .... ....   33.3
```

underlined or highlighted, thereby drawing the crews' attention to it. The flight plans were available for inspection for a considerable period and only collected at the end of the briefing.

Jim Collins took his New Zealand Atlas to this briefing. It is clear that he either kept one of the flight plans which were circulated or that he noted down the co-ordinates, since he subsequently showed his family on maps that the track ran down McMurdo Sound. Furthermore, after the Royal Commission had concluded, I met in Los Angeles a friend of the Collins family, and from him learnt that Jim had explained to him too that he would shortly be taking a flight to Antarctica and that its flight path ran down McMurdo Sound.

I have plotted the co-ordinates from the flight plan which had been circulated at Jim Collins' briefing onto the New Zealand Atlas and Figure 4 shows the flight path which resulted, and which Jim Collins must have believed the aircraft would take when on nav track.

Jim showed his family Antarctic charts he was working on in preparation for the flight. In my opinion, it is more probable he plotted the track onto those, and used his atlas merely to record the Dailey Islands waypoint. But the principle involved is illustrated by the track on the atlas. Jim took his atlas on the flight (it was never recovered) and that suggests he used it for navigation purposes, as well as using his other charts.

ANTARCTICA

Fig. 5 Diagram showing McMurdo route: Exhibit 165

At the RCU briefing the crews were shown exhibit 165, which although not displaying the Air New Zealand route did display a McMurdo Sound route.

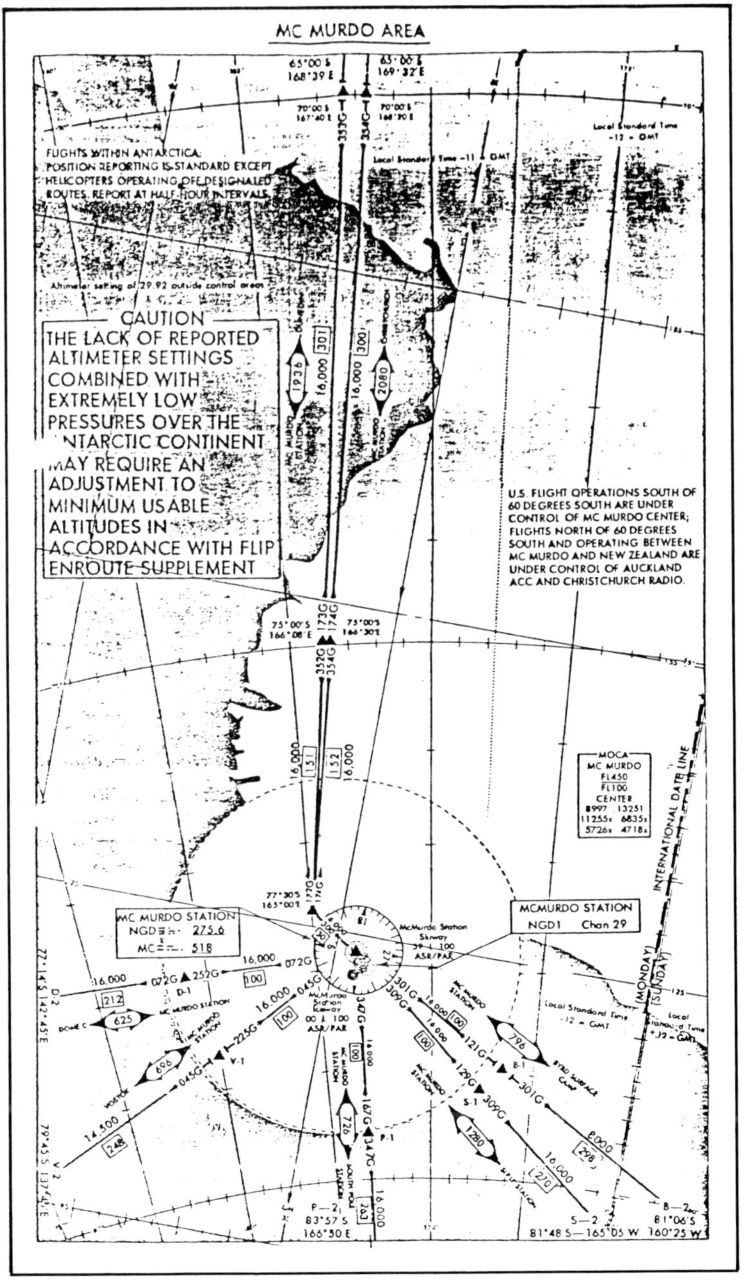

Fig. 6 Map showing McMurdo route: Annex H

The crew's belief that their flight path ran down McMurdo Sound was contributed to by a slide projected at their briefing showing a schematic diagram which displayed a McMurdo Sound route avoiding Mt Erebus. Though an earlier version existed, the slide was produced from artwork prepared by Captain Wilson who briefed the crew on their Antarctic flight.

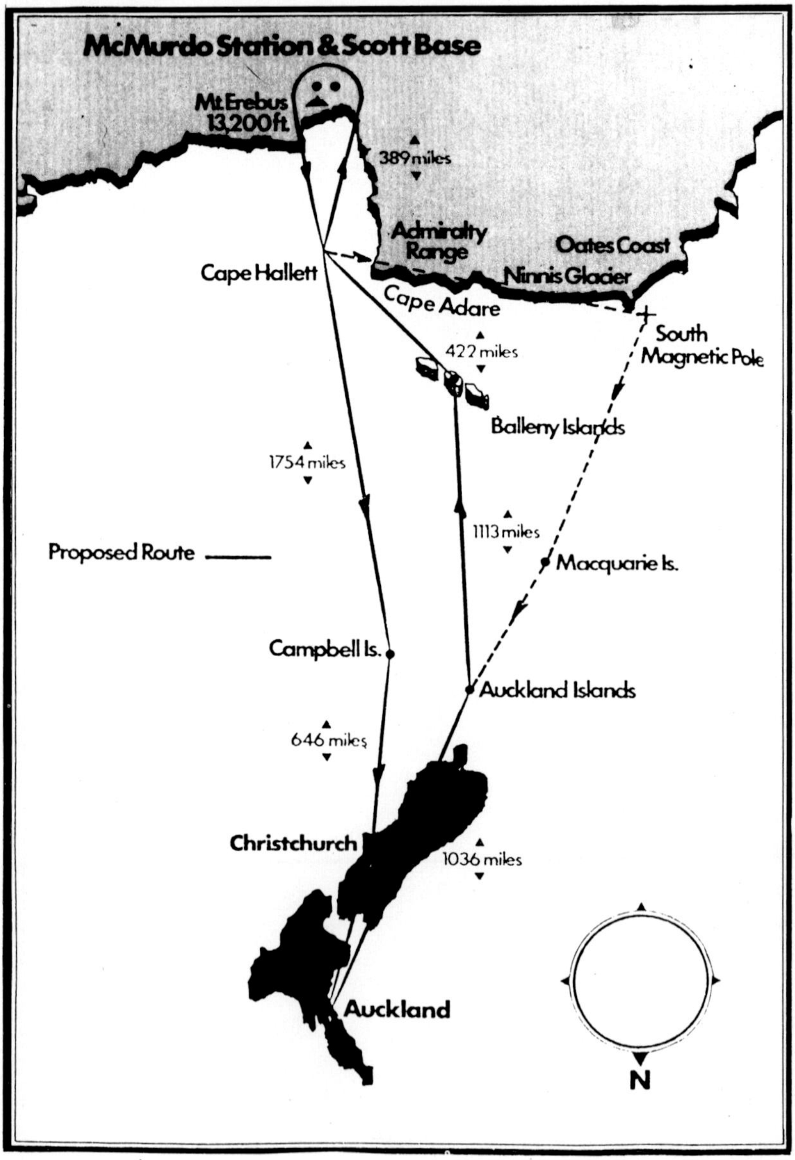

Fig. 7 Map showing McMurdo route: Annex G

The crew were shown at briefing a passenger information map. Fig. 7 is an extract from this exhibit 47, which was produced to the Royal Commission. It was also printed as Annex G to the Chippindale Report. It shows a track down McMurdo Sound, which avoids Mt Erebus on Ross Island. The caption printed on it states the route is to McMurdo Sound and back; no reference is made to a route to Williams Field across Mt Erebus and back.

One could be excused for believing that it was Air New Zealand's intention to have the passengers proceed to McMurdo by a different route from the one they wanted the crew to follow.

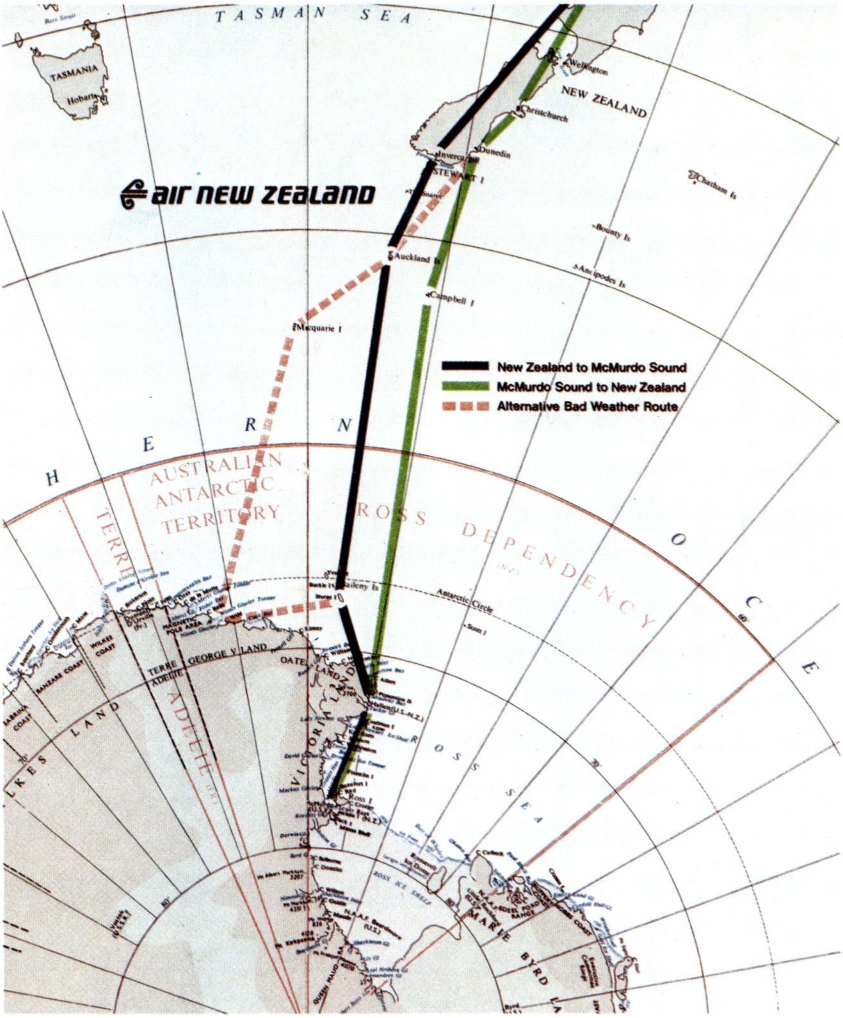

Fig. 8 Antarctic strip chart showing McMurdo route: Annex I

In the Antarctic envelope handed to the crew at their pre-despatch briefing shortly before take-off, was an Antarctic strip chart which depicted two routes between New Zealand and McMurdo. Both routes ran through the Byrd reporting point in the middle of McMurdo Sound.

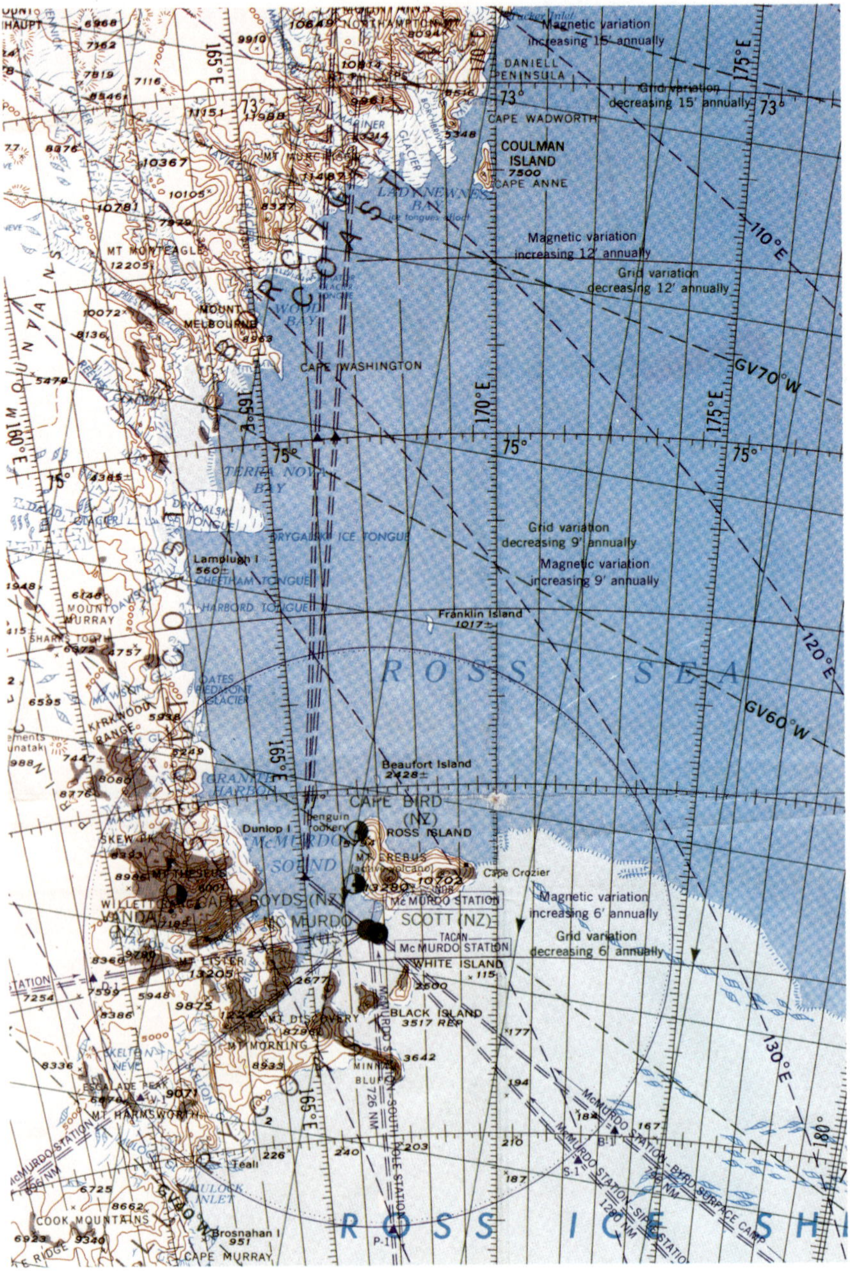

Fig. 9 Radio navigation chart RNC4 showing McMurdo route to Byrd reporting point: Annex F

At the initial RCU briefing this chart was made available to the crew and was referred to during the briefing. Although the route depicted upon it was not the Air New Zealand route, the chart did illustrate a route which over the final leg did run down McMurdo Sound to the same Byrd reporting point shown on the other charts, and it did provide the co-ordinates of that point. Edition 5 is printed below and is the same in relevant respects as edition 1, which was used at the briefing.

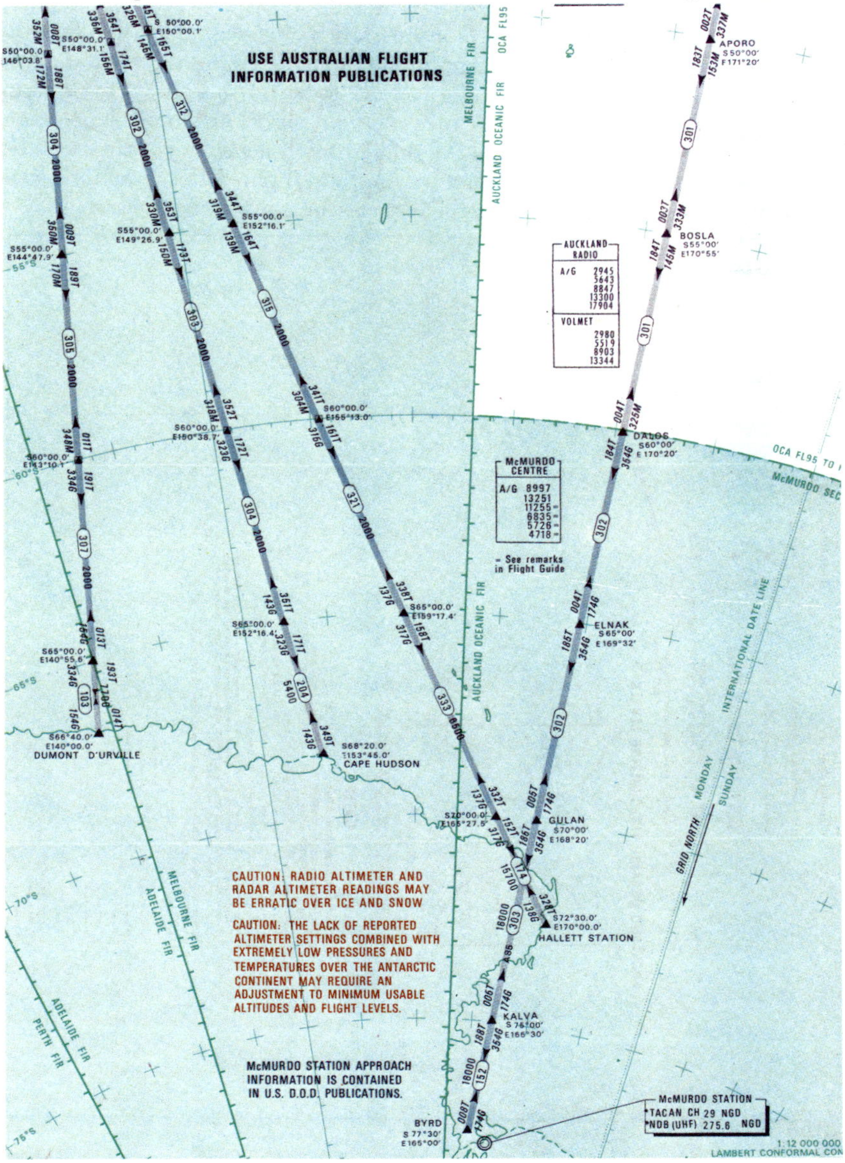

Fig. 10 Track and distance diagram indicating Erebus route: Annex J Not given to crew

Following what the airline told him, Ron Chippindale stated in his report that this diagram, printed as annex J to his report, and here reproduced as fig. 10, had been handed to the crew at the RCU briefing.

Had the crew plotted the co-ordinates for the McMurdo waypoint, which were marked on this diagram, on to a topographical map, they would have noted that the track ran over Mt Erebus and not down McMurdo Sound, since in the bottom right appears the latitude 166° 48'E, which is the latitude for Williams Field lying almost directly behind Mt Erebus on a flight path from Cape Hallett.

Evidence given at the Royal Commission made clear that the diagram was not distributed at that time, and that if it had been handed out at all, it must have been in the Antarctic envelope given to the crew by Captain Johnson shortly before the departure of their flight. Other evidence conflicted and Justice Mahon concluded that Annex J never formed part of the 1979 flight documents and was not on the fatal flight. Consequently it was unknown to the crew and it is irrelevant to the causes of the crash.

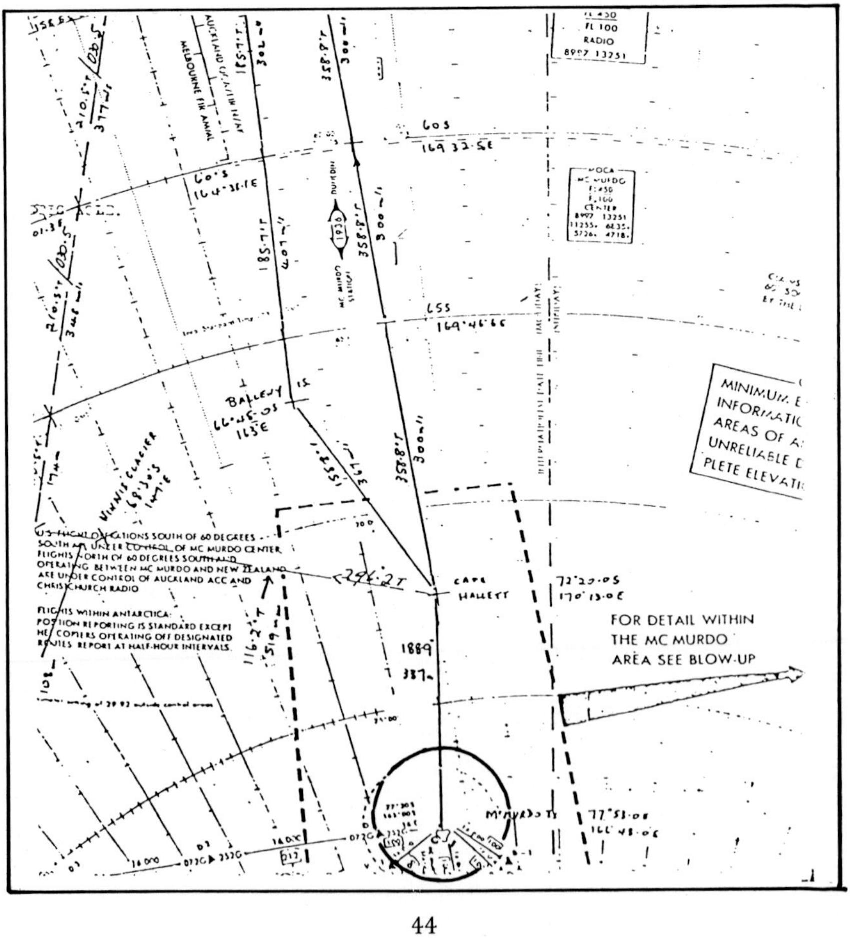

Fig.11 Track and distance diagram showing McMurdo route: Exhibit 164

Exhibit 164 of the Royal Commission Proceedings is here reproduced as fig. 11. At the bottom centre it shows the co-ordinates of the Byrd reporting point in the centre of McMurdo Sound, 165° 00'E 77°30'S. An arrow points south-east leading to another line drawn back to Cape Hallett with an arrow pointing north.

Justice Mahon decided that the evidence given to the Commission suggested that this diagram was included among the flight documents given by the airline to the crew on the fatal flight. It would suggest to them that the flight path ran down McMurdo Sound.

Another chart (fig. 9 on page 43) given to the crew showed the same co-ordinates as being those of the Byrd reporting point, and as being located in the centre of McMurdo Sound. This latter chart was the radio navigation chart RNC4, which was printed as Annex F to the Chippindale Report.

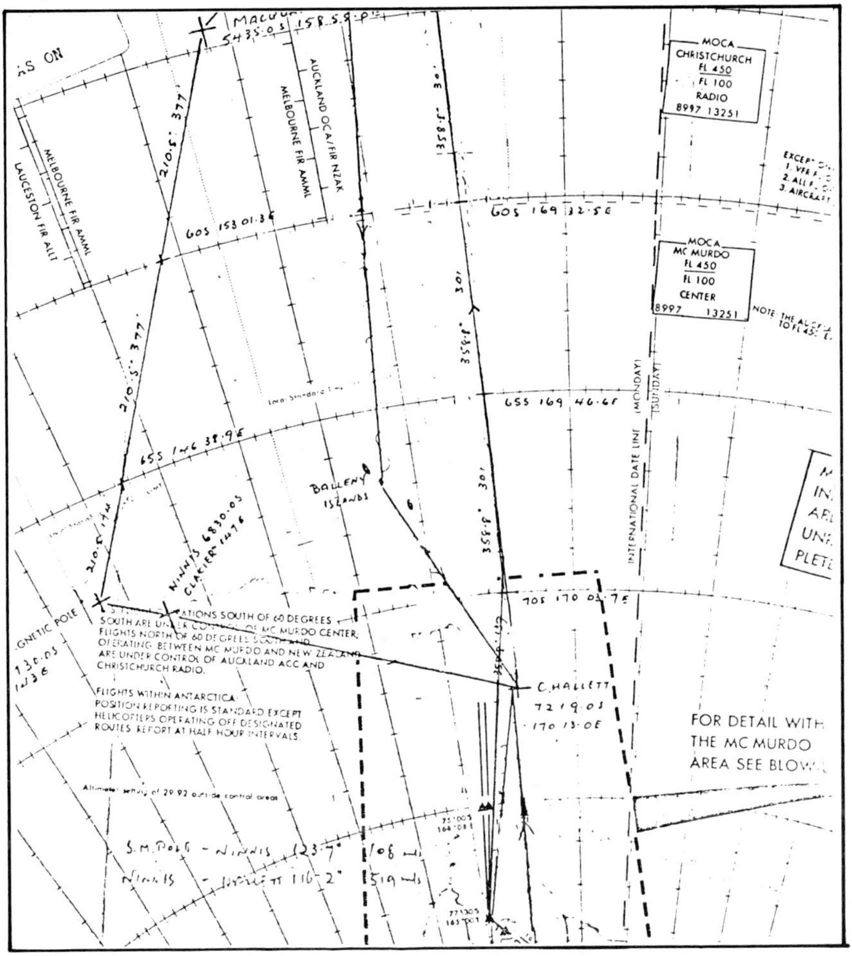

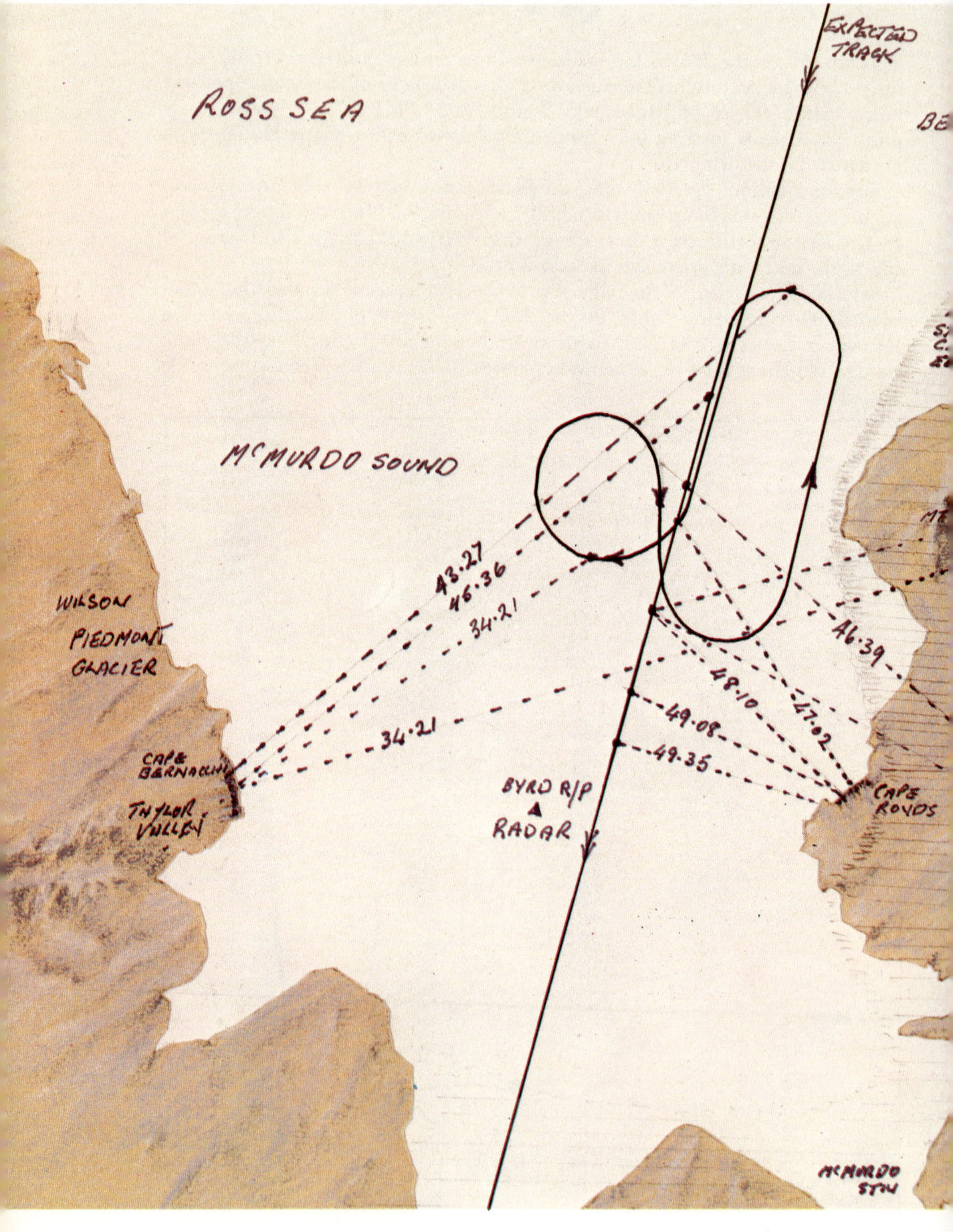

EXPECTED TRACK
ROSS SEA
McMURDO SOUND
WILSON PIEDMONT GLACIER
CAPE BERNACCHI
TAYLOR VALLEY
43.27
46.36
34.21
34.21
46.39
48.10
47.02
49.08
49.35
BYRD R/P
RADAR
CAPE ROYDS
McMURDO STN

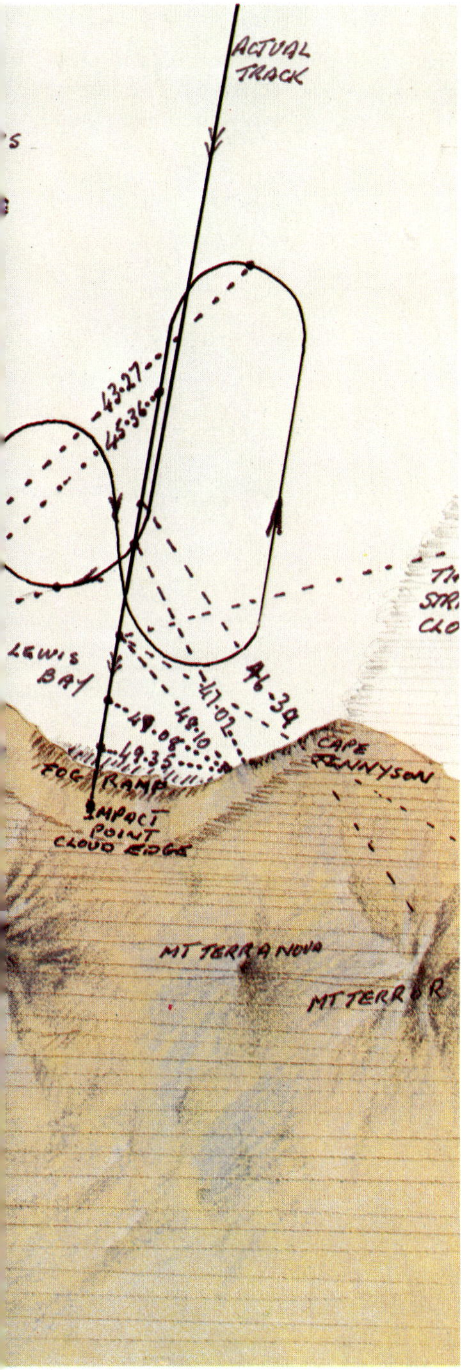

Fig. 12 Map showing bearings of sightings: Azimuth map

The significance of this map is fully explained on pages 49-60.

The technique used here to illustrate the flight path the crew believed they were on is to draw it on the left in McMurdo Sound. The actual flight path is illustrated by drawing it on the right in Lewis Bay.

These same facts are shown in fig. 14 on page 61 by a different technique. That figure shows only one flight path in Lewis Bay, but so as to illustrate the assumed path in McMurdo Sound, a shaded drawing of Ross Island is displaced to the right, to where the crew believed it lay.

For clarity the cloud formation present is not illustrated. A reconstruction of the upper and lower and the lower cloud formation appears overleaf in fig. 13.

Fig. 13 Cloud maps

These two maps are reduced prints of fig. 12 on pages 46 and 47 and they illustrate the cloud formation at the time of the descent orbits. The upper map shows both the high level and low level cloud, while the lower map shows only the lower level cloud.

The same cloud formation, but in cut-away profile form is illustrated on page 62 in fig. 15 headed descent, cloud and terrain profile.

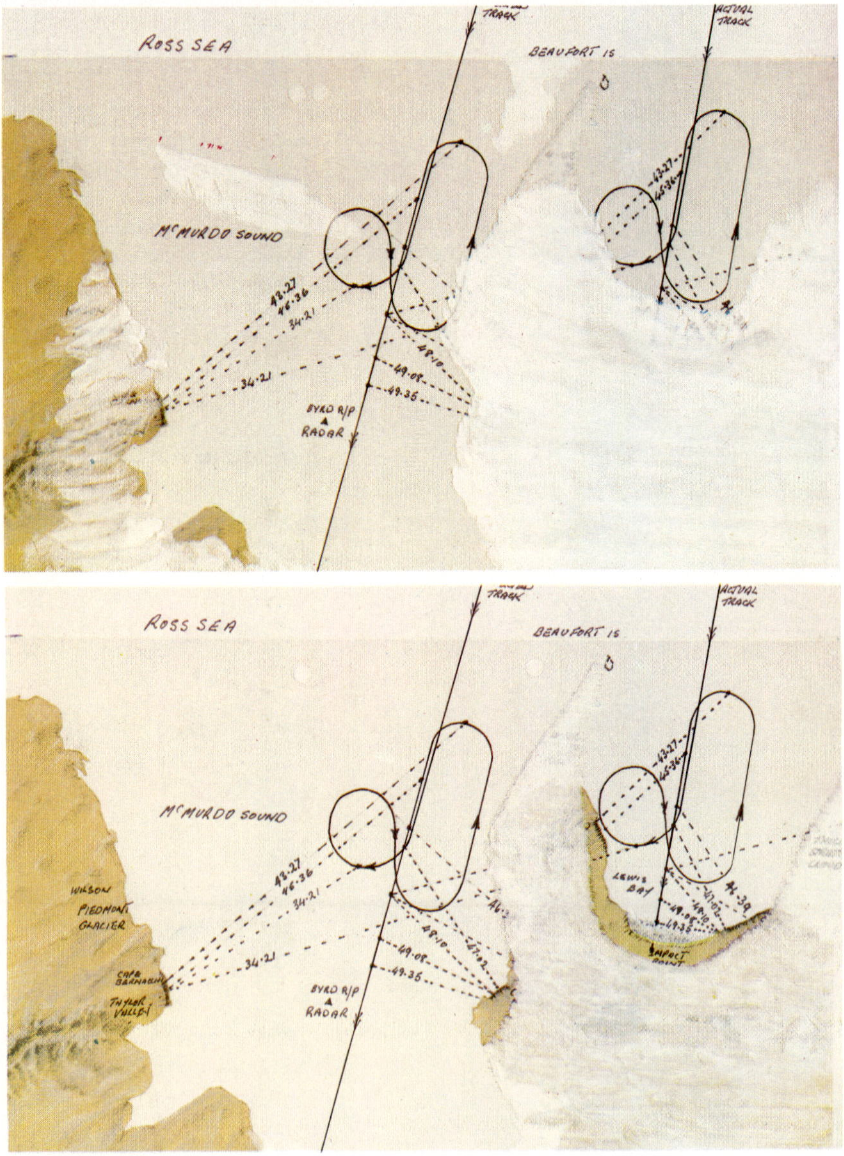

Fig. 12 Map showing bearings of sightings: Azimuth Map

Before one considers the significance of this map, there are some general issues which one should keep in mind.

I believe that Collins in the initial briefing would have gained the same impression that I did regarding the radar at McMurdo. I understood that because the McMurdo air traffic control used it for their own military aircraft, its approach function would be limited to 6000 feet and 40 miles. Military aircraft could feed into that, in the Byrd area, by use of their TACAN which is their equivalent of our VOR-DME set up for civil aircraft. They could commence their let-down below MSA ahead of getting to McMurdo and feed into the primary radar (approach radar) at Point Byrd. That is how the military route lines up.

For Air New Zealand, on the other hand, I believed that the initial approach let-down procedures fed into the NDB, with a cloud break procedure down to 6000 feet within the area of the approach radar for McMurdo ATC to continue on and monitor the approach of the aircraft and their own local traffic, such as helicopters.

It is probably correct to say that Collins would have been genuinely surprised while he was still at safety altitude approaching McMurdo to hear McMurdo ATC offer him a radar let-down at that altitude. (Clearly indicated by his tone of voice on CVR.)

He probably either did not understand their reasons or else he would have been surprised that they had the radar switched into the surveillance mode, which, even though it has blind spots overhead, would allow them to monitor his initial approach and probably allow him to come below MSA on approach.

However, during the next stage of the approach his views would have been altered. Even though he saw that his transponder was being interrogated, he still apparently did not get any radar direction. At that stage he would have reverted to his original belief, namely, probably believing that they were incapable of giving him a radar let-down above 6000 feet and outside a range of 40 miles. So his departure briefing to make a visual descent to 6000 feet was correct. I am sure that this belief precipitated his own decision to make the approach visually, separating himself from obstacles by visual means, while approaching visually to the area where he knew the final approach radar would certainly be able to pick him up. That is in the vicinity of the Byrd reporting point.

I think Collins believed when McMurdo came up with that offer of a radar let-down that he was getting a bonus issue, that the airline obviously did not know about. While he was probably delighted to avail himself of this facility, and he certainly said so in the transcript, the offer did not eventuate, even though he got closer and closer to McMurdo.

For some reason McMurdo ATC were having difficulty in picking him up, so he probably reverted to his original opinion that the radar was limited in its function to an approach radar, limited to 6000 feet and 40 miles, in a fixed direction, i.e. towards the Byrd reporting point.

Another point on the visual identification during the descent phase I think needs to be stressed. They did not treat this leg as being any different from any of the other legs, where the fulcrum or waypoints remained unaltered. As they were flying leg by leg, they were doing precisely as they were doing on their last leg. That is, they were looking out from a position indicated by the HSI (see fig. 39 on page 115) towards known objects, landmarks in each direction.

So they had the advantage of a running fix on the HSI. At the same time, I am quite certain they were monitoring that HSI, and its integrity, with their visual sightings of prominent objects as they were making their way along. Because there were no discrepancies noted right up until the final leg, I am sure they believed the final waypoint would also remain precisely where they believed it to be. The sightings that they made would have been made probably almost as casually as the previous ones at the safety altitude. Because, though they had departed safety altitude, they still believed that they had the full protection of more than the regulation requirements in visibility.

If anything had occurred on any previous leg that caused Collins to doubt the integrity of either his navigation equipment, or the loading of his computer, then there is no question that he would have taken various steps to ensure that the position of the aircraft, in terms of co-ordinates, was re-ascertained and re-plotted ahead for whatever was necessary for the exercise in hand. But, all Collins had on the previous legs was total re-assurance that the navigation equipment was working properly, his HSI was indicating a continuous running fix of complete validity, within a mile or so, and this was all tying in with the land masses that he could see and the continual fixes that he was actually using, even though he did not need them up to that time. So he would be lulled into a sense of false security, in that everything had ticked over up to that time. He had no reason to suspect that anyone would have interfered with the last leg.

As it happened, the co-incidental sightings of the land masses did not, in any way, cause him to believe that the last leg had been interfered with.

My analysis which follows of what was said by the crew shows where the aircraft was as each remark was made and, when the remark concerns a geographic position, it shows the direction the speaker was looking. My purpose is to demonstrate that nothing that we know the pilots saw is inconsistent with what they would be expecting to see during a flight up McMurdo Sound.

If I were proceeding in the aircraft's navigation mode along the navigation track, anywhere along the routes that we fly and I saw, as I expected to see, a physical feature located out to one side of the aircraft, in the expected azimuth (direction or bearing), at what appeared to be the expected distance from the aircraft; and all these points confirmed the location of the aircraft as indicated by the distance to run indicator displayed before me, then I personally would take what I had seen as a positive visual fix of the geographical location of the aircraft.

Jim Collins and his crew were receiving all these indications. If I had been in their position at that time, it is very likely that I would have believed that I had made the same positive visual fix. I would probably have been misled in the same respects as they were, and I myself may well have crashed on Mt Erebus.

I am convinced that it is only with the benefit of hindsight that others now put forward theoretical means which they allege Collins ought to have adopted at the time in order to ascertain that the geographical location of the aircraft was some 26 miles off the course which he believed it was on.

The first factor which deceived him was showing him a printout of the computer flight plan used for the previous 14 months, and then six hours before the flight altering the one critical leg of it by about 26 miles, without telling him the change had been made. His belief in the accuracy of the flight plan was reinforced by the passenger maps given to him, and by the track and distance diagram exhibit 164, which Judge Mahon held (upon the weight of the evidence) had been given to him, and also by the general impression given at the briefing.

The final factor which deceived him was that local conditions in the McMurdo area, which were unknown to him (and indeed unknown to me at the time of my flight), acting in conjunction with an altered flight path, caused him to make what he believed was a positive visual fix.

Unknown to any of them, the features were not of McMurdo Sound but of Lewis Bay, and straight ahead, concealed by the most insidious form of whiteout, namely sector whiteout, lay Mt Erebus.

When making their positive visual fix, how were Collins, Mulgrew, and the other crew in the loop deceived by observing physical features in the directions expected in order to confirm that the aircraft was located in McMurdo Sound?

The answer is provided by fig. 12 which shows on the left the flight path including the orbiting descent which they believed, on the evidence of the computer flight path and their observations, that they were following. On the right is the actual track which, unknown to any on the flight deck, the aircraft was flying.

At every point on their descent orbit they could see texture, that is contrasting terrain, in the direction they would expect to see it had they been in the left orbit pattern.

Confirmation that those on the flight deck were scanning for and observing such visual texture is provided by the transcript of the cockpit voice recorder, which recorded a number of references to geographic features made during the orbits.

The actual position of the aircraft during its orbit at the time when the comment is passed is shown by the intersection of the broken line with the solid line representing the orbit.

Each broken line represents the direction (in all probability) toward which the speaker was looking when he spoke, while the figures show the time the comment was passed.

It is easier to understand the diagram if we commence at 43.27 (43 minutes 27 seconds past midnight Greenwich mean time). Peter Mulgrew was actually looking from a point on the right-hand orbiting path toward Cape Bird. With the aircraft heading true west the texture observed would be in the direction of between ten o'clock and eleven o'clock. Such a direction is a bearing, which in navigation is more commonly called an azimuth.

At the time of 43.27 Mulgrew believed the aircraft was in the orbit on the left in McMurdo Sound. When heading true west and reading from map to terrain, Mulgrew would expect to see Cape Bernacchi lying on the direction of between ten o'clock and eleven o'clock. In fact, he did see texture in the direction he expected and, despite his experience, he identified it as Cape Bernacchi, even though in reality he was looking at Cape Bird from the orbit pattern on the right.

Had Mulgrew correctly identified Cape Bird, he would have known the aircraft was about 26 miles east of where he believed it to be, descending on to a disastrous course which would lead it to an inevitable collision with Mt Erebus.

At 43.27 Mulgrew said: 'There you go. There's some land ahead.'

Initially I thought this may have referred to Beaufort Island, but subsequently I thought about this a lot more, and examined the photographs taken by the passengers. I became convinced that at that point in the orbit they would be following normal airmanship procedures, so that they would all be looking in the direction to which they were turning.

At that stage in the holding pattern, the aircraft would be banked to the left at about 20-25 degrees angle of bank. Mulgrew is sitting directly behind Collins on the left-hand side of the aircraft. The moment the aircraft banks left, Mulgrew loses all vision from out of the right-hand side of the aircraft. His chances of seeing Beaufort Island, assuming that it was visible at that time, would be zero because he could not be seated in his seat and still look through the coaming, since the island would be out on his right. The crew, in operating their normal technique of conducting a turn, would be looking out in the direction that they are turning into. They would all be looking fairly hard around to their left, which is the line of azimuth that I have shown to the texture of Cape Bird. So they would be scanning that area, and they would not be scanning the area where Beaufort Island is located. In most cases the cockpit coaming would be obscuring their vision, though the co-pilot possibly may have been able to see it, but he would not have had any reason to be looking in that direction at that time.

The crew had never been to the area before, and if at any stage they saw Beaufort Island they may well have mistaken it for a break in the ice, or for Dunlop Island. But Mulgrew knew the area well, and had he ever sighted Beaufort Island, it is difficult to believe that he would not have immediately identified it as Beaufort Island. Since he knew the geography of the area so well, he would have known immediately that if they were flying down McMurdo Sound, as he believed, then Beaufort Island must have been to their east throughout. Sighting it to the west would have showed that they must instead be flying into Lewis Bay.

For Mulgrew to have identified Beaufort Island, and not to have immediately told Collins in the strongest manner that he was in Lewis Bay would have been suicidal and knowing Mulgrew as I did, I cannot believe that. I do not find credible the theory of Captain Kippenberger that Mulgrew and the aircrew were simultaneously afflicted by some mental or physical condition throughout the whole relevant period of flight.

Captain Gemmell testified to the Royal Commission that using the DC10 simulator, he reconstructed the stage of the flight where the aircraft turned left to recapture navigation track. He used a visual presentation for Rarotonga to represent Beaufort Island and he concluded that it was readily visible ahead from both the pilots' position and the observer's seat.

I do not think this evidence affects the validity of my interpretation. The point which he is really making refers to the early part of the left-hand turn into the intercept for the navigation track southbound. At that stage, what Gemmell says is quite correct, and an exercise in the simulator would not have proved much at all. At the moment they commence their turn to the left, then they open up the coaming area and expose Beaufort Island to a greater extent than it would have otherwise been visible. But the point I am making is that once they have gone about 90° through the turn, the coaming angle in fact closes off, because Beaufort Island then moves to the right-hand side of the cockpit. The time at which I understand that Mulgrew made his sighting is when they were approximately 90° through that turn. My opinion remains unaltered by Gemmell's evidence. I still believe that at that stage the coaming cut-off angle would have obscured the view, or at the least would have made it difficult for someone in the observer's seat to have seen it.

As the turn progressed toward the navigation capture, the chances of anyone on the flight deck seeing that island would be extremely remote. However, for the passengers on the right-hand side Beaufort Island, would become increasingly visible as it passed towards the rear and gained vertical dimensions (appeared to project higher from the ice) as they descended.

At that stage, without a doubt, all of the crew are going to be looking well out towards the left in the normal practice of airmanship, which is to look into the area towards which the aircraft is proceeding.

When the aircrew was initiating the first part of the left-hand turn, why, if Beaufort Island were unobstructed by cloud, was it not observed? At that time it would have been within the scan pattern of any members on the flight deck. However, it is my contention that either of two things may have prevented them from 'perceiving' Beaufort Island at that stage. The laminations of cloud, scattered patches still remaining below them, could easily have moved across Beaufort Island at precisely the time the crew were scanning that area. I think passenger photograph no. 2 (fig. 21 on page 68) shows the very piece of cloud which may have obscured the island for most of that run north on that leg. The cloud, even though it is over a fixed point on the surface — apart from wind movement, has a great deal of relative movement, as the aircraft moves its own position over the ground. The effect could be that while the piece of cloud was not located vertically above Beaufort Island, the movement of the aircraft during the orbiting pattern could easily have the effect, owing to parallax, of shifting that cloud into a position obscuring Beaufort Island, in terms of being in line with the aircraft. Passenger photograph no. 3 (fig. 22 on page 69) tends to confirm that this probably happened.

I remain of the belief that if the aircrew did obtain an unobscured view at any stage when they may have been scanning that area, then the similarities to the black holes in the ice could have deceived all except Mulgrew. See fig. 21 on page 68 and fig. 23 on page 69. Mulgrew himself must have been proceeding along the passenger cabin at the time so as to reach his observer seat on the flight deck and while doing so would not have had a view of the island. Even once he reached the flight deck, he would have to seat himself, buckle himself in, and put in all of his lanyards for the PA system. Only then would he commence to scan and by then I believe the aircraft would already be banked left through 90° of its turn. If only Mulgrew had arrived on the flight deck a couple of minutes earlier he just may have been able to see and recognise Beaufort Island. I have no doubt at all that had that been the case, then this particular Erebus disaster would not have occurred.

Having clarified the meaning of figure 12 by examining the 43.27 lines, let us revert to the earlier time of 34.21. When interpreting the comment made at that time, one must bear in mind that Captain Arthur Cooper is an airline pilot of great experience and he knew the speakers whose voices were recorded on the CVR, except for Peter Mulgrew. He was one of those who transcribed the CVR tape. I have relied upon his understanding of the 'loop' and his knowledge of the speakers for much of my interpretation of the CVR transcript. For instance, because the tone was confirmatory, I believe that on the balance of probability the correct reading of the words timed at 34.21 GMT is 'There's Wilson'.

On the other hand, Ron Chippindale, without a proper understanding of the loop as an airline pilot, and without experience of the speaker's voice, believed the words supported his opinion that the crew must have been expressing uncertainty, and that what was said was 'Where's Wilson?'. Chippindale included the words in his interim report, but he omitted them from his final report.

At 45.36 Mulgrew says: 'Taylor on the right now'. That is from the American transcript, the accuracy of which both Justice Mahon and I prefer to that of Ron Chippindale's version.

Here Mulgrew believes he is looking toward the Taylor Valley on the right of the aircraft, shown on the left-hand flight path while in fact he is looking at the texture beneath Mt Bird, shown in the actual flight path on the right. Once again, the direction or azimuth is the same that he would expect if he were looking toward the Taylor Valley from the position he believed he was in. He can now see to the right as the aircraft is no longer banked left, so his line of sight is not now blocked by the cockpit coaming.

At 46.39 Brooks, as part of the crew loop, asks: 'Where's Erebus in relation to us at the moment?' Note in particular that the question is asked by the engineer, sitting toward the rear of the cockpit, not by either the pilot or co-pilot.

The reply, as recorded in the American transcript with the mileage added to by Chippindale's transcript is 'Left about 24 or 25 miles... about eleven o'clock... yep'. There is a sound of paper rustling, which realistically can only mean that the crew are handling the map which they are reading from.

Confirmation is provided some 33 seconds later when the explanation is given: 'Didn't want to block the window too completely with it'. The rustling most probably represented the refolding of one of the large maps of the route owned by Collins which he used to plan his flight and which were missing from his home after the accident. The references to rustling paper and not wanting to block the window are recorded in the American transcript, but Ron Chippindale omitted them from his version. Chippindale implies in his press release that Collins had no such maps of his own and that the evidence given that he did possess and did work upon such maps was invented. For details see note 2 on page 327. I am told by Captain Cooper, who was one of those who prepared the American transcript, that these remarks were made in confirmatory tones, indicating no concern or anxiety. Hence I do not attach significance to subsequent comments indicating some uncertainty.

Peter Mulgrew then says: 'Yes - - - no, no, I really don't know.' But if that did represent any momentary uncertainty as to position, that was dispelled seconds later when, at 47.02 he said: 'That's the edge'.

By 'the edge' I believe he is referring to Cape Royds, which is the most threatening edge to their track.

If you take the same direction from the cockpit as Cape Royds lies from their believed track on the left, and trace along the direction or bearing from the true position of the aircraft shown on the right, then you arrive at an actual line of sight toward the ice cliffs west of Cape Tennyson.

To the west of that again lies the fog ramp which creates the illusion to the observer that he is looking along a flat ice plateau. To my mind Mulgrew clearly believed that what he 'perceived' as that plateau was McMurdo Sound, located to the west of Cape Royds, exactly as he would expect to see it from the position he believed he was in within McMurdo Sound.

At 48.10 Mulgrew says: 'I reckon Bird's through here and Ross Island there. Erebus should be here.' Once again, we have it on the authority of Captain Cooper, that he spoke in a confirmatory tone.

Mulgrew is scanning left of the aircraft to locate and identify the high ground of Ross Island. I am sure he was pointing at the map and pointing to the features on the left of the aircraft. At that time the higher features of Ross Island were covered by cloud, so that when Mulgrew is pointing back along the other line marked 48.10, while he is in truth pointing into a massive cloud marked in fig. 12 on page 47 thick stratus cloud from the right-hand flight path, he believes that he is pointing back, along the same bearing, from his believed position on the left-hand flight path, towards where Mt Bird should lie.

Collins has been listening to all this, checking from the map that the landmarks are on the correct azimuths to verify their position, and all the time cross-checking his position on the distance to run indicator in front of him.

I know this from the evidence of his bringing his notebook and atlas to the briefing, from the evidence of his family, from the evidence I obtained from his friend in Los Angeles, and from the evidence that despite flying VMC he continued to bring his aircraft back on to its navigation track. I know what a methodical and thorough pilot he was, and unless he had pre-plotted his flight path so that he would know his position at all times merely by looking at his distance to run, he would never have reverted to navigation track even during a descent orbit. It is clear to me that in accordance with a plan carefully made in advance he was sensibly using navigation track as a back-up system to his flying visually, in effect he was flying with belt and braces caution.

Collins had been told by the airline, wrongly as it turned out, that the only ground-based navigation aid, the non-directional beacon, which would have given him precise bearings, had been withdrawn. So, in order to compensate for this lack, he had substituted the technique of flying visually in conjunction with flying on what was, to his knowledge, the only instrument which was available to him, namely on his pre-plotted navigation track.

I know from Captain Cooper that the first demur made in the crew loop was at 48.55, when Collins asked Cassin: 'What's - um - have we got them on the tower?'

Captain Cooper assures me that this was the first tonal indication from anyone in the crew that concern was starting to be felt.

They have now proceeded beyond the area which they have swept visually and so Collins is giving the first demur. Captain Cooper could tell by the tone in which Collins spoke that this is the first demure into the crew loop. This is the first indication that they are not going to achieve their objective of enabling the passengers to take good photographs. By that I do not mean that the crew are concerned for the safety of the aeroplane; they are not. He is simply saying that this exercise which we are trying to do is not going to be productive. I cannot see the huts out here on the left round the edge of Cape Royds, and I should be able to see some. I have not been picked up by the radar. What do you guys think? They are not the least bit concerned that they are going to hit anything, because they are all looking at this long flat white plateau ahead, which, they know, goes well past the Dailey Islands. There is no question of any obstruction ahead of them.

Although Collins is not concerned about terrain obstruction, he is saying that something is not jelling here, we are not getting some sighting of these buildings, or any of the things which I would have expected would be coming into view.

Cassin replies at 48.59: 'No - - - I'll try again', being his confirmatory demur put into the loop.

Nine seconds after Cassin's reply, Mulgrew says at 49.08: 'Looks like the edge of Ross Island there', said in a confirmatory tone, thus confirming the existence of a position as expected. Now, if Mulgrew had not believed he was pointing to what most likely was Cape Royds, but instead had correctly identified the texture as Cape Tennyson, with my knowledge of Peter Mulgrew I have no doubt at all that his tonal input would have been quite explosive, since he would have instantly recognised that they were all within seconds from death.

Peter Mulgrew had been listening to the conversation in the loop, and following it on the map, while scanning with his extra knowledge of what the terrain looks like. He is looking under the top cloud and clear of the bottom cloud layer. That edge is visible to him, anyway. But in fact, we know that what he is looking at is really the edge of Cape Tennyson. That azimuth, again, is exactly right for what you would have expected. Mulgrew is looking hard around to his left; he sees what he believes is Ross Island going safely by on his left side. So he is virtually reassuring the crew that threatening terrain is no problem. Again, they see the plateau, they know that Ross Island is passing on the left, concealed in a big wad of cloud. When remarking on the edge Mulgrew's confirmatory tone merely indicates that things are as expected.

Reading from map to terrain means that they are following down their own map with what they believe is the proposed track, so they are looking along that with the distance remaining on the HSI indicator to guide them and show them what position along the track line they would be in. They have a pictorial, or plan representation. The HSI indicator shows you are on track by means of a little bar. There is a plan view of the aeroplane and sitting superimposed on it is a floating bar. If you are dead on track, that bar is centred exactly with the miniature aircraft. The distance to run to the next programmed waypoint is displayed upon the same indicator. See pages 114 to 116.

The reference to 26 miles north (at 49.30) reveals they are confirming with the HSI positions which they have established visually. Although he is not depending on that for navigation at all, because he is using visual separation, it happens to be a convenient way to proceed to the Dailey Islands waypoint, and to proceed past, and inside the Byrd radar reporting point. What better way, assuming you possess all the visual protection that you need, is there to come past the Byrd reporting point than to do so on the nav track which you know takes you past that point at a distance to run of 23 miles? So the pilot knows that when he reaches 26 miles he has only three more miles to run in which to pick up the radar. He has several options, he is clear of cloud, he can turn right, because he knows he has plenty of distance between him and the Victoria Land coast, or he can climb straight ahead and report he is overshooting to the controller, or he can continue to run inbound as he is doing remaining clear of cloud, expecting that eventually the radar will pick him up. But that is obviously not his intention. I would think his intention is to turn right, and back track along the same route, and then proceed visually up into the dry valley area.

It is hard to know precisely what his plan was except to stay in visual contact.

Sixteen seconds after Mulgrew's reassurance, engineer officer Brooks at 49.24 says: 'I don't like this'. What he says represents the first verbal demur as opposed to the first tonal demur made by Collins. In other words it is the completion of the total fail safe voting loop initiated by Collins. It is an expression that he shares the concern of his captain; but there is still no suggestion of any urgency, or of any action required.

Brooks is seated behind the pilots, so he has lost sight of the textured cliffs both to right and to left, and in doing so he has passed from sector whiteout into full whiteout. There is no doubt, as I understand the phenomenon of full whiteout, that once you are subject to it, even though you do not know what the phenomenon is, you are likely to realise that something strange is happening.

Being in full whiteout creates, so it is said, the effect of being inside a big milk bottle. I think it is this effect upon him which caused him to say: 'I don't like this'.

But Collins and Cassin, scanning from the pilot's seats, still have the sightings of the textured cliffs on each side. It follows that they have only sector whiteout ahead of them and they are not yet in full whiteout, and thus do not share Brooks' concern at the phenomenon. They are not yet inside the milk bottle with Brooks.

Once the pilots finally lost sight of texture and so entered full whiteout, impact was only seconds away, and no action which they could have taken could have saved the aircraft. Of course, the airline had never explained to them what whiteout really is. Instead, they had been misleadingly told that it was the effect created by snow showers causing loss of transparency or visibility. There were no snow showers involved in this accident.

Immediately Collins asks Cassin: 'Have you got anything from him?' to which Cassin answers 'no', and five seconds later at 49.30, Collins says: 'We're 26 miles north. We'll have to climb out of this'.

This is the captain responding to his own initial tonal demure into the loop and the subsequent crew inputs, including Brooks' remarks, into the crew loop.

Three seconds later, at 49.33 Cassin feeds into the loop that: 'It's clear on the right'. Cassin, who is sitting in the right-hand seat can still see texture, which he still believes is in the distance on the other side of McMurdo Sound. He is indicating that it is clear to execute a right turn pursuant perhaps to pre-descent briefing.

But Collins who is on the left has lost sight of the texture to the right, and on his right, unknown to him, he has entered sector whiteout. Thus his input back into the loop is the question: 'Is it?' Cassin replies into the loop 'Yep.'

While the crew are engaged in their crew loop inputs and deciding what action to take in accordance with possibilities discussed at briefing and in view of visual sightings, Mulgrew is continuing to scan so as to confirm landmarks to the crew and two seconds later, at 49.35, he adds into the loop: 'You can see (Ross Island)'. Still there is no dramatic tone to his voice; he is merely indicating Ross Island which is on the left, it is under his observation, and they are passing it safely.

Chippindale has added in his transcript the words 'Right. Fine!' If these words were spoken at all, and their omission from the American transcript suggests that they were not spoken, then 'Right' can only mean, 'am I right?' If, on the other hand, the phrase were taken to mean that you can see Ross Island to the right of the aircraft, that is, to the west, Mulgrew would be saying Ross Island is visible to the right, therefore we must be in Lewis Bay, at 1500 feet, heading for Erebus, seconds away, fine!

The only other meaning is equally far-fetched, namely that Mulgrew believed they were flying to the east of Cape Crozier, which is on the eastern edge of Ross Island, that is flying toward the Ross Ice Shelf. Such a theory would run contrary to all the geographical references previously made.

Cassin, on the right, is still scanning right to provide visual clearance for a turn right and at 49.38 he says into the loop: 'You're clear to turn right'. But Collins on the left still has no texture remaining on the right, is still in sector whiteout on that side and his definite input into the loop is 'No, negative.' Yet again, Collins is demonstrating his cautious airmanship. His co-pilot on the right is still putting into the loop that it is clear to turn right possibly in order to execute what may have been planned briefing procedure to be taken in the event of failing to make contact by the Byrd reporting point.

But when he looks right himself, Collins can no longer see texture and even though he has never been briefed on true whiteout, let alone sector whiteout, he is not prepared to turn in that direction. Despite his flying on navigation track, he is plainly strictly following visual flight rules, not knowing he is the unwitting victim of the illusions of polar flight.

Just as Cassin sees texture on the right, which he believes is Cape Bernacchi, so Collins on the left can still see texture to the left which he believes is Cape Royds, but which in reality is Cape Tennyson, so that even if he had been briefed on whiteout in general, he would not have suspected he was in sector whiteout.

Cassin again repeats into the loop still at 49.38 (twelve seconds before impact: 'No high ground if you do a 180'.

Captain Gemmell testified that this remark showed that Cassin could not have identified the coastline to the west as being in the Cape Bernacchi region, since the remark indicated Cassin thought the area was clear to turn right. But Gemmell's evidence is contrary to the facts. Supposing Cassin had identified the land to be around Cape Bernacchi, and I believe he had done so, then Collins would have a distance of about 20 miles to the right in which to turn, and this is easily within the capacity of a DC10 and provides a considerable safety margin. Gemmell's evidence to the contrary is absurd.

Both Cassin and Collins are about to lose sight of texture, but before that happens the ground proximity warning indicator comes on six seconds before impact. To their knowledge the pilots are flying 1500 feet above the flat ice of McMurdo Sound, each has been looking at what he had every reason to believe was one side of the Sound. They must have believed the warning was a false alarm.

Nevertheless they went into the correct emergency procedure at least as fast as the average crew being tested in a simulator. And a crew in a simulator is expecting the unexpected.

Two seconds after the warning, that is four seconds before impact, the crew suddenly applied the nose up elevator, which raised the nose up from 5 deg to 10.9 deg some 2.5 seconds before impact, causing the aircraft to be climbing at ten feet per second at impact.

The flight recorder showed that just prior to impact number 1 engine had spooled up to 94% of maximum power. There is no reason to suppose the number 2 and number 3 engines were not similarly spooled up, but limitations of the flight recorder preclude proof.

Collins' request for go round power was an instruction to Brooks to maintain engine power sufficiently below 100% as to avoid possible engine damage.

The case, preposterous in itself, propounded by the Civil Aviation Division that Collins, Cassin, Brooks and Mulgrew had all simultaneously been afflicted by some mental or physical illness during the whole approach to impact is rendered even more absurd by the performance of the aircrew when the warning sounded.

Fig. 14 Ross Island Displaced

This shows Ross Island displaced to where the aircrew believed it was located in relation to their aircraft. The figure combines several presentations.

It is formed firstly by a reprint of fig. 19 on page 67 being the passenger photo location map which shows by numbered broken lines where the passenger photos on pages 68 - 73 were taken and the direction they were facing. Fig. 19 should be studied before this present fig. 14.

Secondly, each solid unbroken arrow marked with a time represents the location of the aircraft on its flight path at that time, while the head of the arrow points to where the speaker who made the comment at that time was probably looking. To that extent fig. 14 is a representation of the azimuth map, fig. 12 on pages 46 and 47.

The difference is that the azimuth map showed Ross Island as it is actually located, the right-hand flight path showed the actual flight path which flight 901 followed in Lewis Bay, and the left-hand flight path showed where the aircrew believed they were located in relation to Ross Island.

This displacement map shows the same facts as the azimuth map but by a different technique. Only one flight path is marked. The unshaded outline of Ross Island indicates where the island was actually positioned in relation to the aircraft's flight path in Lewis Bay.

The shaded outline of Ross Island indicates the island displaced to the position in which the aircrew believed it was located in relation to their flight path.

Thus at 48.10 Mulgrew, sitting on the left of the flight deck behind Collins, is looking left towards what he believes is Ross Island, almost certainly back along the azimuth line marked 48.10 in the direction where Cape Bird would be on the displaced Ross Island.

'...and Ross Island there' probably pointing along the centre line of the three 48.10 azimuth lines to where the centre of the displaced island lies.

'...Erebus should be here' — Mulgrew must surely be pointing along the 48.10 azimuth line towards where Erebus should be on the displaced Ross Island map, but where in reality Mt Terror is located on the actual Ross Island map.

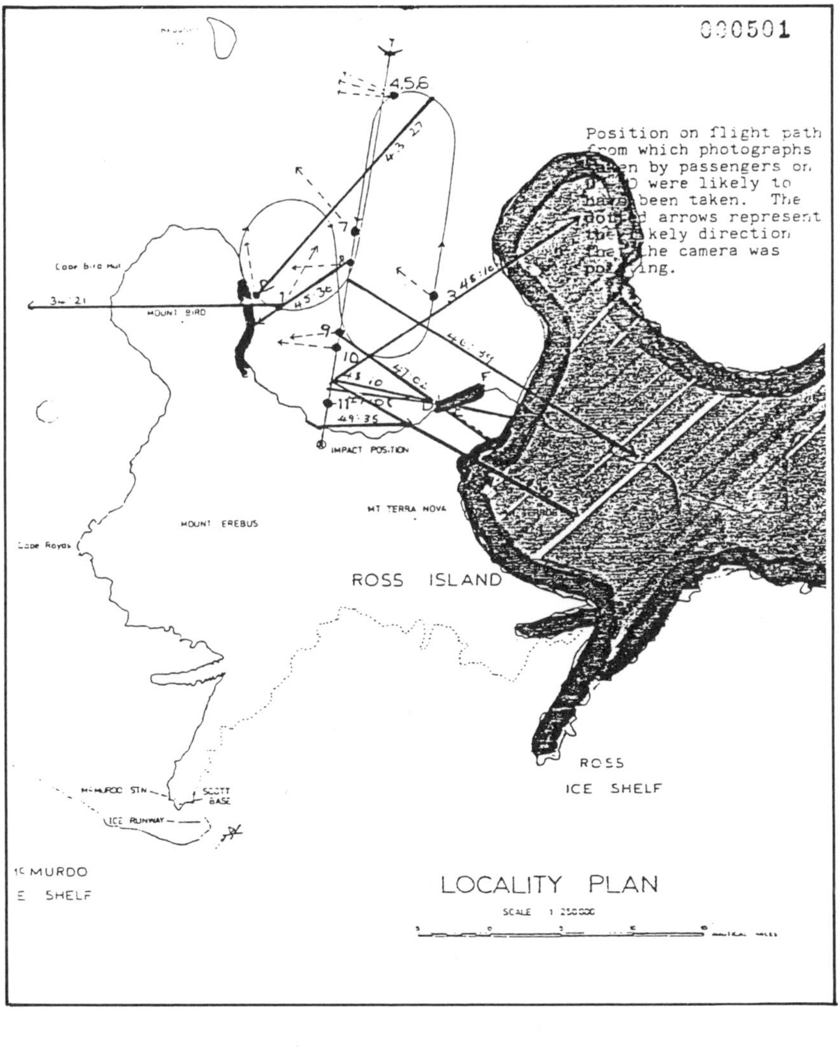

Fig. 15 Descent, cloud and terrain profile

A meteorological reconstruction of the cloud formation at the time of the crash presented to the Royal Commission confirmed that the aircraft flew clear of cloud throughout its flight. The descent, cloud and terrain profile fig. 15 which follows was prepared from that material, which was then imposed upon the descent and terrain profile in the Chippindale Report. It provides a cut away vertical section of heights of cloud.

The figures along the bottom of the diagram — 20, 40, 60 — show the distance in nautical miles from McMurdo Centre. From them one can estimate the position of the aircraft at the time Collins said: 'We're 26 miles north. We'll have to climb out of this.'

The same cloud formation, which is illustrated in fig. 15 in section form, is displayed in map form in fig. 13.

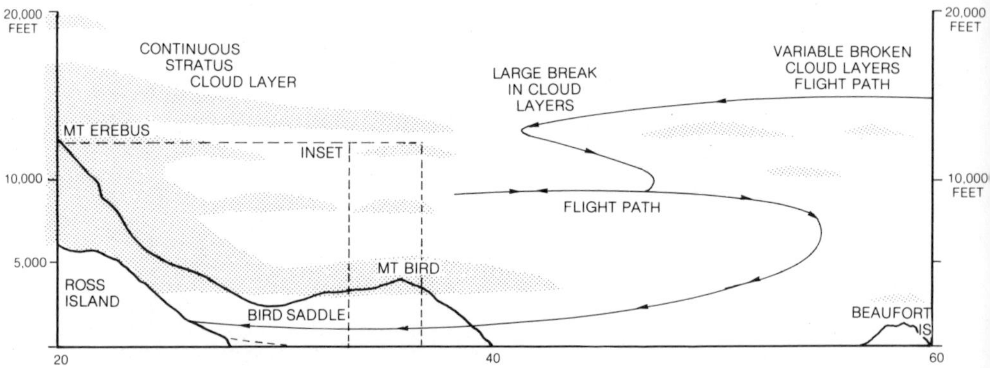

DESCENT, CLOUD AND TERRAIN PROFILE

Fig. 16 Inset showing whiteout factors present

The dotted line marked *inset* on fig. 15 shows the location on that figure of the larger scale fig. 16. The latter figure illustrates the factors which created the optical illusion of polar flight known as *sector* (meaning localised in one area) *whiteout.* These factors are demonstrated by photographs which follow in succeeding pages. Their cumulative effect in transforming the appearance of the entrance to Lewis Bay so as to make it appear to be the entrance to McMurdo Sound is illustrated in fig. 34 and 35 on pages 78 and 79.

The factors which created sector whiteout were firstly the fog ramp which obscured the texture of the ice cliffs of Ross Island in Lewis Bay. The ramp was caused by the back eddy of wind causing sea fog to form and to rise over the ice cliffs. See figs. 31 and 32 on page 75.

The second factor was the white shadowless expanse of the slopes of Mt Erebus with the sun shining on to it from behind the observers in the aircraft to the north.

The final factor which created sector whiteout was the base of the reflective stratus cloud, which caused multiple reflections of light upwards and downwards between it and the white surface beneath the cloud. The same sector whiteout illusion which deceived Collins, Mulgrew and the crew presented itself a year later to Justice Mahon and the series of three photographs which record whiteout developing are printed below on page 77.

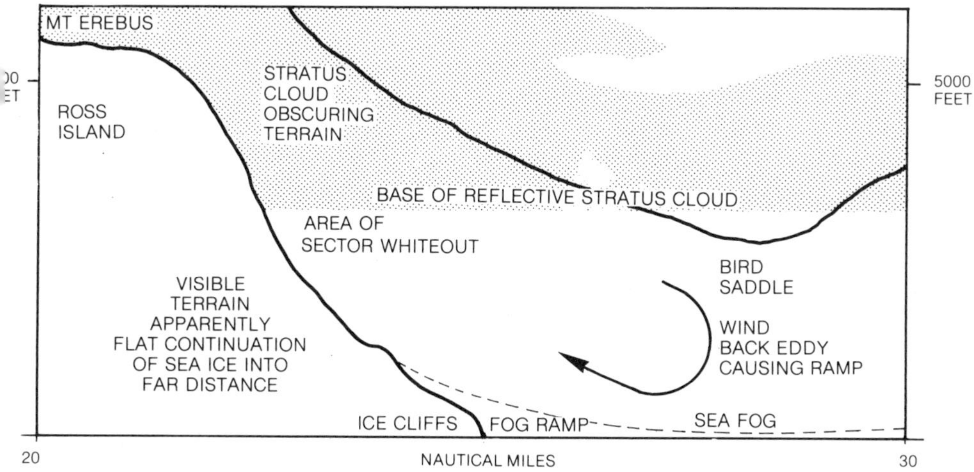

Fig. 17 Aircraft not flying in cloud

It was assumed by some that the aircraft was flying in cloud. Such an assumption is incorrect, passenger photographs, taken shortly before impact, show sun shining in through cabin windows, while other passenger photographs looking out through the windows show clear views unobstructed by cloud. Confirmation that the aircraft was not in cloud is provided by Collins and Cassin making at least 13 recorded comments to the effect that the aircraft was in VMC, and they did not know that anyone would ever play the voice recorder back.

The CAD case was that the crew were not subject to whiteout; instead CAD said that Collins correctly perceived cloud, and possibly fog, ahead to be cloud and fog and deliberately flew into it. Captain Kippenberger, the director of CAD, believed that to have flown in this way, both Collins and Cassin must have been suffering from some physical or mental condition.

Chippindale said he believed that Collins, at the end of his descent to 2000 feet, was confronted with a very vague area in front of him which he may or may not have believed was cloud, and that the aircraft collided with cloud-covered high ground.

Supported by much other evidence, I argued before the Royal Commission that the pilots were flying in clear air and that sector whiteout rendered Erebus invisible. My view was accepted in the Mahon report as correct.

Chippindale said that if indeed the crew were confronted by a whiteout, this situation only existed ahead of them on the coastal slopes of Ross Island.

But Chippindale goes on to add that as the aircraft had completed two orbits, one to the left and one to the right, it would have been apparent that the sea ice and other features produced a pattern of visual cues visible during each turn which were not apparent directly ahead when on the nav track heading. Evidently he means that, though whiteout had not been explained to them, the crew should have been able to detect sector whiteout ahead since the plain white expanse contrasted with the broken sea ice previously encountered in the Ross Sea. Such an argument is patently wrong, since it overlooks the fact that the crew believed the view along the nav track heading represented McMurdo Sound, which they would have expected to present a flat, featureless, unbroken expanse. As nowhere else in the area would produce such a view, what they saw ahead would provide them with further visual confirmation of their location and that McMurdo Sound was clearly visible ahead.

Their belief was mistaken only because the sector whiteout phenomenon made the close slopes of Mt Erebus appear to be McMurdo Sound continuing into the distance and merging with the further flat expanse of the Ross Ice Shelf.

Fig. 18 Descent and terrain profile: below

The letters in green on this profile or section view correspond to the same letters in green on fig. 19. By comparing the two figures the altitude of the aircraft may be estimated on the fig. 19 map at the various points of its descent orbits. Profile fig. 18 also reveals the relationship between fig. 19 on page 67 and the cloud profile fig. 15 on page 62.

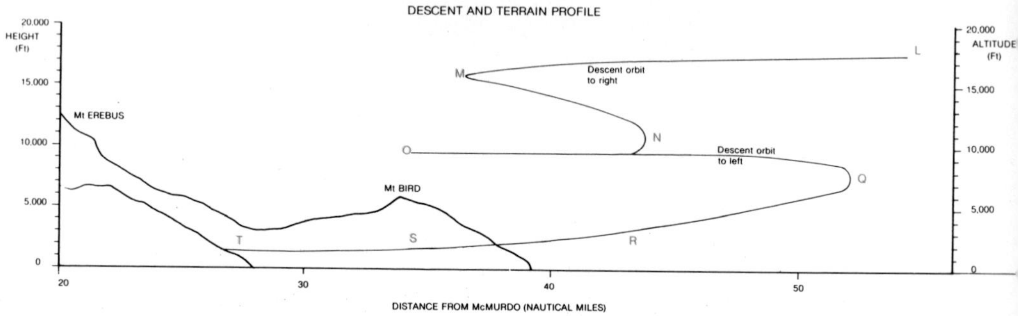

Fig. 19 Passenger photograph map: opposite page

During the figure of eight descent orbit and the subsequent approach to impact on Erebus the passengers remained unconcerned by what they saw out of the windows, and continued taking photographs. A selection of ten of these pictures was produced to the Royal Commission by the meteorologist M.R. Sinclair who, at the time of the accident, was at Vanda Station in the Wright Valley some 130 km to the west of the crash site. He analysed the pictures and fig. 19 being the map facing this page represents his revised opinion as to the location of the aircraft at the time each numbered photograph was taken, while the arrow shows the direction the camera was pointed.

A selection of these photographs follows, each of them carrying the number shown on the map. The photographs without numbers are not from this series, and the latter have been included to illustrate further matters.

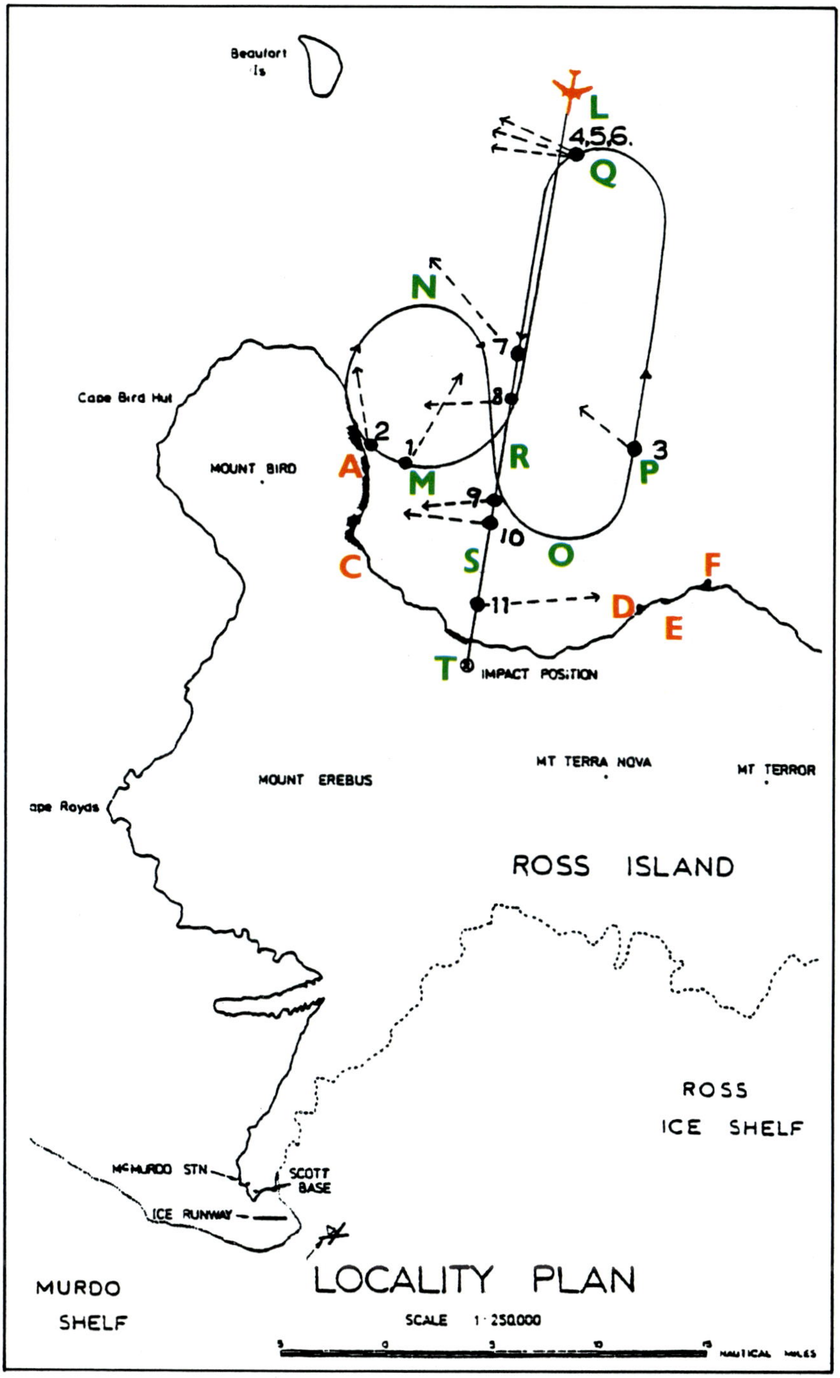
Beaufort Is
L
4,5,6.
Q
N
7
8
Cape Bird Hut
2
1
R
3
P
MOUNT BIRD
A
M
9
10
O
S
C
F
11
D
E
T
IMPACT POSITION
MOUNT EREBUS
MT TERRA NOVA
MT TERROR
ape Royds
ROSS ISLAND
ROSS ICE SHELF
McMURDO STN
SCOTT BASE
ICE RUNWAY
MURDO SHELF
LOCALITY PLAN
SCALE 1 : 250,000
5 0 5 10 15 NAUTICAL MILES

Passenger photos captions

Fig. 20 Photo 1. The aircraft is executing a turn to the right above cloud. The photo is taken out to the right into the clear area that the aircraft is about to descend into. Flimsy whisps of cloud are below and below that again is the clearly defined sea ice.

Fig. 21 Photo 2. A right-hand turn with Beaufort Island visible looking forward. This picture also shows the large breaks in the cloud layers and the illuminated surface of the pack ice visible below together with cloud shadows on the pack ice. Significantly, you can see the finger of cloud extending out from the end of Cape Bird towards Beaufort Island. As the aircraft progresses further to the left, the effect would be for this cloud at the intermediate altitudes to move across and mask off Beaufort Island from their view, until they progressed well to the north into the large cloud break that can be seen at the end of the finger. Another significant feature is the appearance, once you have recognised Beaufort Island, of what appears to be another land mass almost concealed below the edge of the cloud on the bottom left. While it has every appearance of being a land mass, just as much so as Beaufort Island, it is in fact a break in the sea ice. It serves to reinforce my view that the crew, having observed some of these heavy black breaks in the sea ice, could well have believed Beaufort Island to be another such break, or an extension of that one. The aircraft is at an altitude of approximately 16,000 feet.

Fig. 22 Photo 3. Beaufort Island is visible but it is not clear. The aircraft was at 9500 feet, and is likely to have been near position 3 on the north-bound leg. This is the view which the island would present when it was unobscured by the cockpit coaming and within the scan pattern of the crew. Mulgrew was not on the flight deck. The remainder of the crew were not expecting to see Beaufort Island in that location. If they saw anything they would either believe it was another break in the sea ice, which is my belief, or they would identify it as Dunlop Island, as G.W. Shannon suggested.

The aircraft is below the main cloud layer.

Fig. 23 An 'Island' in the Ross Sea?

This 'island' is in fact no island at all. It is simply a break in the sea ice. I include the photograph to demonstrate that should a member of the crew, other than Peter Mulgrew, have seen Beaufort Island, he may well have 'perceived' it as yet another break in the sea ice. The photographs of Beaufort Island which were taken on the flight were photographed from the passenger cabin. If views were visible to the crew, they may have been visible at different times, from different angles and under subtly different lighting conditions. In any case they would be from greater altitude with much less vertical development apparent than in the passenger photos.

Fig. 24. This photograph, taken from the wreckage shows Beaufort Island in the background, to the west, visible above left of the helicopter. The entrance to McMurdo Sound is further west on the far side of Beaufort Island. Peter Mulgrew knew the McMurdo area intimately, and possessed an excellent grasp of geographical location when viewed from the air. Had Mulgrew sighted Beaufort Island he would not have mistaken it for anything but what it was, Beaufort Island. He would have instantly realised that since it was on the right, to the west of the aircraft, the aircraft was not entering McMurdo Sound as it should have been but was instead flying at low level towards Lewis Bay, on a collision course with Mt Erebus. There is no way that Mulgrew, with the experience and intelligence which he possessed, could have made any mistake over this.

Fig. 25 Photo 4. The aircraft is rolling out of its left-hand turn with Beaufort Island starting to pass toward the rear on the right-hand side. The black breaks in the sea ice and Beaufort Island itself show a marked similarity. The crew are looking almost 180 degrees away from Beaufort Island, and the island at this stage is starting to get some considerable vertical elevation because the aircraft has now descended to approximately 7000 feet.

The main cloud base can be seen above the aircraft with the large hole showing up clearly out at the wing tip, and extending right forward. Visibility is excellent. Shadows and highlights on the sea ice indicate a good deal of broken cloud further to the north and in the direction to which they were progressing at that stage.

The aircraft's low altitude makes the island stand out from the sea ice much more prominently than it did from a high altitude when it was within the scan pattern of the pilots. Compare this photo, showing a passenger view of the island with photo 3 fig. 22 on the preceding page which shows how it appeared to the aircrew.

Fig. 26 Photo 8. Taken to the west, towards the Victoria Land coast, showing the well-defined cloud base of the high cloud and some layers of cloud. One layer is apparently at about the same altitude as the aircraft, about 2000 feet. A lower layer can be seen, apparently increasing towards Cape Bird.

Fig. 27 Photo 10. A view taken at the same height and toward the same direction as photo 9, but from about 1 mile south of 9. In the right of both photos 9 and 10 directly above 'A' sunlight can be seen shining on Cape Bird peninsula. Some cloud shadows are visible on the Lewis Bay sea ice. The cloud base above point 'A' of the low level stratus is probably above 2700 feet. It shows that the pass between Mt Erebus and Mt Bird is free of cloud, indicating that the main cloud base is above 2500 feet, and possibly just below 2700 feet.

Very significantly it shows textured cliffs of Cape Bird on the right-hand side with the stratus cutting and forming an artificial horizon across the slopes of Bird. No doubt this line of demarcation between the base of the cloud and the snow slopes continued round in front of the aircraft toward the other textured cliffs of Cape Tennyson, on their left. The Cape Bird cliffs were obviously mistaken for the Cape Bernacchi cliffs on the other side of McMurdo Sound.

This line of demarcation of the cloud meeting the slopes of Cape Bird shown in passenger photo 10 supports my hypothesis of the same effect on the slopes of Erebus ahead giving the impression of a far distant horizon as illustrated in my visual models presented to the Royal Commission in figs. 34 and 35 on page 78 and 79.

Fig. 28 Photo 11. Taken from the left-hand side of the aircraft looking at almost a right angle to the line of flight. Points 'D', 'E', and 'F' which are marked in red on fig. 19 on page 67 are the three dark rock outcrops visible in the photo in order from 'D' on the right to 'F' on the left. Point 'F' is Cape Tennyson, Point 'E' is barely discernible.

In general terms the photo shows the textured cliffs of Cape Tennyson, which the crew probably believed was Cape Royds. Photographs 11 and 10 are considerably damaged by the impact forces.

The aircraft was at 1500 feet and horizontal visibility in the direction of photo 11 did not appear to be reduced, indicating that the aircraft was still below the cloud base. Shadows on the sea ice show that the cloud above was of variable density. There are fewer bright patches on the surface of the right of the photo indicating that there was denser cloud in this area, that is towards the impact area. This would be the direction that Collins and Mulgrew would be examining closely for signs of buildings, etc in the McMurdo-Scott Base area as the edge of what they believed was Cape Royds moved back behind them to give them an unobstructed view of this part of Ross Island.

Fig. 29 Impact Photo. The final photograph of actual impact we believe was showing fluid of some nature on the window. This could be fuel.

Link between Passenger Photos, Times and CVR

In order to illustrate the view visible to passengers at the times when those on the flight deck were passing the remarks which have been recorded, the comment closest in time to when the pictures were taken is shown below. But be particularly warned that the view taken by the passenger will usually be totally different from the view being scanned by those on the flight deck.

Picture of aeroplane. When the aircraft was located about where the picture marked A in red is drawn on the map, Collins says: 'I'll have to do an orbit here, I think.' (32.45 GMT)

Photo 1. Time of photo: before 33.45. Collins 34.21: 'Captain again, ladies and gentlemen. We're carrying out an orbit and circling our present position, and will be descending to an altitude below cloud, so that we can proceed to McMurdo Sound — Thank you.' Note the reference by Collins to McMurdo Sound.

Photo 3. Time of photo: not recorded in evidence. The passenger photo reveals (to those who know what it is) an indistinct view of Beaufort Island as it appeared from 9500 feet.

Photo 4. Time of photo 44.10, distance to impact 25.5 miles. Unidentified speaker 44.06: 'Where are we?'. Brooks: 'About up to here now?' Sound of rustling paper suggesting map being moved to indicate position on map. Cassin 45.00: 'We are now at 6000 feet, descending to 2000 feet, and we're VMC.'

Photo 8. Time of photo between 46.37 and 48.05, distance to impact between 8 and 15 miles. Brooks 46.39: 'Where's Erebus in relation to us at the moment?' Reply: 'Left about (20) or (25) miles.' Mulgrew 47.02: 'That's the edge'. Collins 47.43: 'We might have to drop down to 1500 feet here I think'.

Photo 10. Time of photo 48.18, distance to impact 8 miles.

Photo 11. Time of photo 48.59 (51 seconds to impact), 3.75 miles to impact. Collins 48.55: 'Have we got them on the tower?' Cassin: 'No — I'll try again.' Mulgrew 49.08: 'That looks like the edge of Ross Island there.' Brooks 49.24: 'I don't like this.' Brooks, whose vision was obscured by the cockpit coaming, has lost sight of all textured cliffs and so has passed from sector whiteout into full whiteout.

Collins 49.30: 'We're 26 miles north. We'll have to climb out of this.'

Unidentified 49.35: 'You can see (Ross Island).'

Cassin 49.38: 'No high ground if you do a 180.'

49.44: Ground proximity warning indicator sounds.

Collins 49.48: 'Go round power please.'

Fig. 29. 49.50 Impact photo of cabin window covered in liquid, possibly fuel.

Figs. 30, 31 and 32. The Effect of Sea Fog Ramp

Fig. 30. View of TE 901 wreckage on northern slope of Mt Erebus. Looking north-east towards Cape Tennyson on the northern slope of Mt Terror.

Fig. 31. Same view of wreckage, but now ice fog is coming up from the Ross Sea and beginning to make the slope of Mt Erebus merge invisibly into the Ross Sea.

Fig. 32. Looking up the slope towards the saddle between Mt Erebus and Mt Terror. Some definition of the saddle is visible but its surface is diffuse. The fog which has drifted up the slope from the Ross Sea obscures the ice cliffs on the coast.

Fig. 33 Mt Erebus concealed by whiteout

The photographs in this series were each aimed directly at Mt Erebus. They were taken from the helicopter which was taking Mr Justice Mahon towards Mt Erebus on the same track as flight TE 901. The rising ground which commenced at the ice shelf was clearly apparent in the sunlight, but the mountain itself was becoming enveloped in pale cloud, and in a minute or so it totally disappeared from sight. Even though the mountain slope began only two or three miles ahead of the crash site, no part of the mountain could be seen.

Unknown to themselves, Collins and his crew were confronted with cloud concealing Mt Erebus at the top of their view and fog concealing ice cliffs at the bottom of their view while the sun behind them produced no shadows discernible by them. Multiple light reflection back and forth between snow and cloud created whiteout, rendering Erebus invisible and probably resulting in a false horizon as illustrated in fig. 33 at foot of opposite page.

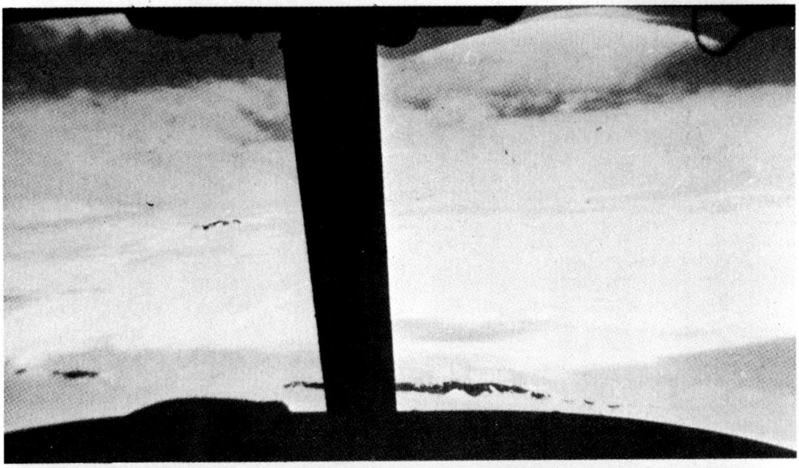

Visual Models Presented to Commission

Fig. 34 Entrance to Lewis Bay.

Cape Tennyson on left with Mt Terror behind it. Cape Bird on right, with Mt Erebus behind it. Mt Bird, further to the right, is out of sight.

Top view represents how entrance to Lewis Bay would appear with no sea fog and no cloud.

Middle view represents appearance of Lewis Bay with sea fog and cloud.

Bottom view same as middle view, but with addition of low overcast. Horizon would probably have been less distinct than shown and in full whiteout the horizon may have disappeared, at the last moment. This is probably what Collins saw and should be compared with what he would have expected to see as he flew up McMurdo Sound. Such an expected view is shown in bottom model on opposite page.

Fig. 35 Entrance to McMurdo Sound.

Cape Royds on left with Mt Erebus behind it. Mt Discovery rear centre. Cape Bernacchi on right.
Top view represents how entrance to McMurdo Sound would appear with no fog or cloud.

Middle view represents appearance of McMurdo Sound as crew could expect to see it without cloud.

Bottom view same as middle view, but with addition of stratus cloud base cutting far distant horizon. — The view Collins would have expected, given the meteorological conditions over McMurdo Sound. Compare with view actually sighted, shown on bottom model on opposite page.

Collins' view of Erebus before impact

Fig. 36. View of Erebus as Collins would have seen it framed by the cockpit coaming according to my reconstruction which I presented to the Royal Commission. This is fig. 34 from the foot of page 78 with the added framing of the coaming.

Fig. 37. On 28 November 1980, a year to the day after the crash, Justice Mahon was flown by helicopter towards Mt Erebus on the same track as the DC10 had taken. Before his eyes, and recorded by him in photographs, the mountain disappeared. Compare his view of the actual scene photographed approximately 10 feet above the crash site with my hypothetical view shown in the preceding fig. 36 above.

1. *Background*

THREE MEN DROPPED from the helicopter and stood shuddering in the Antarctic wind as the clatter of the rotor blades receded.

Approaching the flank of Mt Erebus they had stared down in wonder at the great black scar stretching 600 metres across the pristine whiteness; the path of a rolling fireball dotted with the debris of a DC10 aircraft which had disintegrated under an impact force 200 times greater than that of gravity.

Now the three men felt the full shock of the disaster. Around them the bodies of the 257 passengers and crew members of Flight 901 lay in the snow, some burnt and mangled beyond recognition but many seemingly unscathed by the violence of impact. All the comfortable paraphernalia of life accompanied them — cardigans and cameras, cushions and champagne glasses.

Daryl Thompson, Hugh Logan and Keith Woodsford, experienced mountaineers from Scott Base, began their task in grim silence. Roped together they picked their way through the wreckage, carefully avoiding crevasses. They noted the recognisable fragments of the machine and their trajectory over the ice slope.

The two wing engines had chewed deep into the ice, shearing away the turbine blades so that the spindles became bare shafts. The tail engine, still attached to its fin, had lived on a little longer, urging the ruptured cabin onwards, and finally coming to rest amongst the furthest-flung remains of the aircraft.

At the point of impact the airliner's underside silhouette was punched four metres deep into the ice. From there the machine had bounced, disintegrated, and strewn itself up the mountain.

Soon convinced that nobody had survived the disaster, the mountaineers pitched a small tent on a relatively stable plateau of ice, placed inside it provisions, and then signalled to the hovering helicopter. They flew back to base to confirm the worst fears of those awaiting them. The appalling news was soon relayed to New Zealand, the ill-defined sense of nightmare had become a terrible reality.

Captain Jim Collins' flying roster for the month of November 1979, arrived at his home by post some two weeks in advance. He was thrilled and a little surprised to see that he was assigned to command the Antarctic scenic flight of 28 November.

He had given himself little chance of success when he applied to fly one of the services. Until now, most of the command captains rostered for this had belonged to the small coterie of executive pilots, check and training captains and those who held administrative titles.

Of late however, the absence of executive pilots on scheduled services had enabled the odd senior captain to squeeze past this hierarchical barrier.

His desire to fly south to the ice was prompted by three earlier trips over the North Pole, as a passenger on the London-Los Angeles interchange service flown by British Airways with Air New Zealand DC10s.

Even at 39,000 feet, the vastness and stark clarity of the Arctic enthralled him. While his wife Maria read beside him, he gazed out for hours as the aircraft traversed the ice cap. But the grandeur of the Antarctic, he knew, surpassed that of the northern polar region. It was a vast land mass, ridged and furrowed by mountain chains and glaciers on a scale found nowhere else.

Collins noted with approval the crew he would command on the flight south. There was Greg Cassin, a tall, quiet first officer who had flown for several years with Air Pacific out of Fiji and was now accumulating significant experience on the DC10. The other first officer was 'Brick' Lucas, four years with the airline, but one of the most 'visible' personalities among the pilots; a big, solid young man with a hearty voice and a robust humour.

The two flight engineers, Gordon Brooks and Nick Moloney, were sound linchpins for an operation which would require particularly close monitoring of fuel reserves and engine performance.

The date, too, was right. Warmed by the spring sun, Scott Base and McMurdo Station would be starting to buzz. Excursion parties would be out in dog sleds, and the airborne rush to bring in supplies would be on before McMurdo's ice runway broke up.

Jim Collins had a good mental impression of the flight, even though he had never flown in the Antarctic. Other crews who made the flight discussed it with infectious enthusiasm; the long leg southwards at 35,000 feet, enlivened by the anticipatory thrill emanating from the passenger cabin, and the descent towards Ross Island almost always carried out in clear visual flying conditions.

The American ground controllers at McMurdo would invite the aircraft down for a closer look under guidance of the ground radar and most of the pilots descended to 2000 feet and less. There was the unspoken exhilaration of flying and manoeuvring a 250-tonne airliner in absolute safety over the pure clarity of the Ross Ice Shelf.

His impression of the flight was also sharpened by his addiction to maps and charts, which formed a large section of the family's well-stocked library. He enjoyed mulling over these, often prompted by the queries of four lively daughters. For many years Collins had held a flight navigator's licence, but this lapsed in 1971 because the opportunity to practise disappeared with the advanced technology of the newer jets. Only the airline's smaller DC8s still carried navigators.

Collins also felt lucky about this assignment. The 28 November flight was the last for the calendar year and there was a well-founded rumour it would be the last of the Antarctic scenic trips to be operated by the airline. The Opec suppliers were pushing the price of aviation jet fuel up towards $1 a gallon and each flight consumed about 25,000 gallons of fuel. In airline jargon it was becoming 'marginally economical'.

Air New Zealand was especially well situated for these flights, historically and scenically. A brief look at the map of Antarctica shows it is divided, like segments of an orange, among a number of countries. The narrow New Zealand segment, lying directly south of the country, encloses the western half of the Ross Sea which, like a vast estuary, penetrates the continent to within a few hundred miles of the South Pole. The southern half of this estuary is permanently covered by a great slab of ice, the Ross Ice Shelf. But along a well-defined border, the summer sun breaks and scatters the lighter sea ice to the north, and for a few months the northern Ross Sea becomes a summer seaway — albeit used with great caution and the vanguard of an icebreaker.

Robert Falcon Scott, Roald Amundsen, Ernest Shackleton and other early twentieth century explorers all used the summer access of the Ross Sea to gain a toehold on the ice shelf. Deep within the continent they could then launch their treks to the Pole.

About half-way down the bight of the Ross Sea and only

40 miles from the coast of mainland Antarctica, is Ross Island. It straddles the border of the ice shelf and the summer melt of sea ice. On this strategically placed island, Scott and Shackleton set up their southerly bases (actually prefabricated Australian farm houses insulated against the cold, and preserved intact with contents to this day in the bone-dry atmosphere).

Today there are two much larger bases on the southern shores of Scott Island. The United States Navy's McMurdo Station sits in a cove at the tip of the Hut Point Peninsula, a rugged scatter of blocky buildings and oil tanks, not unlike a transplanted Appalachian coal town. Enshrined nearby is Scott's Discovery Hut.

Two miles east around a small headland is Scott Base, maintained by the New Zealand Antarctic Division. It is a much more modest collection of small, green huts fronting the shore of the ice sea. The New Zealanders tether a string of huskies on the ice sea, and these together with the American transport aircraft operating off the ice runways were popular viewing with the passengers.

But dominating all on the 30-mile wide island is Mt Erebus, the largest of four mountains, and lying almost directly north of Scott and McMurdo. Towering to 12,445 feet, Erebus is an active volcano. A continual plume of smoke and vapour rises from its crater another 3000 to 4000 feet before being whipped into a downstream smear by stratospheric winds.

Scott and Shackleton had approached the southern shores of Ross Island by steaming into the sea ice of McMurdo Sound, the broad channel which separates the island from mainland Antarctica. When Erebus and the bulk of the island had slipped astern, they executed an eastwards turn which rounded them up at the point now known as McMurdo.

It was a logical approach and one still used to this day by American military transport aircraft, the C130 Hercules and C141 Starlifters, approaching the base from New Zealand and Australia. After flying safely over the broad, flat expanse of the McMurdo Sound sea ice, they are lined up directly for the ice airfield once the eastern turn is made.

For the previous 14 months a similar approach had been programmed into the navigation systems of the Air New Zealand DC10 airliners. It meshed in sensibly with the American ground control.

The accepted flight path south from New Zealand took the aircraft down to the bald and desolate Balleny Islands where a slight turn eastwards was made to position the aircraft over Hallett Station. The navigation track then turned slightly south, lining the aircraft up for the final 390-mile leg to Ross Island and the clear approach of McMurdo Sound.

Despite the fact that the Antarctic is the coldest place on earth, in summer the area can be a place of unsurpassed clarity and stillness. Those who work there in the dark winter months know only too well how changeable and ferocious conditions can be at ground level. But the 12 airline crews who had made the previous scenic flights had seen the region at its most beautiful. Cocooned in the warmth of a metal hull they had stared out for 200 and 300 miles over glacier and mountain range. Only the curvature of the earth limited the view.

The airline had nominated an alternative route. Should a crew find Ross Island shrouded in a rare summer weather front, the aircraft would deviate to the South Magnetic Pole some 300 miles westwards. But at the time Jim Collins heard of his rostering no crew had been forced to do this. Only one flight was to make detour and this was just days before the flight of 28 November.

Most pilots unlocked the aircraft from its automatic navigation system long before reaching the environs of Ross Island, descending manually to give passengers the best viewing possible before being taken under guidance by McMurdo radar for low-level circuiting near the twin bases.

This then was the common store of information inherited by pilots approaching their first flight to Antarctica. The finer details would be filled in for Jim Collins at the audio-briefing they would get from the company's Route Clearance Unit (RCU) and the stint in the ground simulator.

It was in Collins' nature to work over all the information, to mentally define and plan for the unexpected.

On the first two Antarctic flights, at the beginning of 1977, the airline had followed the military practice of ensuring that the captains of these flights had earlier flown the same route, and therefore had some familiarity with conditions and the approaches to McMurdo. It was in fact an obligation written into the company's regulations.

The requirements had, however, been dropped from the subsequent nine flights. The airline believed that the experience of the RCU briefing and the simulator was sufficient.

Later, the airline was to describe Collins as a 'meticulous

pilot'. It was a faintly cynical response which seemed to beg the question — why then did he allow this to happen?

Meticulous Collins was, his fellow pilots assent, but by no means rigid as the term can imply. In the immediate post-crash days his colleagues made two points about the man: he was a supremely good pilot who always did his homework, and he was popular — quietly good humoured and stable.

There were no loose ends and untidy matters to be cleaned up when Collins called a taxi to take him from his St Heliers home to the airport early on 28 November — only unfulfilled plans. Fresh blue cod — unobtainable in Auckland — was to be picked up that night at the special shop, known to crews, near Christchurch Airport, and the family had agreed that a decision would have to be made within the next few days on the type of off-the-beach yacht they would acquire to teach the four girls sailing.

Collins' private plans were almost totally enmeshed with his family — his wife Maria, and daughters Kathryn, nearly 16; Elizabeth, 14; Phillipa, ten, and Adrienne, six.

When the girls were very young he had cooperated with a friend, boat designer Frank Pellin, in designing a cabin cruiser for holiday trips on the Hauraki Gulf. It had to be roomy, reliable enough to get him home in time for his flight rosters, and above all safe, he stipulated.

The girls teased their father about his cabin checks before sailing. He would tick off a check list which ended with lifejackets for everybody, and one to spare.

With 24 years of flying behind him, and youthful experience as an Air Force airframe fitter, Collins had a bent for what aircraft designers have come to call 'human engineering'. The idea is not only to put everything within easily accessible reach, but to place them where they follow the natural movements and inclinations of the human body and mind.

At the age of 45, he had changed remarkably little from the youngster who received his pilot's brevet at RNZAF Base, Wigram in 1954. He was a handsome man, a fraction under six feet on a lean 11-stone frame, with a full head of dark hair, craggy eyebrows and a square chin. His most striking feature was his electric blue eyes.

He had no worries about the six-monthly medical check pilots are subject to. These could end a career overnight; put a man formerly earning $50,000 a year onto the unemployed list because his blood pressure was a few points too high or his hearing had lost its acuity in the higher frequencies.

Jim Collins left the Air Force in 1958 as a Flying Officer, and joined the then Tasman Empire Airways Ltd (TEAL was later to become Air New Zealand). From Second Officer on DC6s, he progressed to Electras, DC8s and, in 1973, to the DC10. He earned his first command in 1964, at the early age of 29.

He met Maria in 1961. She was a science graduate working as a biochemist in the clinical laboratory at Auckland Hospital. Maria describes her marriage as 'blissful'.

Like other pilots, Captain Collins had implicit faith in the DC10's navigation system — implicit within the reservation of good airmanship, which means counter-checking where possible and never accepting it as infallible.

The inertial system was introduced to jet airliners ten years earlier, and the specialised job of flight navigator began to disappear from the flight deck. It is self-contained, which means it has no need to refer to anything outside the aircraft, such as radio beams or sun sightings.

The core of the system is the inertial platform, a gyroscopically stabilised platform on which small sensors are placed. The pressures exerted on the inertial platform by the movement of the aircraft are measured by the sensors, and are fed to a sophisticated computer which will interpret them and commit them to its memory. The computer can then compute exactly how far the aircraft moves, in what direction, at what speed, and what deviations occur.

The pilot can feed into the computer the route he wants to follow — a series of navigational positions known as 'waypoints' along a track. The computer, guided by the information relayed continuously to it from the inertial platform, can then work the autopilot and fly the aircraft.

The information is transmitted to a visual display unit, and the pilot can see, and if necessary call up, details about the progress of the flight.

Most big aircraft, including the DC10, have three inertial platforms, and if one malfunctions the other two automatically isolate it. The whole package, inertial platforms and computer, is known as the Area Inertial Navigation System (AINS).

The AINS was in its element on the Antarctic flights. The vast, lonely stretch of southern ocean could offer no surface-based radio aids. Only within a 40 to 50-mile radius of McMurdo at the southernmost extremity of the track, would these aids become available.

The AINS would guide the machine accurately — within a mile each way of the track for every hour flown — until the aircraft descended to a low altitude, then the crew would fine tune its position using the ground aids. The AINS is relied on absolutely for cruising at altitude — it is the passage-maker.

All the pilots who flew south spoke approvingly of the extreme accuracy of the AINS. Like the human mind, the system has one failing. Inaccurate information fed into it is acted on as if it were accurate.

Captain Jim Collins

Greg Cassin

2. *The Loop*

CO-PILOT GREG CASSIN often said he would willingly pay Air New Zealand to let him fly. His wife Anne called him a 'flying fanatic'.

So in fact was Anne. They met through flying, and shared the interest through 13 years of marriage. Anne was one of the few airline wives who took an active and detailed interest in the technicalities of the pilot briefings; what radio aids were available, the diversionary airports, whether they would be monitored by radar on let-down, etc.

At the age of 37, Gregory Mark Cassin had a respectable total of nearly 8000 flying hours in his logbook, gained on a wide variety of transport aircraft. He was a tall, lean man with a slight stoop which, with his quiet voice, gave him a studious appearance.

The couple met at Rotorua Airport in 1965 where Greg was working as an airport communications officer, and doubling in his spare time as a flying instructor at the aero club. Anne was well ahead on her private pilot's licence course when he took over tuition from Anne's father, John Munro, a well-known and popular commercial pilot in the 1960s. By grim irony, Munro was killed when his Piper Cub crashed on take-off from Kaipara Flats on 28 November 1965 — exactly 14 years to the day before his future son-in-law was to be rostered for the Antarctic flight.

After several years instructing at various aero clubs and the Taupo Flying School, Greg joined Fiji Airways (later to become Air Pacific) in 1970. After four years the family returned home and Greg joined Air New Zealand as a first officer on DC8s. There were now three children and their education had to be planned. At the time of the Antarctic flight, Maria was 12, Stephen, nine and Jacinta, four.

Nick Moloney and Gordon Brooks, the two flight engineers rostered for Flight 901, were old hands who had both come up through the system. As lads they joined the airline as apprentice engineers and after wide experience on the ground transferred to aircrew.

The airline selected its flight engineers from among its best engine and airframe fitters. Once they were through the stiff training course, they were expected to upgrade their skills and knowledge continually, as are pilots.

If it is to be flown safely, the modern jet airliner relies not on the competence of its captain, but on a triangular linkage of three minds, the two pilots 'up front' and the flight engineer at his panel just behind the co-pilot.

Airlines sometimes apply the mechanical term 'fail safe' to the fully co-ordinated crew. Vital systems and structures on which the safety of the aircraft depends have backup systems which, for most of the time, are redundant. However, should there be a failure in the main system, these redundant systems continue to operate. An example is the triple channel auto-pilot for automatic landings. Should any one of these channels fail, the others pick up the load and continue on with the task.

The term 'fail safe' explains something of the psychological bonding on the flight deck. But aircrew have a more succinct name — they call it 'the loop'.

Communication on the flight deck involves a continual flow of comment and questioning, observation and confirmation circulating between the three. If one man fails to respond properly or to add any input to the flow, the others take the necessary action, throwing an imaginary switch to isolate him from the business of flying the aircraft.

A dramatic example of the loop is practised in the simulator flying crews periodically undergo. The check and training captain overseeing the exercise may secretly pass a note to one of the pilots saying, 'You've had a heart attack'.

Without showing too many overt signs, the pilot will immediately cease to take any part in flying the machine. If the loop is working well, this should be noticed within seconds by the two other crew.

The engineer will yank the seat withdrawal lever to pull the 'dead' pilot clear of the controls, and make an immediate check of the flight position and instruments with the other pilot before attempting any first aid. The safety of the aircraft and 400 passengers remains paramount. This is of course followed immediately by the engineer carrying out the first aid procedure for a 'witnessed collapse'.

Normally, the loop works to pick up small oversights; perhaps a momentary lapse in concentration and the misreading of an instrument. Another aspect of the loop is the need for each man to be confident in his own observations and deductions. He must be positive and speak his own mind. There is no place for the old fashioned deference which in former times caused a junior crew member to withhold criticism of action taken by seniors. The abnormally timid, the demurrer, has no place on the flight deck.

In striking the psychological bond, the cross-flow of talk between the three crewmen takes on its own peculiarities. A kind of verbal shorthand develops, added to by the use of flight acronyms.

To the eavesdropper, observations can sound like exclamations, questions like interrogations, and a sharp reminder can sound like an ill-tempered outburst. Remarks are intercepted in mid-sentence; queries answered before the verbal question.

One vital, if not so obvious, factor in the loop involves the detection and interpretation of body language — those unvoiced gestures of anticipation, approval or doubt, a response qualified by the quizzical lift of an eyebrow or hesitation in complying with a command. These, as much as dialogue, have a crucial role in the quality of the cockpit loop.

It makes for difficulties for accident investigators when deciphering the contents of cockpit voice recorders — the automatic device which is switched on all the time during flight, and which always retains the last 30 minutes of cockpit conversation. The voice recorder can not detect the subtle interplay of body language. This 'gap' in the record was to be a problem in the aftermath of this flight.

On most scheduled services, the DC10 would have carried just the three technical crew. But on Flight 901, three pilots and two flight engineers were rostered. This was the number stipulated in 1977 when these special flights were approved by the Civil Aviation authority. Actually the authority had approved a crew of two captains and one first officer, plus the two engineers, but the airline in the meantime had dropped the two captain requirement.

The reason for the large crew is fairly obvious. They were flying into an area which was unfamiliar, and some of it would be at low level. The duration was also longer than the average flight. Even with the safeguard of a well-functioning loop, no single crew could be expected to remain at their prime throughout more than 11 hours. Rest periods for individuals were essential if the loop was to remain intact.

Gordon Barrett Brooks was an instructor in fail-safe crew training and also the only one among the technical crew who had made a previous ice flight; on the second trip in 1977.

He could be a firm disciplinarian on the flight deck, quick to come down hard on anybody who seemed to be losing concentration. Even senior captains felt his uncharacteristic 'bite' occasionally, and respected him for it.

Brooks was a fine bass baritone who required little coaxing to exercise this talent. He had the presence to go with the voice. He was more than six feet tall, dark and very handsome.

He and Jim Collins were close friends. In the early 1970s they were based together in Los Angeles and the rapport grew from then on. They shared an engineering background and were both fit and naturally athletic.

Never one for wasting words, Brooks would sit back and listen while an argument developed; then when it reached a climax he would step in and round up the salient points in a few words. The Brooks 'overview' was usually accepted as final in the crewroom.

His wife Margaret was an air hostess with the company for eight years. They had a son, Charles, 18 months old, and were busy renovating an old home in Remuera.

Nicholas John Moloney, 45, was one of the personalities among the airline's flight crews. He was a strong, ruddy-faced man with blue eyes and a shock of fair, gingery hair.

He had mobile features and put them to good use. He was always telling jokes and always ready to strike the amusing pose. Above all, he was known to all the kids who lived near his home in Te Atatu, as an easy confidant, a Dutch uncle who disappeared frequently and always had something new to show them on his return. If the toy stores in Los Angeles were having a run on some gadget, Moloney would have a supply tucked away in his flight bag when he returned home.

Nick and Lorraine Moloney had three sons, all in early adulthood. Tony, 23, is a structural engineer in Auckland; Danny, 22, a civil engineer in Sydney; and Richard, 19, an apprentice motor mechanic in Auckland.

Lorraine recalls his excitement when he came home with news that he was rostered for an Antarctic flight.

'I asked him who the captain would be. He said: "One of the best — Jim Collins".'

Whatever the many ways cabin crew contribute to the comfort of passengers, the main reason for their presence is safety. It is the point all airlines emphasise throughout the initial training of stewards and hostesses, and in the regular refresher courses they are required by law to undertake during their flying careers.

Evacuation procedures, first aid, survival training, the safe stowage of goods, fire fighting, even the ability to detect signs of stress among passengers and other crew members are central to the curriculum.

The night before take-off, Chief purser Roy McPherson spent some hours going over new safety literature for cabin crew. It was more of a habit than a chore, says his wife Elizabeth. 'He took it very seriously.'

Ironically, Roy McPherson was to receive another sheaf of safety literature as he boarded the aircraft next morning. It was Qantas Airways' notes on survival in the Antarctic prepared for crews who flew south.

For several months, Captain John Spence, a senior airline inspector with the Civil Aviation Division, had been trying to persuade the airline to equip its Antarctic aircraft with survival equipment as a precaution against the extremely unlikely possibility of a forced landing or ditching in the region. The airline was not keen to do so because the possibility was so remote, and if it did happen the chance of survival would be very small.

Recently returned from Sydney and talks with Qantas, Spence noted the summer dress of most passengers on Flight 901 — open-necked shirts, even shorts — and handed over the literature to the most appropriate member of the crew. There would be little doubt that Roy McPherson, an approachable and dependable man, studied them in the spare minutes he could catch.

McPherson led a youthful cabin crew, average age for the men 31, and for the hostesses, 26. Yet there were strong family ties with the airline running through the group of 15. Assistant purser Mike Finlay had also married a hostess; Suzanne Marinovic was from a family of four sisters who were all air hostesses, and was married to a purser. Purser Russell Scott and steward Steve Simmons both had brothers in cabin crew. Hostess Katrina Morrison's father was a foreman at the airline's jet engineering base.

There was also another tie which linked them in varying degrees, especially senior cabin crew members Sue Marinovic and Graham Cater. A few months earlier the airline's cabin staff had mounted a huge effort to give disadvantaged children a treat. Months of fund raising and a glittering town hall ball raised enough money to enable them to charter one of the company's jets and fly 230 children to Los Angeles and a visit to Disneyland.

The other eight members of cabin crew were: Purser Martin Collins; assistant purser Dave Bennett; senior cabin crew member Bruce Maxwell; cabin crew members Elizabeth Carr-Smith, Dianne Keenan, Jim Lewis, Dave Sicklemore, Steve Simmons and Marie Wolfert.

3. *RCU Briefing*

ON 9 NOVEMBER 1979, Captain Collins and one of the two first officers rostered for the Antarctic flight, Gregory Cassin, filed into the bleakly functional room which serves as the route clearance unit, and which is situated conveniently next door to the airline's DC10 simulator.

Another crew was present for the Antarctic briefing — Captain Leslie Simpson, the company's training captain for DC10s, was to take the flight down on 14 November. With him were his two co-pilots, First Officers Gabriel (now Captain) and Irvine.

Like Jim Collins, Simpson had also been the holder of a flight navigator's licence, and had logged 1000 hours in this job as well as 14,000 hours as a pilot. He was a quietly-spoken man, a first-class pilot, and a stickler for precision.

Unlike the pre-flight despatch briefing which takes place just before a flight and deals with the immediate problems of weather, communications and navigational aids, the RCU briefing begins as more of a guided tour of the route the pilot will fly, and his approach to the airport at the other end. It is taken at a leisurely pace, with an audio-visual screening and a 'talk-through' session which prompts discussion.

The RCU's role is familiarisation for pilots about to fly to an airport new to them. Earlier, a captain about to fly to an unfamiliar airport would be required to make the flight under supervision of a pilot experienced with this flight. An RCU briefing did away with this stipulation provided the director of CAD gave an exemption — and of course made savings for the airline.

There was to be no intended landing in the Antarctic. But the approach, the let-down and the low flying in the McMurdo area was considered at least as important as a landing approach to an airport.

The airline had expended some theatrical flair in making its Antarctic briefing as concise and interesting as possible. There were twin projectors to beam colour slides in unison onto twin screens. A taped narrative alternated information on terrain, communications and navigational aids with occasional sparks of dry wit.

Captain John Wilson, the supervisor, was a man experienced for the job. A lifetime of flying had begun as a wartime pilot in the Pacific (he was awarded the Air Force Cross), and ended at retiring age as a DC10 captain with the company. In between, he had pioneered routes in the East Indies, written operational manuals for Guyana Airways, and had accredited instrument ratings to other pilots. He was a lean, alert man with the command captain's knack of engaging attention. There were few areas in the world he had not flown, but until two days before the briefing he would have numbered the Antarctic among them.

Wilson's trip as a flight deck observer was on the only Antarctic flight that flew the alternative route to the South Magnetic Pole. Low cloud closed in around Ross Island, so the aircraft diverted some 500 miles before reaching mainland Antarctica. Nevertheless, he had taken notes and brought along a copy of the flight plan.

As he took his place before the pilot group, he noticed that Jim Collins had brought along a black-bound volume of the *New Zealand Atlas,* and that First Officer Gabriel had a *National Geographic* pull-out map of the Antarctic. Interest was high; it was almost unknown for pilots to bring along their own maps.

The exact details of the discussion at the RCU briefing will never be known because later accounts were to differ markedly. But the audio-visual screening remains as a record. Two flight plans had been prepared. Both routes would follow a common track south from New Zealand via the Balleny Islands to Cape Hallett. Should a crew learn that weather conditions were unsuitable in the McMurdo area, they would take a secondary route from Cape Hallett to the South Magnetic Pole, via the impressive Ninnis Glacier.

The primary track was a straight leg south to McMurdo from Cape Hallett. A close watch on fuel would be necessary and the crew would need to ensure they did not linger at low altitudes in the viewing area where the fuel burn would be excessive. The aircraft would leave Auckland at almost maximum all-up weight, 250 tonnes including 109 tonnes of turbine fuel.

At 60 degrees south they would be required to change to grid navigation. Grid is used in both polar regions of the world, and a quick glance at an atlas will show why. Towards the polar regions the longitude lines on the globe converge increasingly until they all come together at one point, the north or south pole, as the case may be.

In former times this converging caused difficulties for navigators because of the progressively diminishing distance between longitudinal points. The nearer the pole the greater was the margin for error as the minutes in each longitudinal degree diminished to a few miles and eventually to a matter of yards. Apart from the difficulty of accurately plotting converging longitudes, the magnetic compass spins uselessly in these regions and cannot be used in the normal way for pinpointing the direction of a radio beacon.

The grid can be likened to a giant chessboard grid replacing the normal lines of longitude and latitude over the polar areas. It is aligned at its centre on the Greenwich date line. At 60 degrees south, the crews would align their compasses with grid north to fit in with the navigational information which would be beamed from McMurdo.

Within the aircraft's inertial navigation system, however, the computer would retain memory of navigational waypoints based on the normal lines of longitude, and continue to fly the machine according to these.

Discussion on the grid changeover took time at the briefing and simulator session because it was an important part of the route definition. However, it is of little importance in understanding the train of events which overtook Captain Collins' flight.

Meanwhile, the series of colour slides flashed onto the screen depicted the view from the aircraft as it flew south — and the pilots noted that the track of the flight lay over the sea ice of McMurdo Sound, with the high terrain of Ross Island out to the left, and the mountains of the Antarctic mainland to the right. It was an impression they expected, and had no need to question.

Erebus, as it loomed into view, was also to the left in its correct place (actually this slide, it was later learnt, depicted the mountain from the south, but the oversight was not discovered until later).

'Now approaching Erebus at 16,000 feet, the minimum sector altitude,' the taped voice said.

'In VMC (visual flying) a descent to this minimum altitude up to 50 miles before McMurdo will be found advantageous for viewing. . .'

The presentation noted that the non-directional beacon (NDB) positioned on the McMurdo ice airfield had become 'not available'. In former times this radio aid could have been used by pilots for locating the position of McMurdo should the area be obscured by cloud. Homing in on the NDB, they could have descended through the higher cloud until picked up by the McMurdo radar and safely guided into the vicinity south of Ross Island.

To overcome the absence of the NDB, the descent procedure was now amended. If the aircraft was to be taken below the minimum safe altitude of 16,000 feet, it must be under strictly visual flying conditions (VMC) and with authority from the McMurdo controllers.

Visibility had to be at least 20 kilometres, and there should be no snow showers in the area. The descent should be coordinated with the local radar control because there might be other aircraft flying in the area.

Finally, using the McMurdo TACAN aid as a centre point, the aircraft could operate at 6000 feet in a closely defined arc over the sea ice and ice shelf. The TACAN (Tactical Air Navigation Station) is a military aid not normally used for civilian aircraft, but — as in the case of the New Zealand DC10s — pilots could tune into it as a distance measuring aid. Like the civilian Distance Measuring Equipment (DME), it would tell them how many miles away they were from the installation, and tick off the miles as they approached the installation.

Homeward bound, it was permissible to climb the aircraft to 10,000 feet for viewing the Dry Valleys on the Antarctic mainland, and the stretch of coast north to Cape Hallett. If visual flying conditions were not present, the company's minimum safe altitude of Flight Level 160 should be maintained (a flight level, as distinct from a height, is the barometric reading of height on the cockpit instrument which has not been adjusted for local variations in pressure. It is abbreviated to hundredths — hence flight level 160 is 16,000 feet).

In retrospect there were two omissions from this audio-visual text, and the talk-through plus simulator sessions which followed. But these were not to become apparent until much later. . .

Firstly, what was the track the aircraft would follow southwards from Cape Hallett to McMurdo?

The audio-visual had spoken simply of McMurdo as the southern destination, and if a ruler is placed on the map between Cape Hallett and the McMurdo Station, the direct line between them runs over the top of Mt Erebus.

But the five pilots present were well aware that on previous flights (certainly those within the past 14 months), the aircraft navigation systems had been programmed to fly to a southerly destination, a final waypoint west of the Dailey Islands, some 25-26 nautical miles to the west of McMurdo Station. It was similar to the route used by US Navy aircraft, and it ran safely down the centre of McMurdo Sound over low-lying sea ice. McMurdo Sound, except for two low-lying promontaries half-way along, is a stretch of flat ice 40 miles wide. Even the tips of the promontaries are nearly 30 miles apart.

On reaching this final waypoint, the aircraft would have been expected to turn left, and under guidance of McMurdo controllers complete the few miles to go to the ice airfield.

Few, if any, previous pilots had followed this navigational track exactly. In most cases the weather had been clear and they had disconnected the aircraft from its automatic navigation system, and flown manually.

Captain Wilson was later to testify that he believed the track ran direct from Cape Hallett to McMurdo Sound, over Mt Erebus. It was a misconception held by a number of other senior people in the airline's flight operations section.

The track over Mt Erebus was certainly the one used when the first flights were made in 1977. It had been changed in the meantime, and the newer track was much more sensible and the only one the McMurdo authorities were actually allowed to accept for their own aircraft.

Captain Wilson was also to claim later that he indicated to the five pilots, in the talk-through at the RCU briefing, the route over Mt Erebus by running his pen down a navigational chart of the area. It was a claim that came to be disputed by Captain Simpson and other pilots who were there.

Among a number of maps and documents placed on the table before the pilots at the RCU briefing were four copies of an actual flight plan, one used only two days previously on Captain Wilson's Antarctic flight. Flight plans are difficult documents from which to obtain a quick impression of a flight because they consist almost wholly of figures and code words.

Down the left side is a list of navigational waypoints or coordinates which will be punched into the aircraft's navigational computer. On the right is a table showing the aircraft's fuel burn-off as it moves along the route.

Flying as they do at 30,000 feet and above, modern pilots accept these waypoints at face value, well-used points along well-trodden tracks. Most waypoints are code words to be interpreted by the computer. On the Antarctic flight plan, however, the waypoints were written out as navigational co-ordinates — a latitude and longitude position, accurate to the minute.

It later became clear that Captain Collins must have learnt at the briefing precisely where that flight path was programmed to take the aircraft. Several explanations are possible. He may have retained one of the four flight plans which were distributed. He may have copied the list of co-ordinates down in a small ring-binder he kept for such purposes, and which he took to the briefing. He may have memorised the waypoint or noted it on the atlas he took to the briefing.

Whatever Collins did, the destination waypoint he correctly noted was: 77.53 south, 164.48 east. The co-ordinate 164.48 east was eventually to take on a momentous significance. It is the longitudinal co-ordinate for the 'safe' destination waypoint; the one at the end of the track flying down over the McMurdo sea ice. A track which would place Mt Erebus 25 miles to the left of the aircraft.

The other most serious omission at the RCU briefing was the almost total lack of discussion about so-called polar 'whiteout'. The pilots were told to avoid snow showers and blown snow. No warning was given of the danger in attempting to proceed visually under any overcast.

The military services who flew into the area took special precautions to alert and train crews to cope with the various forms of visual phenomena encompassed by the term 'whiteout'. The captain of every aircraft flown into the McMurdo area by the US Navy or the New Zealand and Australian Air Forces had to have made at least three previous flights into the area under the command of an experienced polar flyer.

Yet here was Air New Zealand sending its DC10 into the Antarctic under the command of pilots who had never even ventured into the area before!

Apparently it was prompted by the belief that since no landing in the Antarctic was contemplated, they were unlikely to be troubled with polar visibility problems. It was a theory to be tragically disproven.

The question of what height the airliner should descend to was discussed in the talk-through. Should conditions remain clear, pilots might well descend lower than 6000 feet at the

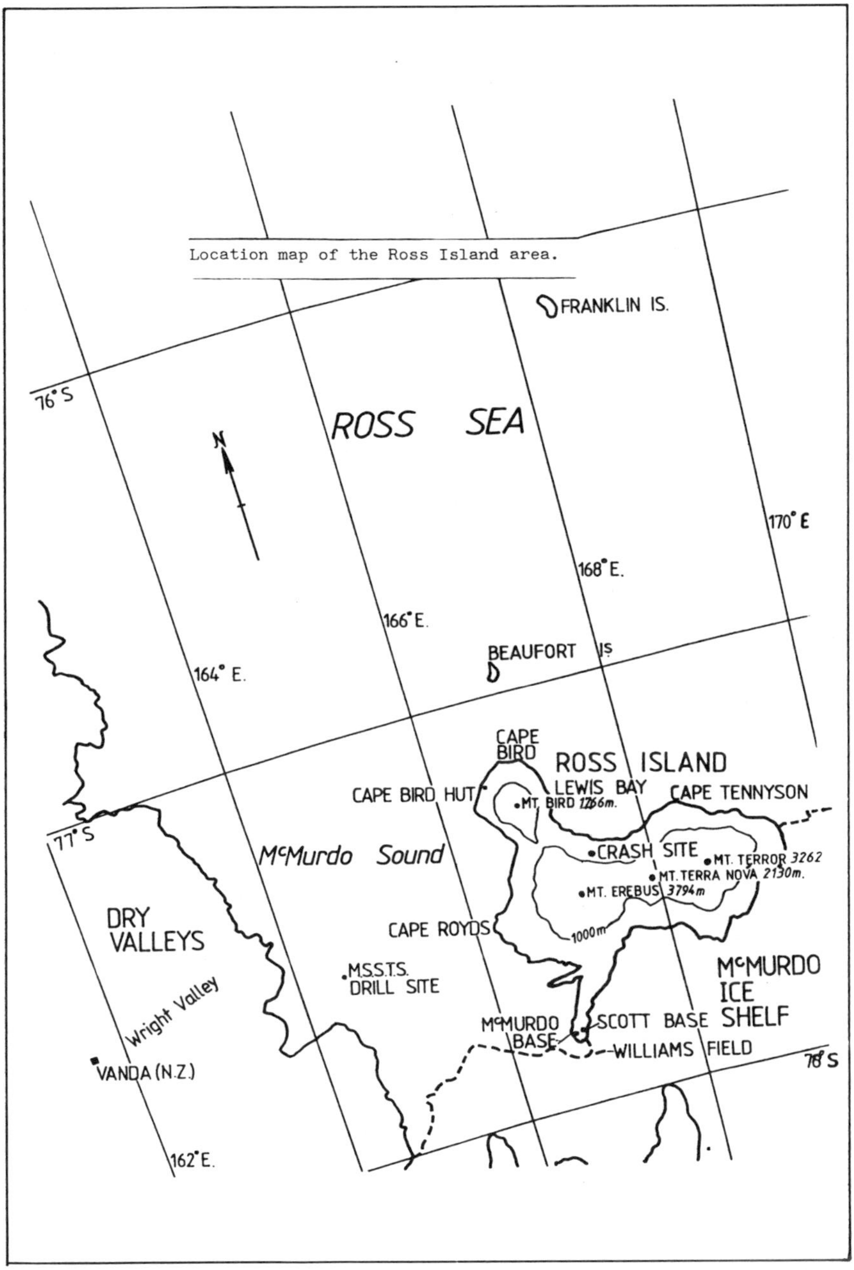

Fig. 38. This illustrates the latitudes and longitudes applicable to the Ross Island and McMurdo Sound areas.

invitation of the American controllers at McMurdo (Mac Centre). In discussion they arrived at the conclusion that the approach radar could only pick them up within 40 miles and below 6000 feet.

It was a point that needed little elaboration. Most of the previous pilots had taken their machines down to 2000 feet, or less, and the cheerful willingness of the American controllers to oblige was well known. The base inhabitants enjoyed the intimacy of viewing the stately form of the proudly marked DC10 as it circled the base at circuit height, viewing and being viewed. A welcome visitor who seemed, even for a short time, to remove the feeling of remoteness.

It was a practice which had apparently received tacit approval. Low flying in the area had been described in a number of publications including the airline's own newspaper, and depicted on film and discussed on radio talkback programmes.

Coming to the end of the RCU briefing, there seemed to be no outstanding matters to be clarified. They went on to lunch and the simulator briefing.

What failed to surface at the briefing also passed unnoticed in the simulator. Captain Ross Johnson, the airline Flight Manager Line Operations DC10 — DC8, was later to testify that he accepted the correct track south as the one lying over Mt Erebus, and that he had indicated this to the crews by marking a large map.

All the surviving pilots who had attended the briefing were to disagree that he indicated a track over Erebus. Nor, said Captain Simpson, was the simulator ever placed in a position 50 miles north of McMurdo so that the crew could practise the approach let down from height, as claimed by Captain Johnson.

The simulator sessions for both crews were in fact largely concerned with the changeover to grid navigation, and the NDB let down to fly into the altitude and range capability of the radar for monitoring the visual sightseeing around McMurdo (6000 feet and 40 miles).

On 14 November, Captain Simpson and crew made their flight to McMurdo. In fine weather they were able to depart early from the flight track having passed the Cape Hallett tip. Simpson manually flew the aircraft close in to the coast of Victoria Land to enable the passengers to view the glaciers and mountains of the Antarctic mainland. He was some 30 to 35 miles westwards of the navigational track when McMurdo controllers gave him clearance to descend to 2000 feet and fly visually towards the station across the sound.

Flying from west to east across the track he was surprised to notice that the distance from the track to the McMurdo TACAN aid was rather greater than he expected. He had thought it would be about ten miles but his instruments showed the distance to be about 27 miles. Over the TACAN he checked the accuracy of his navigation system and found it to be correct. He checked again on the homeward leg and again there was no inaccuracy. It was obvious to Captain Simpson that nothing was wrong with the navigational system, but that he simply had a misconception about the distance between McMurdo Base and the track to the final southerly waypoint situated in McMurdo Sound, which was called 'McMurdo' on the flight plan.

The day following the flight, Simpson telephoned Captain Johnson of Flight Operations and during the course of a brief conversation, mentioned his surprise at the distance between the destination waypoint and McMurdo Base, adding that he had made two updatings of his navigation computer just to check that there had not been a mistake.

As other crews might be surprised by the cross distance, it would be just as well to warn them about the position of the destination waypoint in relation to McMurdo Base so that they would not be tempted to update their navigation systems unnecessarily, he said.

Later, Johnson's recollection of the same conversation was to differ markedly from Simpson's account. Johnson claimed his fellow pilot asked him to get the Navigation Section to look at the position of the McMurdo waypoint on the computerised flight plan, as it would be better situated on the TACAN aid.

He did not recall Simpson specifically mentioning a distance of some 26 nautical miles between the two points. He was aware that in the early flights the waypoint had been positioned on the now defunct NDB aid.

'I had in mind that this position may still have been on the flight plan, and that Captain Simpson was suggesting it would be better changed to the co-ordinates of the TACAN,' Johnson later testified.

He passed these comments on to the Navigation Section, asking them to check the McMurdo position and report back.

There exist two significantly different versions of the same telephone conversation. Simpson was to deny completely Johnson's recollection, and it was the former's version which was eventually upheld in the Royal Commissioner's report.

The misunderstanding was about to set in train the final mistake in a long string of oversights and blunders (See mistake 45 on page 333 from note 3).

4. *Flight to the Ice*

THE EVENING OF 27 November, the night before the Antarctic flight, found Jim Collins relaxing in one of the ways he enjoyed most — at the dining-room table surrounded by maps, charts, his favourite *New Zealand Atlas,* ruler, dividers and pencils. He was marking out on a large chart the path the aircraft would be programmed to fly the next day, using the navigational co-ordinates noted down from the sample flight plan produced at the RCU briefing 19 days earlier.

His two elder daughters, Kathryn, 16, and Elizabeth, 14, were attentive observers. In reply to their queries, he laid the big chart out on the floor; Kathryn particularly remembers her father indicating the path the aircraft would follow — down the coast of Victoria Land and on to McMurdo Sound. It would fly back up the same coast on the homeward leg, he told her.

After working on the maps for nearly two hours, he folded them and, with instruments, diary and black ring binder, packed the lot neatly into his flight bag. It was an early start tomorrow and there was just time to watch the 10 pm television news before bed. (See note 2 on page 327 headed *Did Jim Collins plot track before flight?*)

Six miles away in Newton, a suburb near central Auckland, another man was at work that night on the navigation route for Flight 901.

It was a Tuesday, the night of the week the airline updated information stored in the company's central computer. Among the data to be updated by the operator at the keyboard were the navigational co-ordinates of the flight plan for the Antarctic service.

Following Captain Simpson's telephone conversation, Captain Johnson spoke to the superintendent of the Navigation Section, Laurence Lawton, and passed on his mistaken belief that Simpson thought there was something wrong with the McMurdo terminal waypoint position.

Lawton and his chief navigator, Brian Hewitt, retrieved what they believed was a copy of the current computer printout of the track south to McMurdo, and noted that the final waypoint given for McMurdo was 77.53 south and 166.48 east. But in fact it was not the current flight path, it was one last used 14 months previously (see mistakes 46-48 on page 334 from note 3).

The reader will recall that the eastern position Captain Collins had taken from the flight plan shown at the briefing was 164.48 east — just one digit difference but the equivalent of 26 nautical miles in this region. (For maps which show longitudes, see Fig. 8 on page 42, Fig. 38 on page 101, and Fig. 42 on page 134.)

Lawton and Hewitt were not aware that when the flight plan had first been computerised 14 months earlier, the operator had made a simple keyboard typing error. He had punched a 4 instead of a 6 into the machine, and the error had never been noticed.

The reason why the error was never discovered was that it rerouted the aircraft safely over the sea ice to the waypoint in McMurdo Sound. It was a fortuitous mistake, and the pilots who flew the route never had any reason to suspect it was wrong, or anything other than a sensible revision.

Now Lawton and Hewitt were examining the flight plan used in the 18 months before it was computerised. The final waypoint was the intersection of the two runways on the Williams Field (a permanent runway used by ski-equipped planes). The track lay over Mt Erebus.

On checking this they found that in recent times a very minor correction had been made to this navigational position of the Americans, a matter of 10 minutes, or some 2.1 nautical miles.

They pointed this out to Captain Johnson, and after some discussion decided to place the final waypoint on the nearby TACAN aid, using the updated information from the American charts. The new position was to be 166.58 east. This waypoint was printed into Captain Collins' flight plan just eight hours before take-off.

So in three years the final waypoint came to be changed three times: 1977/78 — 166.48 east, seven flights routed over the top of Erebus to Williams Field; 1978/79 — 164.48 east, seven flights routed close to the military route down McMurdo Sound; 28 November, 1979 — 166.58 east, the last TACAN flight routed back again over Erebus.

Mistakes which meshed with each other like wheels within wheels, until a final wedge jammed the machinery. A simplified train of events would be as follows:

(1) A stupid decision is made: the airline decides to route its flights over the active volcano, Mt Erebus to Williams Field on the other side.
(2) A fortuitous typing error: the destination is shifted 26 nautical miles to a more sensible site and the system

functions well for 14 months.

(3) A message is garbled: a comment in passing is interpreted as a complaint to be rectified.

(4) A mistake within a mistake: earlier 'finger trouble' passes unnoticed and a fatal shift is made.

Within hours, early on 28 November, the airline was to make a final mistake. It was to neglect to tell Captain Collins and his crew that there had been a change made to the final destination. The change was thought to be so small it did not warrant alerting the crew. (see note 3 on page 329 headed *The 54 mistakes concerning flight path*)

Preparations for the flight would have followed predictable lines. Jim Collins' springy step taking him into the Flight Despatch office to be greeted with smiles from his crew. One would expect the voice of Brick Lucas to boom out a flippant greeting.

Ten years ago Lucas' approach to his seniors would have prompted a curt put-down. But attitudes had changed; there was value in the spontaneous reaction, and from a man of Lucas' big frame it carried some impact.

Check and training engineer, Gordon Brooks, must have given the arrivals his usual courteous nod, and maybe remarked that the forecast over the ice looked even better than for his last trip. He was the only member of the five-man flight deck crew to have flown an Antarctic trip previously.

His fellow flight engineer, Nick Moloney, was possibly behind the counter fitting batteries to an inspection torch. It is easy to imagine him remarking enthusiastically on the flight about having the experienced and highly regarded Peter Mulgrew as commentator. Mulgrew would move among the passengers until they reached the pack ice. Then, if it was acceptable to the skipper, he would plug his microphone into the flight deck and take up a position in the jump seat behind the pilots.

One can envisage Greg Cassin move with his usual quiet efficiency over to examine the file of Notices to Airmen, the 'Notams' promulgated by civil aviation authorities to keep airmen up-to-date with recent amendments to regulatory details.

The background of cheerful remarks and snatches of conversation would continue as Collins reached into his navigator's bag for the company briefing notes. These he would lay beside the big pile of flight planning material on the counter.

Glancing at the additional notes he had prepared the night before while mulling over the flight, he would begin tying together the strings of assorted information garnered not only from briefing intelligence, but also from his own reading and the latest reports from the Meteorological Office and Mac Centre.

Command responsibility first came to Collins as a youngster in the Air Force. His ability was quickly recognised when he joined the airline. In the more fleeting years of his late twenties and early thirties, it took him just six short years to move from the co-pilot's right-hand seat to the command position on the left. Now, 15 years later, he had the respect not only of his fellow technical crews, but also of the cabin staff.

The latter may sound odd, but it is highly valued by pilots. While most of the cabin staff know little about flying an airliner, they have an instinctive feel for the worth of a pilot. Maybe ten or a dozen captains in the company are rated in the top stratum.

It is a match of pilot ability and confidence — should there be a mismatch with the polarities too great, sooner or later there will be trouble. A first-class pilot who is over or under-confident is a poor crew member. But an average pilot whose confidence is pitched at middle level is likely to be a very safe pilot.

It is a subtlety which can be partially gauged from a man's everyday demeanour. Minor crises are not infrequent even on modern airliners and, though passengers are rarely aware, cabin crew hooked into the machine's communication and mental bond quickly notice an occasional indecisiveness, slowness in response, or fluster.

On the flight deck fear can be sensed; not the fear of imminent disaster, but of making a bad decision, of being caught unawares, of slipping outside the psychological linkage known as the loop and having your fellow crew join up to exclude you from the immediate function of flying. It happens in the safety of the ground simulator as well as in the air.

But it never happened with Collins. He could snap short of the catchline of a joke to renew control, clamp power levers and flick switches. When the situation returned to normal he would switch back to the story and everybody would relax.

They had noticed this trait some years earlier when, flying out of Los Angeles, a bolt of lightning blew off two compartment doors on the DC10 Collins captained. There was a deafening bang and the aircraft began to buffet violently. The passengers were very concerned but apart from the incredible noise few realised how critical the situation was until they read about it the next day in the *Los Angeles Times*.

After stabilising the aircraft, dumping fuel and turning her back, he explained over the PA what had happened and what he was doing. There were no lies and no fuss.

At the check-in hall in Auckland's international terminal on 28 November 1979, the 237 passengers — New Zealanders, Japanese, American, British, Canadian and a lone Australian — would have chatted as they awaited the boarding call. They were dressed casually in cardigans and open-necked shirts, they carried shoulder bags and cameras. It was a crowd slightly weighted with the middle-aged and elderly, perhaps a little more prosperous than average; on the whole well-read and strongly motivated.

Among the party of 24 Japanese were two doctors, two optometrists, a dentist, a professor emeritus from Waseda University, several company directors — and two honeymooning couples.

Some had scrimped for the trip. A few, like Trevor Maskelyne, a policeman from New Plymouth, had won tickets in raffles. Seventy-six-year-old Valgria Rawlins of Auckland's Mt Eden had decided on the trip in spite of being confined to a wheelchair. There were extended family groups — Tauranga manufacturer Edward Palmer had with him his two married sons and his son-in-law.

Coincidentally, two girls from England, both exchange teachers posted to opposite ends of the North Island, had tickets because they were determined not to go home without seeing the Great White Continent.

For several, the ticket was a gift from husband, wife or parents. For 17-year-old Mark Mitchel of Lower Hutt, it was a bonus from his maternal grandparents for a successful final year at school — he joined them in Auckland for the trip.

A welcome visitor to the flight deck during the trip south was Christine Nicholson, a Christchurch primary teacher, who was studying Antarctica with her Standard Three class. She had just completed a 10-week extension course at Canterbury University and was rounding out her studies with this trip.

There was Lady Helen Robb, widow of an eminent heart surgeon, and Bev Price, New Zealand's most experienced woman mountaineer and one of five survivors from the disastrous Hardeol Peak expedition in the Himalayas in 1974.

Young and old, national and foreigner, the passengers mingled easily and shared one denominator: a sense of adventure, following in the tracks of Amundsen, Scott, Hillary, Fuchs, Byrd and others. For the well-travelled it was a unique break from the regulated routes of international air travel.

The versatile DC10 was a highly suitable machine for the job. By leaving just ten central seats vacant, the drift of passengers could be loosened up for easy viewing.

From the technical crew's viewpoint it was perhaps the best of the wide-bodied jets to handle — smooth, responsive and 'long-legged' enough to provide ample fuel reserves for the lengthy flight. Above all, it was built around a navigation system more highly refined than that of any other big jet.

The company had worked hard to add value to the trip. Brochures outlining the history of Antarctica awaited the passengers. Sir Edmund Hillary, fellow explorer and mountaineer Peter Mulgrew, and the knowledgeable director of the Antarctic Division, Bob Thomson, frequently flew as commentators.

There were inflight films and menus ('Peach Erebus') which fitted the theme. After a documentary film — 'The Big Ice' — the window blinds would be recessed as the pack ice appeared beneath the aircraft. Suddenly, as the brochures said, 'you were seeing . . . you were feeling . . . you were believing.'

Well, perhaps not physically feeling, standing there in shirt sleeves with a glass of champagne.

The flight had exceptional dramatic attractions. While there was a gradual growth in scenic interest, the Southern Alps and the lonely islands of the Southern Ocean, Antarctica — unlike the Atlantic approach to the Arctic — gives little warning of approach in the way of drifting bergs and ice floes. Suddenly there is the ice sea, and then a long time later, the extraordinary mountains and glaciers of the polar cap.

But it would be Erebus and its smoking volcanic bowl, just 28 miles north of Scott Base, which would hold the tourists in awe. Erebus, entrance to the underworld, so named by

Captain Cook, was taller than New Zealand's highest peak, Mt Cook, and three times as vast. Beneath the smoke-choked crater lay a glowing lava lake which could occasionally be glimpsed. Erebus, as the writers said, put man in his place.

Outside the bronzed viewing glass, DC10 ZK-NZP, awaited at Gate Number Two.

Jim Collins and his crew exchanged light banter with flight despatcher Johnny Johnson as they filed out of his office and past the genial security check people.

Johnson is an old and knowing hand; a navigator who has flown with the RNZAF and the RAF and latterly with the airline. He had taken up his present post when navigators were being phased out because of the impending arrival of the Tens. He sent the crew on their way with a colourful cussing, but his salty approach was appreciated.

Brian Wilson, in charge of the passenger handling and a 'trout fisherman supreme', often brought smiles to the faces of the waiting passengers as he made a quick friendly quip to the passing crew.

On the flight deck Collins and Brooks would have discussed the engineering state of the aircraft and insertions in the maintenance log. There was nothing to query; the ten maintenance concessions were minor — the head off a Hi Lock fastener, a trial period for an unmodified generator control unit; minor points which in no way could compromise safety. The machine was going into the hangar for a major C Inspection on its return.

As always, the two co-pilots would probably have moved smoothly into their pre-flight checks. The navigation system primed with the computer-generated flight plan with Lucas, who had not been present at the RCU briefing 19 days before, watching carefully as each way-point was transposed onto the aircraft's display screen.

What they did not realise was that they were most diligently programming their own deaths. The company computer-generated flight plan could not be automatically loaded into the onboard computer, but had to be transposed manually. There is an obvious danger in this procedure. The company guarded against this by the company pre-flight procedure of that time. This procedure required one pilot to load the plan and the other to check that the computer-generated plan and the onboard plan co-ordinates matched. The pilots did match the co-ordinates in accordance with instructions. But unknown to them, and unguarded against by the company pre-flight technique in force at that time, they had been issued with a computer-generated flight plan which differed from the previous ones and which differed from the one used at their briefing.

They believed that by accurately transposing their computer-generated plan into the onboard AINS computer they were programming an automatic navigation track down the clear 40-mile-wide pathway of McMurdo Sound but in fact, because of the altered plan given to them, they had programmed a collision course with Mt Erebus.

But even having the automatic navigation system unexpectedly aim them at Mt Erebus would not inevitably cause a crash, would it? Because, if they were flying in reduced visibility, for instance in cloud, they would be operating under instrument meteorological conditions, consequently they would know that they could not see, so in order to avoid obstacles, they would be flying at safety altitude above Erebus.

This flight, however, had been dispatched with the understanding that they would make a visual approach to McMurdo. Under the ICAO rules 8 km visibility would be sufficient for them to see and avoid any obstacle, but in this case the company had increased this requirement to 20 km. Obviously, so long as they obeyed visual flight rules, should Erebus appear unexpectedly ahead of them, they would see it at least 20 km away — or would they?

The waypoints are geographic reference points which the system relates to in flight. On scheduled services, code numbers are fed into the computer, but in this case the co-pilots typed on the keyboard (oddly known as the 'scratchpad') the co-ordinates listed on the company's flight plan. The first waypoint — necessarily fixed within metres — was that of Number Two Gate at Auckland International Airport; the final position inserted was the new one printed out by the computer a few hours earlier giving a latitude of 166.58 east and establishing a course, unknown to the crew, over the top of Mt Erebus.

At the later inquiry, barrister Roger MacLaren likened this fix for the most southerly point of the flight to the false light beacons used by Cornish plunderers to lure ships with rich cargoes onto the rocks.

In pilot terminology the aircraft was 'up to full navigation status'. One of the pilots would have checked the other's take-off calculations, then placed the tabulations on the pedestal and set his speed bugs (moveable white peripheral markers on the dial of the airspeed indicators).

Collins meanwhile moving into the command seat on the left would make sure his oxygen mask was functioning, an elementary safeguard since the safety of the aircraft would depend on the mask should the pressurisation fail. He would have double-checked the work of the co-pilots with his own percentage check.

Together studying the take-off data, then calling up the route co-ordinates already fed into the navigation system, Collins would satisfy himself they matched the figures on the flight plan.

As the DC10 began to taxi out, Lucas would have excused himself from the flight deck. Redundant for the time being, he would be back later to give one of the others a spell.

On the flight deck the crew would visibly relax after the intensity of the take-off phases. The aircraft climbing steadily, in clean form with undercarriage, high-lift flaps and slats retracted, and swinging gently onto its planned track southwards.

If left alone, the auto-pilot following the dictates of the navigation system would take the aircraft in a straight line over the top of Mt Egmont to Invercargill, and then make a slight turn left to put them in line for the Auckland and Balleny Islands, deep in the southern ocean.

There was, however, no need to stick rigidly to this track if viewing could be enhanced while flying over the New Zealand high country. The Area Navigation System is infinitely flexible. The machine could be detoured off at an angle, or it could fly parallel to, but several miles off, the inertial track.

But the track would remain indelibly printed on the computer's memory and on the horizontal situation indicator (the HSI, mounted directly in front of the pilot, presents him with a continuous running fix on a display which also indicates any deviation from the flight plan. See page 114).

The maker of the navigation system guaranteed that it would not deviate more than approximately one mile during every flying hour. The pilots knew in fact that it was much more accurate — that a deviation of more than two or three miles in a six-hour flight was rare.

Over the PA system, Jim Collins would have greeted the passengers, introducing each member of the flight crew in-

Figs. 39 and 40 Monitoring Position with HSI

Collins had been given a McMurdo Sound flight plan at his briefing and he plotted his navigation track from that on to a topographical map.

Reconstructing the navigation technique which he followed, it is clear he must also have plotted onto the map, at least over the critical area, the distances to run to his next waypoint.

He was flying visually, but was simultaneously to monitor his position by using his HSI indicator. Fig. 39 on page 115 shows this instrument directly in front of the pilot. A close up of the HSI indicator is provided in fig. 40 on page 115.

So long as Collins remained on his navigation track he could follow his position on his map merely by glancing at the distance to run displayed in the top left (in fig. 40 indicating 40 miles to run) and comparing it with the distances to run he had marked on his map.

The instrument shows him that he is on track, or off track by the position of the bar in the centre of the dial.

The illustrations are the only ones which are available to me and are those of a 747, but the system is similar in a DC10.

Certain witnesses called on behalf of the airline, such as Captain Gemmell, suggested that Collins should not have relied upon visual flight in conjunction with that HSI technique; instead it was urged that Collins should have been monitoring the continuous readout of the aircraft's latitude and longitude and comparing them with the positions on a topographical map.

In my opinion this is a totally unrealistic suggestion. The display in question is situated in the roof of the flight deck at above eye level, and to carry out such a procedure would have precluded the pilot concerned from maintaining the proper lookout required for visual flight. Indeed, elsewhere Captain Gemmell suggests that because the crew were somewhat preoccupied with communications difficulties they seemingly did not observe Beaufort Island. The fact is that the attention devoted to communications was minimal compared with the distraction that would be caused by monitoring the latitude and longitude display in conjunction with a map.

The minimal work involved in communications is analysed in Captain Cooper's appendix of evidence. I can imagine the criticism the crew would have attracted had it been proved that the crash occurred when they were attempting to monitor their position by the latitude-longitude readout when flying visually.

The latitude-longitude readout is intended for testing purposes, not for navigation purposes. The crew would have had no cause to consider referring to it in the circumstances. It is only with hindsight that one can cast round for a means whereby the crew would conceivably have

Fig. 39

Fig. 40

ascertained that their flight plan had been reprogrammed.

In short, using the HSI, as Collins clearly was, provided the correct technique for the circumstances, and would have given the aircraft's position accurately, had the airline not altered the Antarctic flight plan without telling him.

Back at the time when he was planning his flight, Collins would have had no more reason to suppose that the HSI would display an incorrect position than to suppose that some instrument failure would cause the latitude-longitude readout to display an incorrect position.

As Captain Gemmell testified, the DC10 carried plotting equipment for use in case of a breakdown of the navigation system. But should the airline have intended all visual flights beneath 16,000 feet in Antarctica to be monitored as he suggested, then crews should have been briefed accordingly, and a navigator added to the crew. In any event, the technique which it was suggested that Collins should have adopted, provides a record of where the aircraft was in the past, not where it is when the plotting is complete, or where it is about to fly towards.

Explanation of reference to distances north

Collins had been given a McMurdo waypoint the latitude of which was 77 deg 52.7' south, and he must have known this was the same as the latitude for McMurdo Station, as may be seen in fig. 42 on page 134.

When Cassin tells McMurdo at 32.07 their position is approximately 43 miles north, and again when at 38.38 he tells them the aircraft is 34 miles to the north of McMurdo, he knows that by giving McMurdo ATC the distance north from the Dailey Islands waypoint shown above the HSI (see figs. 39 and 40 on page 115), he is giving them the distance north of McMurdo ATC. Since he would know the flight plans were transmitted by the airline direct to McMurdo ATC he would also believe they were aware that the co-ordinates locating the aircraft's waypoint by the Dailey Islands gave the same distance north. In reality, the airline, because of a mistake, did not reveal to McMurdo ATC that the waypoint had been shifted from the Dailey Islands to the TACAN at McMurdo Station. Thus McMurdo ATC, as well as the aircrew, believed the waypoint of the computer flight plan was still by the Dailey Islands in McMurdo Sound.

dividually, and Peter Mulgrew. They would have been invited to relax and enjoy the hospitality of the airline. Perhaps he would have pointed out that they represented .00001 per cent of the world's population to have sighted Antarctica emphasising that was only one-tenth of one-thousandth of one per cent.

He would have followed customary routine in pointing out that the aircraft had not been designed for photography trips, so it was important for those allocated window seats to share with others. Invariably they did, earlier crews had found. Long before the pack ice was reached, the mix and mingle had forged a warm camaraderie.

Inevitably Captain Collins would have pointed out that cockpit visits would be allowed until the start of the descent into the polar area, and that Peter Mulgrew would move among them and provide a live commentary on points of interest.

With the aircraft climbing on track over the long black-sand beaches of the North Island's west coast, soon it would be at its allocated cruise height of 35,000 feet. Collins may have placed a 10-mile starboard offset to enable passengers to get a good view from the left-hand windows of Mt Egmont, the first of the scenic attractions. Behind them the cabin would be beginning to buzz, the first bottles of champagne being broached and a group of excited Japanese probably comparing the conical slopes of Egmont with Mt Fuji.

Egmont behind and the offset removed, they flew on towards the South Island whose saw-toothed ranges — the Kaikouras to the left and the snow-covered Southern Alps to the right — were now in sight, their feet still beneath the horizon.

Oceanic Control at Auckland Airport radioed the latest update on terminal conditions at McMurdo, and the fact that it was good was relayed to the passengers. Jim Collins, at this stage, very likely took the opportunity of cautioning that weather conditions could change quickly in Antarctica, and that a decision would be made on whether to proceed to McMurdo when they reached the Balleny Islands. Should the outlook not be good they would divert to the less spectacular but significant South Magnetic Pole.

With interest rising in the cabin as the spiny ridges of the Southern Alps slipped past on one side and the emerald apron of the Canterbury Plains on the other, the captain would point out the white spills of the Fox, Tasman and Franz Josef

Glaciers, and the unique turquoise water of Lakes Tekapo and Pukaki. The undercut chisel top of Mt Cook, with its deep-blue ice face and the hairline crack of a threatening avalanche, shows up, as does Mt Tasman, with its spectacular windblown plume. Miles below the Swiss huddle of the Hermitage; breakfast would be on time, but the cabin crew reluctant to disturb the viewers until the coast was well astern.

On the flight deck the relaxed rhythm of a well-attuned team would be felt. But the crew would be conscious of the difference of this flight, that it would tax their skills more than a routine service.

Left alone to fly at 35,000 feet, the three inertial boxes and their computer would take them unerringly to McMurdo, turn around and bring them home. But this was certainly not the trip the passengers had paid for. The machine would need to descend soon after it passed the Cape Hallett corner, and continue descending near high land masses until it reached the McMurdo waypoint at the head of the sound.

The high altitude reliance they placed on the Area Navigation would revert to more complex and time-consuming methods; radio beacons including Distance Measuring Equipment, gyro compass, grid navigation and most importantly, visual flight.

There might have been an uncomfortable moment when they had discovered that McMurdo Non Directional Beacon could not be used. Despite its name, this beacon could normally be used by crews to show the heading of the aircraft while tracking to or from the beacon.

On reflection though there was really very little to cause concern. The terminal waypoint was some 27 nautical miles west of McMurdo and Ross Island's mountainous slopes, and long before reaching this area they would be flying visually and within the radar orbit.

The ground controllers would have an accurate fix on the aircraft, providing it was below 6000 feet and within the radar range of 40 miles. They would be able to clear the aircraft down, providing separation from terrain and other aircraft, until it was over the ice airfield. There was the additional safeguard of the Navy's TACAN which could be intercepted by the DC10's Distance Measuring Equipment (DME) to provide an accurate read-off of the aircraft's mileage from McMurdo.

Like any good pilot, Collins would certainly have made up his mind that unless visual conditions were good, there was no way he would take the aircraft below Minimum Safe Altitude (MSA) of 16,000 feet. MSA constitutes a platform for pilots. Once below it they leave behind the self-sufficiency of the inertial navigation system and commit themselves once more to guidance from aids on mother earth, and to visual flying — 'eyeballing' it in pilot jargon.

Nevertheless, they still keep a very close eye on the all important HSI, maintaining the running fix this instrument feeds them — the bird's-eye view of their progress along a pre-planned track.

What Jim Collins — and at that time probably all the other captains in the two airlines operating the Antarctic scenic flights — did not know was that in white surface flying and with an overcast, the normal protection written into Visual Meteorological Conditions (VMC) would not provide a safeguard. In other words, the human eye was not designed to cope with these unusual conditions. It could be readily deceived by at least a dozen undetectable visual illusions.

The International Civil Aviation Organisation (ICAO) sets world guidelines for flying conditions. But its definition of visibility fails to take into account the polar phenomenon in which texture is lost — when the eye can see for miles, but the brain fails to perceive what the eye actually sees.

ICAO defines visibility as: 'The distance at which a trained observer can distinguish a prominent obstacle, unlit during daylight hours, and lighted at night.'

Under ICAO-influenced regulations of most countries this distance must be eight kilometres (five miles) for VMC flying. In these flights, however, Air New Zealand had actually increased the VMC distance to 20 kilometres (about 13 miles) as an added precaution; a change which appeared to provide a comfortable margin of time in which to see, recognise and avoid obstacles.

The vital element which is not taken cognisance of in this conception of visibility is texture — the behaviour of reflected light which interprets contrasts, colour, shading, and shape and enables judgments to be made of relative size, distance and depth perception. Texture information, however, varies enormously in quality. For instance, a rocky outcrop in polar regions may be clearly visible at 20 kilometres or more, but a much closer snow-covered outcrop may not — because there is no differentiation in its texture from the surrounding snow plain.

Similarly, lack of terrain texture may conceal from a pilot the horizon line, or even create a false horizon as low-lying cloud meets rising terrain.

Pilots are trained to make texture interpretations; indeed, the ability to do so is vital to anyone wishing to orientate themselves in their surroundings. Very few people, however, realise the insidious dangers in the loss of visible texture. Those who do have learnt by first-hand experience — the 'white surface' pilots operating in polar regions. Jim Collins and his crew lacked that experience. In descending to 1500 feet over the McMurdo area, Collins was aware that he would be doing a type of visual flying rarely experienced in high speed aircraft.

Approaches to airports are made with high lift slats and flaps deployed. In essence the aircraft is artificially slowed down in varying degrees by these devices extended from the wings. In the Antarctic, however, the low flying would have to be done clean. No chances could be taken by deploying these high lift devices because if they malfunctioned in the extreme cold, the aircraft would be unable to return to cruising altitude and speed, and would just not have enough fuel to return home.

Clean flying is high-speed flying. Drop much below 230 knots in a machine such as the DC10, and its fined-down, streamlined mainplanes lose the ability to generate enough lift to keep it airborne. It is in danger of stalling and plunging to the ground.

So the McMurdo viewing would be an exercise which no normal pilot could fail to relish; keyed as it was to the delight of pure flight, a beautifully articulated machine in skilled hands, flown in relation to the surrounding terrain. Skilled flying, certainly, but nowhere near as demanding as, say, an approach to a difficult airport such as Hong Kong.

In the semi-psychological term which has become familiar to all commercial pilots in recent years, it would be a complete 'interfacing' between pilot and machine.

As the modern airliner has become more automated, so the job of the pilots has become more cerebral. In much military flying, and even well into the jet airliner era, interfacing between pilot and machine was a much more direct exercise. An energy circle was established. Firstly the visual clues were transmitted to the brain, processed and with the precise co-ordination of the skilled pilot, transmitted to the tactile sensors, the fingers and wrists. As pressures were applied so back pressures responded, and the flow reversed.

So too with flying the DC10 visually. But the type of skill required for transition from instrument to visual is an experience known only to pilots. The pilot must learn to reset automatically his mental processes, as well as his focal lengths. The confidence he places in his instruments must be carried through, but tempered with the knowledge that his eyes are fallible, tricks can be played under variable meteorological or geographic conditions.

A typical example is the 'black hole' syndrome, experienced occasionally when going visual from instruments while making a low approach over water. While the water may be only 100 feet below, there is the sensation of a bottomless cavity beneath the aircraft.

Jim Collins had made thousands of decisions at altitudes of 100 or 200 feet of whether to overshoot or land. He had the right to feel as confident as any of the pilots who had previously taken the machine visually over the ice. Providing the weather was good, he knew he would be cleared to take the aircraft down to 2000 feet and possibly lower. Most of the others who had made the Antarctic flight, including the executive pilots, had done just that.

As the man in command, he also knew that the final decision on what height the aircraft should fly rested with him. He accepted the somewhat ambiguous company briefing in the same way his fellow pilots accepted the 6000 feet height stipulation — as the cloud-break procedure limit, an intermediate approach altitude in which further descent had to be cleared by McMurdo control, and made in visual conditions with radar monitoring where possible and limited to 6000 feet and below and within forty miles of the busy McMurdo Base. The radar, of course, could not monitor them on their sightseeing around the dry valleys and in the vicinity of the Ninnis Glacier.

The company later attempted to claim it had set a total prohibition on descent below 6000 feet. It failed when it became apparent at the Royal Commission that their own briefing officers accepted 6000 as cloud-break altitude, not as a minimum. (see note 4 on page 335 headed *The 147 mistakes concerning height*)

Fig. 41. This shows the contours of Ross Island in metres, the actual flight marked by an arrow, and the point of impact at 1500 feet. Had an aircraft been flying at 6000 feet in visual meterological conditions, on the flight plan re-programmed by the airline, and encountered sector whiteout ahead, it would have struck the mountain on about the 2000 metre contour line. Had it been precisely on computer flight plan, the aircraft would have been 1.2 miles to the left, headed almost directly towards the peak of the mountain.

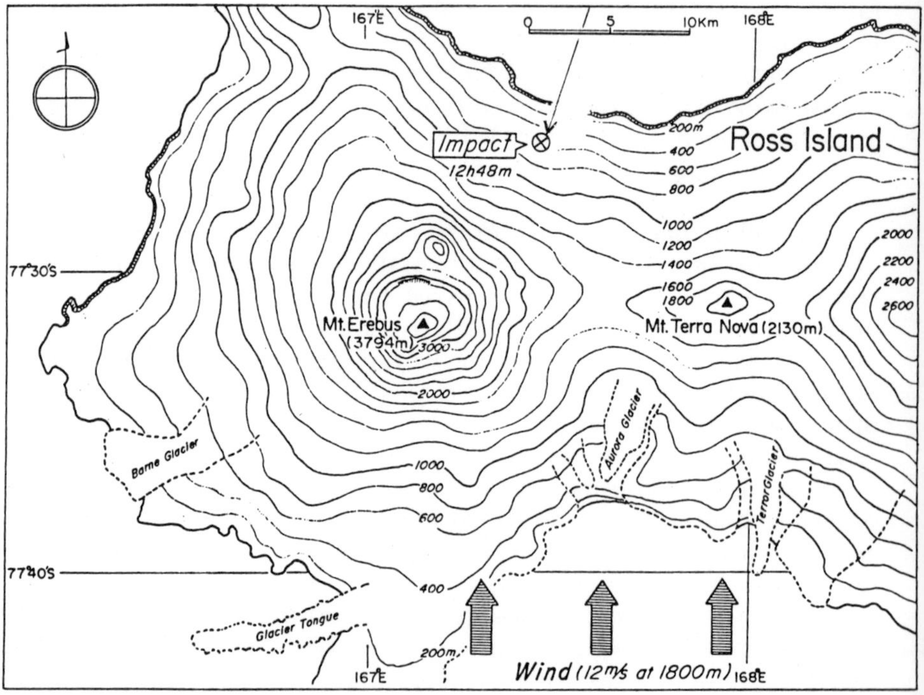

Leaving behind the New Zealand coast, the aircraft pushed out over the southern ocean, the loneliest and most ferocious stretch of sea in the world.

It was a stage in the flight, as earlier pilots had noticed, when innermost qualms emerged; where a feeling of remoteness surfaced. Below, the winds of the Roaring Forties whipped the sea into permanent grey striations. The variable climate and mild turbulence of the land had given way to steady and unremitting ocean patterns.

Back in the cabin it was warm and festive. Chief purser Roy McPherson and senior hostess Sue Marinovic would have set about breaking any last inhibitions with a sumptuous champagne breakfast. Preparations were being made for the first film screening.

The flight commentator Peter Mulgrew was a quiet man but he had a compelling magnetism. After serving in the New Zealand Navy during the Second World War, and later in Japan and Korea, he travelled overland with Edmund Hillary to the South Pole in 1958. Two years later he joined Hillary in the Himalayas. There Mulgrew lost his legs and fingers after suffering pulmonary thrombosis and frostbite. He settled into business but was never content to stay at home for long. He raced yachts with conspicuous success, led an expedition to sail a gaff-rigged 40-footer around Cape Horn in 1973, and was planning an attempt on the North West Passage across the top of Canada.

Mulgrew's sense of adventure was compelling, his intimate knowledge of regions remote from civilisation an asset on the Antarctic flights. Despite his artificial limbs he was accustomed to moving nimbly along the aisle of the DC10, patiently answering a stream of questions and pointing out the fascinating details which floated beneath the aircraft's passage.

The Auckland Islands, some 450 miles south of New Zealand, came into view. They have always been a surprise to everybody, but especially New Zealanders. The main island, much larger than the dot on the chart suggests, is a rolling plateau of what looks like excellent grazing turf. Collins would have banked the aircraft and pointed to the outline of the North West Reef beneath the deep blue water, explaining that the wreck of the ship *General Grant* lay close, yet to be located by salvage experts, and containing a nineteenth century fortune in gold. Perhaps it was the entrepreneurial Japanese, calculating the logarithmic ascendency in bullion prices, whose cameras clicked fastest.

With the flight deck crew checking the final navigation system updates and reviewing the procedure for the imminent changeover to grid, the cross-flow of conversation would establish that they were prepared for the impending mental gymnastics.

They had imprinted the fact that while still travelling nearly due south, they were prepared to accept that they would soon be heading grid north, and azimuths relative to it. They would be also ensuring the practical application from the same briefing, of balancing and checking fuel reserves against the revised wind forecasts.

Flying on the 165° eastern meridian, about four hours out of Auckland and 15 minutes before the Balleny Islands, the first pack ice is sighted.

Scanning ahead from a height of six miles, the ice is a slow dawning revelation. At first it appears to be a light area of stratus ahead. As it nears, the edges sharpen and it takes on the look of translucent rind on blue jelly. Progressively the Antarctic Circle is being broached. The rind becomes multilayered, whiter in patches, cracked by hairlines and flecked with embedded chips.

As Peter Mulgrew often explained, this vast sheet intrudes much further into warmer waters than one would expect. The massive gravitation force of billions of tons of ice on the Antarctic continental dome acts like a squeezed toothpaste tube to extrude the ice sheet hundreds of miles beyond the coast.

This is the release of one of the world's greatest reservoirs of water, a drip feed of ice which, if speeded up by a temperature rise of only a degree or two, would flood the greatest continents.

From the perspective every jet pilot acquires, Jim Collins must have been instinctively aware of the real nature of the gentle pattern he looked down on. These hairline cracks are in fact as broad as estuaries, the white flecks are large icebergs broken adrift from distant glaciers. Warmed by a steady but oblique summer sun, the edges of the continent erode. And the movement is massive.

Flight 901 approached the Balleny Islands, those forbidding footholds on the rim of the continent. Here the autopilot, acting on a minute impulse from the computer, swings the aircraft a few degrees to the left and on track for the next waypoint, Cape Hallett. Collins would have carefully noted the line of constant bearing as the island turning point slid towards the nose of the aircraft, indicating the navigation computers were as usual holding track very accurately and moving the aircraft precisely along the path indicated on his map.

Flying south on this route there is a marked change in

colour and clarity. The atmosphere loses the grey and green softness of the habitable latitudes and takes on the pure clarity of blue and white.

But the Antarctic's third colour is black, the stark black of cliff facings, rock outcrops and ridges. For the Antarctic pilot, the black serves two purposes. It brings home to him the remoteness and inability of the terrain to support any plant life, even the smallest lichen. It also provides the visual ingredient for the human eye to function properly.

The trained Antarctic pilot looks to these dark contrasts for orientation, but knows that they can also be the fangs of death. In isolation, a black basalt face can be a large mountain, maybe 100 miles away. But it can also be a cliff face at point-blank range.

The black calligraphy of the Antarctic was later to ensure the catastrophic end to this flight. But its earlier effect was to add an edge to the gut feeling of caution that Jim Collins and his crew would have shared with the crews of preceding flights. Pressing deep into unfamiliar territory, by unspoken but mutual recognition, all Antarctic flight crews increased the tempo of mental scanning they knew would be part of their existence for the rest of the trip. It manifested itself in the insistent checking, double and treble checking of fuel flows, of numerous performance graphs, of related instruments, of the relationship between the data and the instrument readings.

There was a buzz of quiet assurances, the prick of exclamations. The fail-safe concept was swung into overdrive.

The Balleny Islands, as well as being a waypoint on the route, was the point at which the crew had to prepare for a decision. McMurdo had to be contacted for the latest weather forecast. If this proved to be unacceptable for a safe descent into the area, the airliner would be required to divert to its secondary scenic route, the South Magnetic Pole.

'There is some cloud in the local area, base 3000 feet, visibility 40 miles,' the American controller said.

Collins and Mulgrew probably conferred, and decided to proceed on to McMurdo. They could now make out the Cape Adare coastline of the Antarctic continent, and it was clear. Even from this height the bulk and breadth of the land forms is awesome. Yet this is only a tiny corner of the continent which they would cut across to locate on Cape Hallett, and then turn on to the 400-mile run, down over the low-lying pack ice of the Ross Sea, to Ross Island and McMurdo on its southern tip.

Below, the great 10,000 feet ridges of the Admiralty Ranges unfurl. As far as the eye can see — some 200 miles in this dry and brilliant atmosphere — the continent is a great, buckled, heaving mass, up to the great central plateau, with the South Pole itself at 9,100 feet.

Nearing Hallett crews were able to make out the vast glacier tongues, formed over millenniums, spilling off the Prince Albert Mountains to form solid rafts, the size of Switzerland, in the Ross Sea. Mulgrew would add insights, explaining that the average height of the Antarctic landmass is 6000 feet above sea level; that the mile-thick crust of ice which presses down on the continent, constitutes the world's greatest mass of fresh water.

Collins and crew would have received further confirmation of the accuracy of the navigation system as they passed over Cape Hallett.

The distance from Hallett to McMurdo is 389 miles. There is time enough for passengers to take in and photograph the grandeur of the Victoria Land coastline as it passes on the right-hand side.

For Collins and his crew scanning ahead, the tops of Mt Erebus and Mt Terror should now be appearing over the horizon. But all they could see was a big patch of cloud covering the area up to a height of about 15,000 feet.

Greg Cassin responded to his captain's request to try to reach McMurdo on the VHF radio, but there was no reply. Collins meanwhile was re-checking the frequency of McMurdo's DME. It should be starting to show some distance now on the aircraft's instrument.

The weather radar showed a useless, scattered return of echoes — as expected over the ice. (see note 5 on page 340 headed *What did the on-board aircraft radar show?*) The glowing green printout on the display screen of the area navigation computer was showing 'on track', and about 80 miles to run to McMurdo, confirming the running fix shown on the HSI.

It looked to be reasonably clear over the Royal Society Mountains, to the south-west and out to the left of the Ross Ice Shelf. Collins decided to start his descent down to his safety altitude of 16,000 feet, and to proceed on track. This would still leave the aircraft above the highest cloud, and he could anticipate to get an accurate weather report from McMurdo soon.

The last report had not been very promising, but it did appear that it might be improving even though Ross Island seemed to be completely obscured. Large breaks in the clouds were revealing the black basalt faces on mountains in the Royal Society Range.

There was still no contact with the VHF radio, so Greg Cassin sent off a message on the HF set to McMurdo telling of the 16,000 feet decision, and asking for present weather, and surface visibility. Mac Centre replied that cloud base was above 3000 feet and visibility more than 40 miles. The controller offered to provide the crew with a radar approach.

Collins, hearing the offer, nodded approval. His co-pilot told the controller that radar would be appreciated, and then asked for the frequency of the base's TACAN set. The controller confirmed that the frequency their equipment was already set at was correct. Cassin replied that they were not receiving on the DME, but would call again when the aircraft reached 16,000 feet.

Collins noted the time at 00.10 Greenwich Mean Time (G.M.T.) on the instrument panel chronometer. It was the custom for the captain to do the descent briefing about this point, including all the technical crew and the commentator whose previous experience and recognition of terrain is particularly valuable in visual descents.

The line on the chart showed the nav. track taking the DC10 to the head of McMurdo Sound, past the radar pick-up point, Pt Byrd. It would be here that they would break off to follow the radar or to fly visually around the local area. Most of the important sightseeing is in this area.

The crystal-clear sky and apparently limitless visibility indicated that cloud base was well above the normal circuit height at the airfield enabling a run in to be made from McMurdo Sound across the base area providing a clearance could be obtained and radar monitoring was available for separation from other aircraft. Cassin and Brooks had made entries in the radio log of incoming calls.

'Visibility is forecast to be down from 40 miles to five miles in occasional snow showers, with some broken cloud at 4000 feet.'

Reading back, Collins would have noted the previous forecast gave a base of 2000 feet broken cloud, and 40 miles visibility. The situation looked likely to change quickly, necessitating an update closer in.

The overhead speaker burst into life.

'Nine Zero One, this is McMurdo Centre on High Frequency. I've got the forecaster here to give you your conditions.'

The forecaster came on: 'We have low overcast in the area at about 2000 feet, and right now we are having some light snow, but our visibility is still about 40 miles. *(0018.05 GMT)*

Collins commented: 'The cloud's come down a bit, so we may not be able to get to McMurdo. Very hard to tell the difference between cloud and ice from here.' *(0018.11 GMT).*

McMurdo Centre: 'I can give you an update on where the clear areas are around the local area.'

Cassin responded: 'That would be handy. We'd like a clearance to Flight Level 160.' *(0018.29 GMT)*

McMurdo radio came back with permission to descend and to maintain Flight Level 180. *(0018.47 GMT)* There was a 2000 feet difference in the request and the response. Flight Level 160, an altimeter setting of 16,000, is a standard used in New Zealand air space. But though legally still in New Zealand air space, they were now under guidance of American air traffic controllers and the Americans maintained a standard of FL 180 (18,000 feet altimeter setting). It was a noticeable but minor point which was to have no influence on the flight's final outcome.

A few seconds later the McMurdo forecaster came on: 'It looks like the clear areas are approximately between 75 and 100 miles to the north and west of us. But right now over McMurdo we have pretty extensive low overcast.' *(0018.52 GMT)*

Cassin: 'Roger . . . New Zealand 901. Thanks.' *(0019.14 GMT)*

It has not been with any idea of turning this account into a work of fiction that I have at this stage felt a need to inject a detailed reconstruction of events as I think they may have taken place. Normal standard operating procedure is an amalgam of the personalities and their interactions as crew members, the psychological sets and the visual and perceptual aspects of operating in the loop continually monitoring not only the performance of the machine but also of each other. This model is, I believe, a more accurate portrayal than snatches of conversation quoted out of context. All direct speech is derived from voice and flight data information. Where words have been inserted in the gaps, gestures mentioned, these would of necessity have taken place in order for the recorded remarks to have been made. I have attempted to credit unclaimed remarks to the persons I feel most likely to have made them. It has been my aim to round off what my earlier evidence attempted to achieve, namely that the checks and safeguards built into normal procedure were totally inappropriate for the circumstances encountered. I have used my long-term exposure to the behaviour patterns and operating techniques of the individuals concerned (covering numerous flights and up to almost 30 years in some cases), together with my knowledge of the crew loop procedures, and my detailed study of Antarctic conditions and the various illusory conditions that can be produced. These I have superimposed on the memorable details of my own flight to the Antarctic with Peter Mulgrew. I have attempted to use all of these factors to take the reader inside the cockpit, and reconstruct in detail to the very best of my ability the flight as it may have progressed.

Collins looked at his map on the pedestal on which Moloney was pointing to the area at the mouth of McMurdo Sound. The engineer nodded at the cloud forms ahead and slightly to the right of the aircraft.

'That would be round about Cape Bird, wouldn't it?' he remarked.

The two pilots scanned ahead. 'Right, right,' Cassin agreed.

Collins meanwhile reached forward and armed the altitude alert for the Flight Level 180. Looking ahead, he could see what appeared to be a solid wad of cloud over the McMurdo area, with two fingers of lighter laminations extending out on either side, as predicted by the forecaster.

There was no cause to doubt the integrity of the navigation system, so Collins left it coupled to take the machine up McMurdo Sound, as marked out in the briefing information.

Obviously there was no point in going over McMurdo Base at altitude as they would be in solid cloud with Erebus unseen beneath, and no let-down procedure available for them at this height. The best course would be to obtain clearance to let-down out here where conditions were perfect, over the low-lying sea ice. They could then take a look under the cloud base to see what it was like. Then, liaising with the tower, they could move in beneath the cloud to the radar pick-up point.

Moloney, reading Collins' expression, commented: 'A low overcast over McMurdo.'

'Doesn't look very promising does it?' the captain replied. The others responded negatively. They recognised the door was open for them to express their slightest concerns. *(0019.39 GMT)*

Their optimism rose as the radio from McMurdo came on with its offer: 'Within a range of 40 miles of McMurdo we have radar that will, if you desire, let you down to 1500 feet on radar vectors. Over.' *(0019.56 GMT)*

'Roger, New Zealand nine zero one. That is acceptable,' replied Cassin.

Collins expressed his relief. 'Crikey, that's what we want to hear.'

He eased back slightly on the three power levers, thinking maybe the briefing was wrong and they could pick up the aircraft above 6000 feet, and beyond 40 miles.

They were now passing over a thinning belt of stratus through which the fragmented pack ice showed. But only a short way ahead, a big break of some 40 by 20 miles, broken only by a few fluffy clouds, was showing up.

The safe course was to descend into this space in an orbit which would bring them around and back onto the navigation track again.

From where they were the base of the major cloud areas appeared to be about 10,000 feet. Once they got below this, they could be in the clear with miles of visibility to continue the run up McMurdo Sound until intercepted by radar.

If conditions proved unsatisfactory once they descended into the clear opening, then there would be no problem climbing back to 18,000 feet, and flying on to McMurdo waypoint, before doubling back to do a run over the Dry Valley on the Victoria Coast mainland — which they could still see was in sunshine.

The aircraft, locked on to 18,000 feet, was still flying in crystal sunshine. Elusive glimpses of pack ice showed through the scattered cloud below, and the tantalising knowledge that the land was close at hand kept the crew on visual tenterhooks, scanning keenly for the white glow of sun on snow, and the darker patterns of rock. It was some consolation for the crew that the radar was operating, and that the Americans were happy for them to fly over the ice airfield at 1500 feet.

There was a nagging doubt at the back of Jim Collins' mind, possibly associated with the briefing, that the approach radar would be unable to pick them up above 6000 feet. On previous flights the approach had been largely clear of cloud and, as a sure aid, the feed in had been attained by using McMurdo's NDB (Non-Directional Beacon) to bring them into the orbit of the precision approach radar, at 6000 feet.

Now the NDB was not available, so this let-down would have to be achieved in clear air space beneath cloud. And once contact was made with the radar, the ground controllers could use its vectors to ensure safe separation from other aircraft which might possibly be flying in the same area, as they guided the DC10 towards the huddle of insulated boxes which marked McMurdo's airfield.

Sue Marinovic peeped around the flight deck door to see if the crew needed any more coffee, but was waved off with a grin from Moloney. 'We're going down.' *(0020.17 GMT)*

Cassin, observing the luminous green display of the AINS, told his captain: 'Looks like 114 miles to run.'

Collins took up the PA microphone.

'Ladies and gentlemen, we are going down initially to 18,000 feet. The cloud cover over McMurdo has increased although the visibility is 40 kilometres, so visibility is good.'

He reached forward and adjusted the rate of descent slightly.

'We will be taking advantage of the radar facilities at McMurdo for let-down, which should take us below cloud and give us a view of the McMurdo area.

'That is always likely to change, of course, depending on any variations in the weather. But we are hopeful we will be able to give you a look at McMurdo today.' *(0020.30 GMT)*

As the aircraft continued to sink he leaned forward to read the fuel totaliser gauge. 'What's the weight here?'

Gordon Brooks, who had now relieved Moloney in the flight engineer's chair told him: 'One hundred and ninety-nine and a half.' (199,500 kilograms.) *(0022.29 GMT)*

Collins made a quick calculation on the landing speed graph and told the engineer: 'We'll have a minimum of 260 knots.'

'OK, beauty.' *(0023.07 GMT)*

This was a necessary calculation which now gave the crew a cruising speed for their low altitude work. Low flying requires a necessary speed margin over the stalling speed when the aircraft is flying 'clean' — without the aid of high-lift devices used for landing approaches. This margin is set at 50 percent above stalling speed, and the weight is a major factor when working out the stalling speed.

Actually, as Collins noted, the safe speed would have been

252.5 knots. The 260 brought it up to a rounded figure, and gave a good balance to the rate of progress.

Collins had noticed that his co-pilot was having trouble informing McMurdo of the aircraft's position. The machine's inertial set showed they were 90 miles out from the waypoint, but confirmation was needed from the DME, and this was not coming through because there was no sign of a lock-on with the radio beam from McMurdo's TACAN set.

Cassin was using the HF radio, but it is not the clearest reception, and is easily distorted by atmospheric disturbances, though it has the benefit of not being unduly affected by obstructions such as ridges and mountains.

So Collins took over and was given the TACAN frequency by McMurdo. He then asked to try the short-range but very clear VHF radio. When all attempts to raise VHF failed, they resorted back to HF.

To the two pilots it appeared that McMurdo's VHF transmitter must be of a fairly low power output, and this suspicion was confirmed by the controller's advice to try making VHF contact when the aircraft was 80 miles out. Normally, Collins knew, VHF contact should come on well before this, at high altitude.

What was actually happening at this stage was that the aircraft, now approaching the height of 18,000 feet, and being, unknown to the pilots, on a direct track to McMurdo Base, not to the waypoint in McMurdo Sound, which Collins had plotted, had Mt Erebus interposed between it and the transmitters.

VHF and the TACAN beams only travel on a direct line of sight. They were low enough now for the mountain to block off this line of sight, although the more flexible HF transmissions could get through.

The navigational track heading was being maintained, and Collins noted the mileage to run. The display unit showed 80 miles, and the airspeed indicator 295 knots — the mileage tumbler jerked every ten seconds. A quick mental reckoning put him about ten minutes before the McMurdo Sound waypoint near the Dailey Islands if he continued on the same heading, holding the same height.

But there was no point in this exercise. The edge of the main cloud front was drawing closer and he would have to get beneath it before he could make the dog's-leg turn to hook in south of the island, and fly visually under radar guidance to the ice runway and the TACAN. He tried twice more to raise McMurdo on the VHF, but the radio remained mute. 'I'll have to do an orbit here I think,' he announced. *(0031.01 GMT)*

Cassin and Moloney looked out of the right side of the aircraft and noted the layer of cloud about 6000 feet below. 'It's not clear on the right-hand side here', Moloney said. *(0031.08 GMT)*

He was backed by Cassin who pointed out that there was an area totally clear of cloud to the left. Just beneath the edge of this cloud, indistinct but promising, they noted a faint tracery of dark. It was where they expected to see the western coast of Ross Island, in the vicinity of Cape Royds.

What they were seeing was in reality the basalt cliff faces of Cape Tennyson, running roughly parallel to Cape Bird, but 22 miles to the east.

Ahead was the deep bight of Lewis Bay and beyond that, Mt Erebus, which was invisible. The time was 31 minutes 8 seconds past midnight, Greenwich Mean Time, and 18 minutes, 42 seconds before impact.

Scanning around, Collins prepared for the descent. He told his co-pilot to raise McMurdo on the HF radio. 'Tell him we would like further descent. We have contact with the ground and we could, if necessary, descend doing an orbit.'

Cassin confirmed with McMurdo Centre that there was still no contact with them on the two VHF channels. They were about 43 miles 'north' and descending VMC, he told the Americans.

'Roger Kiwi New Zealand nine zero one. VMC descent is approved and keep Mac Centre advised of your altitude,' the reply came back. *(0032.08 GMT)*

The friendly tone of the controller, his obvious acceptance of the lack of VHF communication, and of the visual approach, indicated that the descent would pose no difficulty — nor concern — for those overseeing the operation from the ground.

All they needed was this orbit; then a run in on the navigation track, up the sound and under the cloud for a further ten miles. If McMurdo was not visible, or they were not showing on the station's radar, then there would be ample room to double back and head north.

Brooks had missed the last part of the conversation with the controller because he had taken off his headset as he

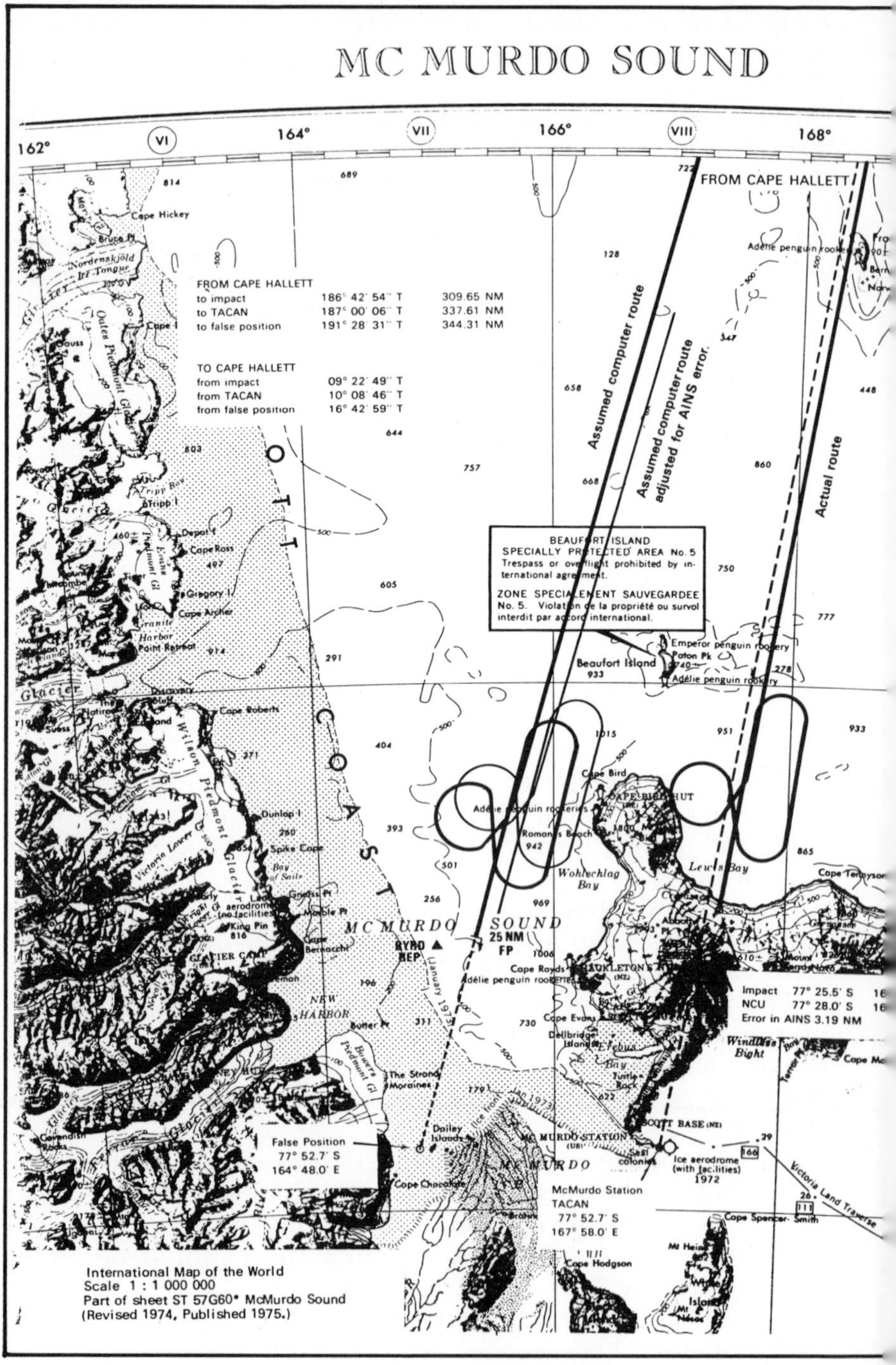
MC MURDO SOUND
FROM CAPE HALLETT
to impact 186° 42' 54" T 309.65 NM
to TACAN 187° 00' 06" T 337.61 NM
to false position 191° 28' 31" T 344.31 NM
TO CAPE HALLETT
from impact 09° 22' 49" T
from TACAN 10° 08' 46" T
from false position 16° 42' 59" T
FROM CAPE HALLETT
Assumed computer route
Assumed computer route adjusted for AINS error.
Actual route
BEAUFORT ISLAND
SPECIALLY PROTECTED AREA No. 5
Trespass or overflight prohibited by international agreement.
ZONE SPECIALEMENT SAUVEGARDEE No. 5. Violation de la propriété ou survol interdit par accord international.
Beaufort Island
MC MURDO SOUND
25 NM
FP
Impact 77° 25.5' S
NCU 77° 28.0' S
Error in AINS 3.19 NM
False Position
77° 52.7' S
164° 48.0' E
McMurdo Station
TACAN
77° 52.7' S
167° 58.0' E
Ice aerodrome (with facilities) 1972
International Map of the World
Scale 1 : 1 000 000
Part of sheet ST 57G60* McMurdo Sound
(Revised 1974, Published 1975.)

Fig. 42: Navigation track given to Collins at RCU briefing is shown on far left in thick outline. Dotted line represents track as re-programmed by airline eight hours before flight. Crew were not told of the change. The actual flight path followed by flight 901 is shown by thick solid line on far right.

At his briefing Collins was shown the Antarctic computer flight plan, consisting of a series of waypoints, each located by its latitude and longitude. The final waypoint before the return was 77°52.7'S 164°48.0'E. That waypoint was underlined on the flight plans which were shown to the crews that were being briefed. Collins believed that so long as he kept his aircraft on its navigation track, it would follow the path shown to that waypoint to the west of the Dailey Islands. He departed from navigation track on this leg only to carry out the figure of eight descent orbit to let down through a large clear area in the cloud cover which was developing. It is an escapable conclusion that Collins believed that there was some inherent protection or safety provided to the aircraft by its adherence to navigation track. There is no other explanation for adhering to it while flying visually.

The route on the left which he believed he was following is marked by a dotted line, 25 miles from the waypoint, being where Collins believed the aircraft was located when he said at 0049.30 GMT: 'We're 26 miles north. We'll have to climb out of this.'

The map was exhibit 250 supplied to the Royal Commission by the Department of Land and Survey, and the route calculations were made by their Geodetic Computing Branch from data supplied by the Civil Aviation Division, Ministry of Transport.

obligingly moved back to allow the passengers through to the view up front.

Cassin radioed McMurdo: 'We are vacating One Eight Zero. Will advise level.' *(0032.10 GMT)*

Collins tilted the right wing down so that they could see how much clearance they had on that side. He saw that the cloud was about 8000 feet below. 'I'll come around there and take her down to 10,000,' he said.

'OK,' Cassin responded as he set the auto-pilot to control the rate of the aircraft's descent.

Meanwhile, Brooks was returning to his seat and replacing the headphones when he noticed Cassin's manipulations. 'They gave us 10,000 feet?' he asked warily.

Moloney explained that the Americans had just given approval for visual descent, and not a height.

'Got to keep them advised,' the co-pilot chipped in.

Brooks had missed part of this cross-flow as he adjusted his earphones. He pressed the point. 'No, see if we can descend VMC.' *(0032.33 GMT)*

He was reassured as both Cassin and Moloney nodded, and realised he had missed clearance from the Americans. He quickly caught up with the play, scanning the new settings made on the instrument panel — which now made sense. 'I see alt. heading select and vert. speed.'

As the aircraft swung around on its right-hand turn, he leaned forward to explain to one of the Japanese passengers that while there was not much to see at the moment, they would get good photographs when they got below cloud base.

Collins and Moloney were discussing the speed reduction to 1.5 of the stall speed for descent below 6000 feet.

Mac Centre was again trying to raise them on VHF, but without success. However, the tape recorder at Mac Centre was getting it all down. The orbit pattern, which the crew believed (because of Collins noting the flight plan given at briefing) the aircraft was following, is the left-hand orbit on Fig. 42 on page 134. Unknown to them, because of the unnotified change in flight plans, the actual orbit was the right-hand orbit on page 134.

Collins picked up his PA handset again. 'Captain again ladies and gentlemen. We're carrying out an orbit and circling our present position, and will be descending to an altitude below cloud so that we can proceed to McMurdo Sound.' *(0034.21 GMT)*

Brooks leaned across from the observer's seat and looked out of the right side of the cockpit. 'There's Wilson,' he said — meaning Wilson Piedmont Glacier. (see note 6 on page 342 headed *Could 'Wilson' mean Wilson Piedmont Glacier?*)

Collins had his finger on the map about 15 miles short of the Byrd reporting point. That tied in nicely with the brief sighting he had of the high basalt face he believed was on the slopes of Erebus far out to the left, and the horizontal line of a cliff face which lined up with the edge of Ross Island around Cape Royds.

Actually, the basalt face was on Mt Terror, to the east of Erebus, and the cliff he assumed was Cape Royds was really Cape Tennyson.

One of the passengers interrupted their viewing to ask if it would be difficult to get down to see Scott Base.

'They'll have no problem once they get down clear of this cloud layer,' Brooks responded. *(0034.21 GMT)*

As the aircraft came around on the full 180 degrees of the orbit, Cassin noted that the long-awaited contact with McMurdo's surveillance radar had been made. The aircraft's transponder was working, its blue light indicating that the ground radar was being intercepted, amplified and returned to the Mac Centre controllers.

In fact, this return would have placed a clear image on the radar during the full time their aircraft was between radials 186 and 180. A map showing this is printed in The Mahon Report, page 129.

It was a reassuring sight to the crew, and Collins responded with a brief 'OK'. He reflected on the limited range of 40 miles. It seemed all McMurdo's VHF equipment suffered this limitation.

At this stage, the aircraft's orbit had taken it close to the western tip of the Bird Peninsula. The radar could reach them over the low-lying coastal slopes of Mt Bird. But contact was not to last as they turned onto an easterly heading, back towards the navigation track. However, a radar trace detected or otherwise would have been faithfully produced on the radar screen at McMurdo for an appreciable time.

For the crew, confirmation of their belief, if they needed any, that they were orbiting over the sea ice, mid-way between Ross Island and the Antarctic mainland, was piling up. They had received no suggestion from McMurdo that their location as revealed on radar was other than normal and safe. Their psychological set was being reinforced.

The turn over the top of the finger of stratus cloud gave Cassin and Moloney a good chance to observe the large clear hole in the cloud which lay ahead as they came back onto an easterly heading. It appeared to coincide nicely with the navigation track running up the sound. Unfortunately though, a finger of cloud spread across the track and joined up with a great slug of almost unbroken cloud which appeared to be sitting over Ross Island to the east. (See passenger photos 1 and 2 on page 68.)

Clearly they would not be able to descend visually on this orbit alone, but would have to find a further orbit which would bring them low enough to continue the entry through the clear area.

Once lined up in this position, it should be no problem to bring the aircraft down over the vicinity of 'the Byrd' — the Pt Byrd radar reporting position which lay almost centrally in the rough triangle of sea ice encompassed between Ross Island and the mainland. Then there would be the clear run up the sound under radar guidance, and clear skies ahead over the Ross Ice Shelf. The cloud appeared to be localised over the Ross Island land mass.

As Collins swung in over the navigation track they lost VHF contact again. He resolved to press south along the track, but still above cloud, for the few miles which would put them over the Byrd reporting point area. This should enable them to regain VHF contact; it was only 23 nautical miles from the McMurdo terminal waypoint.

If contact was not made, they could turn north again until they could position themselves for the visual descent through the clear area. This run would give them the best chance of making an early assessment of conditions for photography.

He knew the regulations would allow him to take the aircraft down to 1000 feet above any obstacle which could be seen from a five-mile radius. But running alongside the Victoria coastline at 1000 feet would not be the best vantage-point for the photographers. A height of at least 1500 feet would be a minimum from which to obtain a good perspective. This also appeared to be the lowest altitude McMurdo approved for these scenic flights. So he would use his lower limit of 1500 and try something a little higher — bearing in mind the need to stay well clear of cloud.

Collins swung the aircraft over to the left, out of the south grid and navigation track, to begin the turn which would have them backtracking on a brief northerly course. He eased the power levers back, squeezing the auto throttle disconnect and simultaneously pressing the indicated airspeed button.

The big airliner swung easily onto the new navigation track, the nose lowered into a descent attitude and automatically assumed the correct angle to hold the airspeed commanded by the pilot.

As he finished, Collins glanced casually around seeking the brief eye contact with each crew member and the note of affirmation it would carry. The new track looked good, set to pass clear of the large fingers of stratus on either side, particularly the one on the right whose ragged edges occasionally parted to throw up glimpses of threatening black basalt faces at high altitude on Ross Island.

From the navigational position they believed themselves to be in, these escarpments appeared to be on the western face of Mt Bird.

The escarpments were however on the slopes of Mt Terror — almost exactly on the same bearing and the same distance as their assumed position would have placed Mt Bird.

The descent profile of the machine was obviously going to get them down below the base of broken cloud before they reached the continuous cloud band stretching across the track forming the northerly border of the clear hole. This was only, as pilots say, a 'whisper layer', an insubstantial smear of haze. But Collins had to ensure, by turning or increasing the rate of descent, that he would stay visual.

He must have turned around to see Lucas leaning into the flight deck to look at Brooks' map. He was obviously wondering why the aircraft was turning north. Collins explained: 'We are just going VMC to McMurdo and then coming back in.' *(0042.27 GMT)*

The meaning in the pilot's verbal shorthand was clear. The aircraft was backtracking and maintaining visual conditions. When it was below cloud the crew would turn in again and fly towards McMurdo Station in visual conditions.

The aircraft was now sinking between the 10,000 feet and 9000 feet levels, maintaining a steady 273 knots. The time was 42 minutes 27 seconds past midnight, Greenwich Mean Time — seven minutes

23 seconds before impact.

Collins reached up to set the initial altitude warning for 2000 feet. Cassin noted and confirmed the action. As he was leaving the flight deck Lucas heard the voice of the McMurdo controller, still coming over the HF channel. The controller asked the crew to report to him when their distance measuring equipment showed them to be ten miles from McMurdo.

To Lucas, this clearly indicated that Mac Centre was aware of the reason for the manoeuvre, and of the position of the aircraft, and was standing by to welcome them into the sight-seeing area. He moved back into the cabin to make way for Peter Mulgrew who was returning with briefcase and headset.

'Right,' Collins said. 'Are your headings OK?' His co-pilot and the two flight engineers nodded as they checked to see the grid headings matched.

Collins: 'IAS hold, IAS hold.' The three confirmed that the indicated airspeed was at the level programmed. *(0042.46 GMT)*

Positively, but imperceptibly, the taut lines of communication and monitoring were tightening as the aircraft continued its gentle descent. On either side, the cloud laminations rose above them like walls.

Collins was scanning carefully along the left side where he knew the invisible thread of the flight path tracked. When he re-engaged the automatic navigation mode, the aircraft would return to this magnet like a fleck of iron.

As he expected, the view to the left was of clear sea ice with lots of black holes punched into it. Below, visibility was excellent, and by now they could see the bottom hem of the cloud curtain — visibility appeared to stretch for hundreds of miles between cloud and sea.

Now was the time to make the turn towards McMurdo. He could be totally confident that the landmarks would show up clear and unobstructed. His careful airmanship would be followed by the others as he turned his head to the left to check before the turn.

'We're VMC around this way so I'm going to do another turn in,' he said. The others affirmed with nods as he set the heading selector knob for a 40-degree intercept of the in-bound navigation track. Their confidence was reinforced as the machine came around on the 25 degrees bank, allowing them all a clear view to the left of the aircraft. *(0042.49 GMT)*

Even though Collins knew that a clear path between cloud and sea ice lay ahead, and that he would have to decide whether it was suitable for sightseeing, he felt the same stirring within him which happened on an approach to an airport lying below the surrounding terrain.

Adrenalin would flow because of the blind trust one had to put in the proven skills of electronic engineers — their ability to aim a narrow radio beam clear of obstructions so that a pilot flying blind could still manipulate the controls to avoid a collision.

It is an anxiety, basic to man, which even the most experienced pilot never fully comes to terms with. But allied to this, the body copes by subconsciously changing gear — 'hyping up', they call it.

It speeds up the sampling rate from all the available visual cues, intensifying the receptiveness of that limited capacity single-channel computer, the mind. Whether the approach is made on instruments or visually, it is the human eye which provides the inputs.

Collins would notice in a confirmatory glance that he shared the process with his co-pilot. They were lucky to belong to a relatively small airline. To fly together often was to know each other's mannerisms and inflections. This could be vital in providing early warning of discomfort and dissatisfaction with the way the job was being done.

The power levers moved in response to a nerve flicker from the computer, a reminder of the vast amount of horsepower these infinitesimal pulses of energy commanded. Power paid for at the rate of eight tonnes of fossil fuel an hour for the three engines — and commanded by a minute electrical signal.

Collins had often reflected with his fellow pilots on the vastly disparate energy forces which combined man, machine and navigation in an unbroken circle. At its most massive this energy could shred 300 tonnes of machine and its occupants in a fraction of a second; an extension of the energy which reached the brain from the eye, and which amounted to one-twentieth of the voltage of a torch battery.

It was an abstract exercise, yet it held immutable lessons for the modern jet pilot. Should the flow of energy falter at its slightest point it could bring disaster. Conversely, harmony could be restored to rampaging brute energy by a flicker of the same mini-voltage.

He knew these small energy units were to be their main

protection against collision with other aircraft, or the ground. Yet now they would have at least 40 miles of visibility, and an auto-pilot maintaining a heading with extreme accuracy up the clear airspace of McMurdo Sound. But he was still happy to have five sets of human eyes, and he felt still more assurance as Peter Mulgrew took his seat behind.

'Sorry I haven't got time to talk, but. . .' Collins was leaning forward to look at the troublesome VHF selections.

'You can't talk if you can't see anything,' Mulgrew responded before he had time to finish. There's no sense in describing an endless panorama of sea ice, he thought.

Collins took up a discussion with Cassin on the best VHF frequency to try, and Brooks offered to relieve Moloney in the engineer's seat.

'There you go — there's some land ahead,' Mulgrew exclaimed enthusiastically. *(0043.27 GMT)*

They were all looking to the left, the direction the aircraft was turning. There was the dark rock-face of a coastal cliff to the south-west. It lay almost exactly on the azimuth for the part of the mainland coast known as Cape Bernacchi. They confirmed his sighting.

'Where's Mt Erebus?' Brooks asked, lifting up his seat to look over the left-side cockpit coaming.

Armed with the new-found sighting of what they confidently thought was Bernacchi, Mulgrew was able to indicate the spot on the map east of the track line. 'North of McMurdo.'

At this stage the aircraft was almost directly 35 miles north of Mt Erebus. The cliff face Mulgrew pointed out was not Bernacchi but a similar escarpment lying along the eastern shore of the Bird Peninsula. The azimuth map, fig. 12 on page 46, shows where the crew were looking, and the remarks are analysed on pages 49 to 60.

The crew were well aware that Ross Island, out of sight to the east, was their most threatening terrain since they had the Victoria coastline in view.

Their immediate concern was a good visual sweep of the area into which they were turning. As they peered out they could see a long white plateau of sea ice extending more than 40 miles before it merged into the horizon. The merging point was obviously about the area in which the McMurdo terminal waypoint by the Dailey Islands lay.

Certainly there was now no problem about tracking up the

sound on the navigation track. Collins' interrogative glance revealed no qualms.

'I'll arm the navigation again,' he said, putting the aircraft into its final sweep to link up with the navigation track. *(0043.35 GMT)*

The high basalt rock he had noticed earlier, and which he assumed could only be on the slopes of Mt Bird, was now masked by cloud hanging like a curtain to the left and ahead. The Ross Island coast was obviously going to pass about 20 miles away to the left, and it would give the passengers something to see as it revealed itself while approaching McMurdo. Peter Mulgrew would have something to comment on.

Besides, he could see the fairly even base of the stratus was going to become almost continuous ahead, and would shelve down gradually to about 2000 feet — as the McMurdo forecaster had estimated. He could safely maintain a level beneath it.

He knew from the simple formula learnt in his basic training days that at 1500 feet, the aircraft should be able to intercept the VHF over the flat terrain from a distance of a little more than 40 miles. But this could be considerably reduced by a low-powered transmitter. Cassin, who seemed to be reading his thoughts, affirmed his satisfaction with a thumbs up, and 'OK'. *(0043.35 GMT)*

The machine was now coming around almost fully on its 180-degree turn, and as it regained the unseen rails of the navigational track leading south, Collins noted: 'Alt. Nav. cap. IAS hold.' (altitude noted, navigational track captured, hold the same indicated airspeed.)

The others nodded. They were all looking carefully for features which tied their progress into their position on the map. The irritant of the perverse VHF still dogged the two pilots as various channels were suggested and rejected.

'Keeps you busy on comm. doesn't it,' Collins sympathised.

'Well, you know it's ah. . .a bit of a nuisance,' Cassin responded with mock exasperation. They both knew that on this final run they would almost certainly regain VHF contact; and visual conditions were excellent. *(0044.06 GMT)*

McMurdo came on the air. Again it was the HF channel. They asked for a report of the cloud base in the area. Cassin replied: 'Roger New Zealand Nine Zero One. Fifty miles north, the base was one zero thousand, 10,000. We are now at 6000 feet descending to 2000 and we're VMC.'

As he spoke he noted that their optimism was not misplaced. The shelving layer of cloud they were now beneath seemed to lie right over McMurdo Sound. On both sides, but particularly to the left, it merged into a much thicker overcast. There was no hard line of an horizon, but this was not to be expected beneath the feathering base and on a downward approach.

Mulgrew, constantly scanning outside, had failed to spot any more terrain beyond the supposed Cape Bernacchi which had now drifted around to the right. The aircraft was descending in a clear corridor hemmed by cloud.

Collins turned around to the explorer and lifted his right headset earmuff off, pointing to the west. 'We had a message from the Wright Valley and it's clear out there.'

'Oh, good,' Mulgrew smiled. *(0045.26 GMT)*

Collins: 'So if you can get us out over that way. . .?'

'No trouble,' Mulgrew replied.

'Right,' nodded Collins, sweeping his eyes across the cockpit to where Mulgrew was now looking. 'Taylor on the right now.' (see note 7 on page 344 headed *Location of Taylor Valley*)

Mulgrew nodded his agreement. Although he was no pilot, he knew it made good sense to establish firmly their position with the aid of McMurdo's radar before heading off on a new track. Besides, the passengers would at least get a sighting of part of Ross Island. The cliffs of what must be Cape Royds, on Ross Island, were already emerging from under the overcast on the left.

What Mulgrew was observing were the cliffs of Cape Tennyson on the eastern flanks of Lewis Bay. The time was 45.36 GMT; the aircraft was to continue on a straight track to the point of impact, four minutes 14 seconds later. The crew believed that the view ahead was the entrance to McMurdo Sound, Fig. 35 on page 79. The actual view was the entrance to Lewis Bay, its appearance altered by sector whiteout, Fig. 34 on page 78.

Mulgrew decided it was time to update the passengers. 'Peter Mulgrew speaking again folks. I still can't see very much at the moment. I'll keep you informed as soon as I can see something that gives me a clue to where we are — we are going down in altitude now and it won't be long before we get quite a good view.' *(0046.02 GMT)*

There was nothing to achieve by telling them they were still over sea ice, and if they strained they might see Cape

Royds. They were now descending through a gap in the clouds, and clearly these clouds were obscuring Mulgrew's long range view to the east and west of the aircraft.

Collins scanned methodically through an arc of 100 degrees, noting the surface of the sea ice was smeared here and there by shallow pools of surface fog.

The cloud base seemed to be lowering at the right rate to bring it down to 2000 feet overhead McMurdo, now some 40 miles ahead. For the first time he could see land on either side, confirming the navigation system was doing its job.

The flat sea ice stretched reassuringly ahead, and now there was a clear horizon line across the snow at the head of McMurdo Sound. Soon, once the western rump of Ross Island was passed, there would be radar and visual contact with the base.

'A thousand to go,' Cassin said, as the machine slipped gently past the 3000 feet level. 'Alt., Nav. track. Vert. speed.'

Collins was engaging the automatic throttle having rechecked the speed was 252 knots. He responded, 'Speed.'

'Speed I see,' Cassin came back, confirming that the aircraft was now completely configured for the run to the radar pick-up. *(0046.28 GMT)*

The two pilots and Brooks noted that the altitude warning showed they were approaching 2000 feet, which had been set on the counter. Collins decided there was still plenty of clearance between the aircraft and the overhanging cloud base at this stage. He waited for the throttles to move forward automatically locking the machine at 2000 feet.

Looking out to the right, Cassin saw the well-defined cloud base cutting across the hills behind the Victoria coastline. Should it be necessary, there was plenty of space to turn right and head back towards the Taylor and Wright Valleys. Brooks had lost sight of the coastlines as they slipped back beyond the cockpit coaming, and raised himself for a better view.

'Where's Erebus in relation to us at the moment,' he asked Mulgrew. *(0046.39 GMT)*

'Left, about 20 or 25 miles,' Mulgrew replied, 'about 11 o'clock.'

Collins picked up the discussion, noted it tied in with his assessment of 15 miles to run to Pt Byrd, and nodded agreement.

'Yep, yep,' Cassin said, noting the distance to run to the head of McMurdo.

'I'm just thinking of any high ground in the area, that's all,' said Brooks agreeably, easing back into the seat and running a practised glance over his dials and switches. *(0046.48 GMT)*

At this point the aircraft had entered the confines of Lewis Bay. The western arm, Mt Bird Peninsula, was beginning to pass astern. To the crew, the flanking coastline looked exactly as they would have expected had they been flying down McMurdo Sound, 28 miles to the west. They were just three minutes two seconds, and about 15 miles, from the point of impact.

Realising he had a better seat position for viewing Ross Island, Mulgrew leaned forward to point out the position he believed they were in, on the map. 'I think it will be left.'

'Yes, I reckon about here,' Moloney said.

Mulgrew's finger was on a central spot on the northern approach to the sound. He scanned the wad of heavy cloud backing up from the cliffs of what had to be Cape Royds, but he failed to sight the basalt face they earlier noticed on the higher slopes.

'No, no. I really don't know. That's the edge,' he said with the clear confidence of a visual sighting, and he pointed to the coastal cliffs to the left. *(0047.02 GMT)*

None of them could have any doubt that Mt Erebus was well masked by the thick cloud beyond the coastline, and they were happy to see it passing many miles away to the left. They knew the captain would not consider turning left until they were well past the hidden hump, and under guidance from the radar.

Collins noticed the movement ahead of a spur of vapour hanging from the otherwise even cloud base. He reached for the pitch wheel. The aircraft was beginning to go into hold, and the slight pitch down which he turned on the wheel cancelled the altitude hold.

'We might have to drop down to 1500 feet here, I think,' he said. *(0047.43 GMT)*

'Probably see further anyway,' the co-pilot concurred.

The clean sweeping view ahead of the aircraft contrasted with the cloud build-up to the left.

As they descended through 1700 feet, the overhanging spur passed above them. 'It's not too bad,' Cassin said.

Collins nodded and decided he would level out again at 1500 feet. 'I see vertical speed for 1500.' *(0047.55 GMT)*

The blue and brown altitude spheres of the four artificial horizons in front of the pilots were showing about six degrees of nose-up altitude as the big silver wings held a constant angle of attack to the airflow — the slight nose-up was a characteristic of the DC10, flying clean and below its accustomed cruising height in the stratosphere.

Although the two pilots and Mulgrew could feel wholly confident that what they were seeing was what they would have expected to see, both were becoming aware of the strong contrast in the light intensities from above to below this relatively thin and occasionally broken canopy of cloud.

Their distance to run on the HSI showed they had yet to reach the point they reached directly above them on their former run south — the point where they broke off to begin orbiting in order to get below the overcast. Above the cloud layer they had not experienced this peculiar diffusion of the sun's rays. Now they attributed it — correctly — to reflection off the ice. The peculiar thing was that it seemed to give them more light, rather than less as they would have expected.

The trap was now closing on the unsuspecting crew. Not only had a unique set of circumstances invalidated all the careful preflight map preparation by shifting a 13,000-foot mountain in front of them without their knowledge, but the layer of cloud they were so carefully avoiding — as they were instructed to, because the reduction in transparency would reduce their visual protection — was now insidiously rendering their eyes oblivious to the physical nature of the area they were flying into. *(See caption to Fig. 16 on page 63)*

Peter Mulgrew had been looking carefully at the western edge of Ross Island. Placing the map on the pedestal, he pointed well back to the left. 'I reckon Bird's through here.' *(0048.05 GMT)*

His tone was positive as he swung his finger into the cloud to the north of the cliffs. Collins noted this, along with the terrain warning light which he cancelled simultaneously as Cassin called: 'Terrain 1500. Alt. hold.' (confirming they were 1500 feet over the ice.)

Collins detected a slight change in Brooks' voice as the engineer called: 'Hold on both Nav. track.' He sensed his old friend was telling him that from where he was sitting he couldn't see anything because of the cockpit coaming, and felt he was slipping out of the loop. *(0048.23 GMT)*

Now Brooks and Cassin were discussing the TACAN frequency, and Collins and Mulgrew resumed their keen surveillance of the cliffs, passing to the left. They were looking towards the southernmost end of the cliff face, waiting for a glimpse of the block buildings and fuel tanks of McMurdo Station.

From this position they could then make an immediate decision on whether the viewing would be worthwhile over the area. Their scanning was not being rewarded — no sign of the expected scatter of huts.

'Actually, these conditions don't look very good at all'. Collins said thoughtfully.

'No, they don't.' There was disappointment in Mulgrew's response. The explorer venerated this part of the world, and was disappointed he might not be able to pass his enthusiasm on to the passengers. He wondered whether it would be worthwhile drawing their attention to Ross Island before they turned away.

Collins turned again to his co-pilot. 'What's um. . .have we got them on the tower?' *(0048.55 GMT)* (see note 8 on page 344 headed *When did aircrew first express anxiety?*)

His tone indicated he was considering abandoning the run over McMurdo.

'No, I'll try again,' Cassin said. He did not sound hopeful, and he noted that they only had about five more miles to go before they were abeam of Pt Byrd.

Mulgrew heard these remarks, and chipped in with a confirmatory remark. 'Looks like the edge of Ross Island there.'

He was pointing to the cliffs they had been watching to the left. Now they appeared to be edging slightly towards the aircraft's flight track. As they expected, the western coast of the island seemed to have a large hump projecting into the sound.

Brooks had noted the change in tone of the two pilots, and added his own verbal vote to the consideration: 'I don't like this.' *(0049.24 GMT)*

From his position, the cockpit coaming blocked all visual references to the terrain texture and surface. Without this, the strange, almost eerie light was registering rather more in his consciousness than it was with the pilots. He had no knowledge that this in fact was a feature of whiteout conditions.

A passenger seated on the left had just shot a 35mm colour slide of the coastline the crew presumed was the western side of the island. Other passenger photographs recovered from the wreckage showed the airliner's interior bathed in sunlight. The photographs suggest that passengers were in a festive mood and taking photographs from both sides of the aircraft.

They were still well clear of cloud, and visibility was good. They could see the flat sea ice stretching endlessly ahead.

Collins had now made up his mind to head back. They still had no radar or VHF contact, which they should have picked up some minutes earlier, unless McMurdo had some power supply problems. There was certainly not much to photograph, except sea ice. 'We are 26 miles north. I'll have to climb out of this,' he said casually, but with regret. *(0049.30 GMT)*

But first there was another option to be weighed up. He knew he still had a lot of clear sea out over to the right over which he could turn and back-track on a parallel to the navigation track. Cassin noticed Collins looking out to the right in an effort to gauge the distance to the mainland coastline, but the cockpit coaming was masking his view. The co-pilot assured him it was 'clear on the right'.

Mulgrew indicated that in his opinion, straight ahead or round to the right was fine. The Ross Island rump was now passing safely by on the left, but he thought it would be helpful to fix in the captain's mind their position before a manoeuvre was undertaken. 'You see Ross Island?' *(0049.35 GMT)*

Collins was looking out to the left at the cliffs. He decided to reject the right turn, and responded: 'No, Negative.' But as he pulled the heading selector knob, the aircraft commenced a right turn onto the previously accepted intercept heading. Collins casually reselected the inbound heading and the aircraft rolled back to wings level, prior to its left turn to regain heading.

Cassin could see the clear sea ice stretching for about 25 miles across to Cape Bernacchi on the right. He reinforced the right turn option by saying, 'No high ground if you do a 180.' *(0049.38 GMT)*

Collins was about to tell Cassin to call up McMurdo on HF to report that they would be climbing ahead on grid north if they could get clearance. The cloud base appeared to be quite high above them now. Failing clearance, he would certainly stay visual and make the right turn.

The cloud base seemed to be higher now. It would, he thought, be good sense to climb towards it at this stage — they should certainly get the radar then. The flat sea ice still stretched reassuringly ahead for miles.

The apparent increase of the cloud base above them was, he was not to know, an illusionary effect of sector whiteout.

Suddenly the radar altimeter dropped to about 700 feet. Almost immediately, the 500 feet terrain warning light blinked on and an urgent electronic voice called: 'Whoop, whoop, pull up, pull up!' *(0049.44 GMT)*

The hands of Collins and Brooks flew almost simultaneously to the power levers. 'Go around power, please,' Collins called.

This is a command to the flight engineer after the captain has advanced the power levers for the flight engineer to line up the power settings at a value a bit below the full thrust available and avoids unnecessarily overboosting the engines.

The parameters on the flight recorders showed exactly the 'conditioned reflex' rate and actions one would expect to see in the simulator during the times when an unexpected ground proximity warning is triggered by the simulator instructor who is checking to see that the crew immediately carry out the 'go around procedure' and decide if it is a false warning afterwards.

The reflex 'go around procedure' the captain initiates by rotating the aircraft to approximately fifteen degrees nose up, at the same time, he moves the power levers towards the full power position calling 'go around power' to the flight engineer who follows up on the power levers and lines all of the engines up at the correct power setting, a little below full power, so that the engines do not overboost.

The crew must not attempt to decide if it is a spurious warning prior to initiating this action, even if they are inclined to believe it is a false warning. ***The HSI is explained on page 114 to 116.***

He peered through the windshield in disbelief at the unchanging flat surface ahead, and heaved hard on the control yoke.

In a calm but urgent voice Brooks called off the ground clearance: 'Five hundred feet. . . four hundred feet. . .'

Even in these last few seconds of their lives they did not know their eyes had been deceiving them. They must still have believed that they were getting a false warning signal. This is clear, since they made no remark about an impending crash, Collins' command for 'go around power please' was made in a firm but unworried tone, he did not 'firewall' his engines as he would have done had he seen Erebus ahead.

The actions of the crew provide further proof that the aircraft at impact was in clear air, not cloud. If they had been flying in cloud when the ground proximity warning indicator sounded, the aircrew would have genuinely believed the alert, been warned of an impending crash and would have responded to it as an extreme emergency, as they would have done if they had seen the mountain. In his final seconds Collins probably checked his HSI and saw he was exactly on track, with 24 miles of unobstructed distance to run over the sea ice, and he was still only seven miles beyond the point which he had visually swept from above in his earlier orbits.

5. *Impact*

THE TWO UNDERWING engines and the rear underside of the fuselage slammed into the rock-hard ice slope with massive force, leaving a deep indentation in the shape of the aircraft.

As the lower panels of the fuselage shredded and carved deep into the ice, the wing tanks ruptured. Seventy tonnes of fuel, retaining the inertia of 252 knots, broke free like a hellish rainstorm. It ignited almost instantaneously and balled up the slope engulfing the exploding machine like a shower of napalm bombs.

Catapulted from the inferno, the nose section, part of the top of the fuselage, and the tail plane continued on, spinning, rolling and finally coming to a shuddering halt.

The last piece of debris to settle was the tailplane fin containing the rear engine. Even after impact this engine continued for some seconds to deliver full power, rocketing through the inferno for hundreds of metres. The searing heat of the fireball had melted and blackened the ice of Erebus' slopes. It refroze within seconds, but left a marker which was easily picked up nine hours later by a searching US Navy C130 Hercules.

Those aboard felt nothing. Death came to the 257 occupants in milliseconds — far quicker than turning off an electric light. Mercifully, they had no conscious inkling of its arrival.

There was a slight jiggle in the movement of the aircraft in the last half-second, accepted by the passengers as mild turbulence, not even enough to slop a drink. This was caused by the slightly compressed cushion of air which built up under the machine as it swept low over the mountain's flank.

In response to the captain's call for go-around power, the nose of the aircraft tilted upwards slightly. Three feet of soft, powder snow that lay between the keel surface of Flight 901 and the permanent ice cap had no escape from the undeniable forces of physics. The immense compression set up the process of instant sublimation; under this heating effect snow and ice hardly had time to transpose to water before becoming steam and water vapour.

There was a counter-action — Newton's law of equal and opposite reaction. The impact force was returned to the structure of the aircraft whose body angle was now raised by nearly 10 degrees, and increasing as the more fragile underbelly panels yielded to the impact.

This left the main support bulkheads and frames to stand the full force of the concussive shocks which they, in turn, transmitted through the structure even before the pressures began to buckle and tear their rivets.

The shock waves, transmitted at about the speed of sound under water (approximately 600 mph), travelled through the floors of the cockpit and cabin, and up through the legs or buttocks of everybody on board. A millisecond camera would have shown a small ripple or bulge of tissue rising up from the part of the person in contact with the floor, and moving towards the head.

The initial wave would be detected by the nerve bundles, but before their signals could travel at their leisurely 90 mph to report to the cortex, the shock wave would have shattered the intricate network of electro-chemical interconnections. The brain would be an inert mass of biochemicals, no longer organised or responding.

From that point they were clinically dead. The effect was not unlike the infamous Armalite rifle, which can kill by shock waves even when the bullet does not hit the victim. They felt nothing. Life was snuffed out in a fraction of time so small it is difficult to comprehend, and they were enjoying themselves when the end came.

The photos presented in evidence to the commission taken within the cabin in the last minutes of flight show the cabin bathed in sunlight, and everyone smiling and enjoying themselves.

The human misery Erebus exacted was from those they left behind. It also left a legacy for other airmen, and the airlines which operate the big jets. The crew loop which Jim Collins and his crew so diligently and exactly maintained could only be as good as the wider loop which exists within the airline and the industry.

At 6.45 on the evening of 28 November, nearly five hours after the crash, the telephone rang in the Collins home. Maria answered it and the caller announced himself as Captain David Eden, Air New Zealand's director of flight operations. He apologised for disturbing the family, but he thought he should let her know that Jim's aircraft had been out of radio communication for some hours. It was nothing to be concerned about, and he would let her know when communication was re-established.

'I have had similar calls before over delays and technical problems, so I was not unduly worried,' Maria said. 'But then he said something which sent a chill through me. He asked if I was alone, and advised me to call someone in to keep me company.'

After Eden rang off, Maria's worries began to build. She realised Jim's aircraft was not due to land at Christchurch for another 30 minutes, and that Captain Eden had no reason to tell her about a technical problem until the aircraft missed its scheduled arrival time.

She called a friend, a fellow pilot of Jim's, Captain Ross Gordon. He was reassuring. There were a score of reasons why an aircraft could lose communication, especially on an Antarctic flight. There were sunspots which could make radio communications impossible; tricks could be caused by the influence of the South Magnetic Pole . . .

Nevertheless, Maria asked her parents over to keep her company.

At 9 pm, she knew the DC10 would be coming to the end of its fuel, so she rang Captain Eden at his office. His tone sounded gloomier. He confirmed that the aircraft would be reaching the end of its fuel reserves, but he advised her not to be pessimistic.

At 10 pm, her parents heard a newsflash announcing that the aircraft would have run out of fuel by now. Minutes later Captain Eden was on the telephone to confirm this. The aircraft may well have ditched, he told her. There was a chance.

The house began to fill with friends and neighbours. Some did not know for sure whether Jim was on the missing aircraft, but they called because they knew he had talked excitedly about being rostered on an Antarctic flight.

'Like most pilots' wives I had occasionally thought of this sort of thing happening,' Maria said. 'But it was not in the least like I expected. I was too benumbed for immediate grief. One thing kept going through my mind as the night drew on very slowly.

'It was the desperate hope that if anything had happened to the aircraft, it had been very quick. If there were survivors, with Jim among them, I knew they would probably be injured and in frightful conditions of cold and wet.

'I knew Jim would be trying to organise things. He had been good at survival training in the air force, and if they were on the ice, or in liferafts, he would be trying to organise a rescue, apportioning whatever food and warm clothing there was, and trying to keep spirits up.'

The midnight news brought the worst: wreckage believed to be that of the DC10 had been sighted on the slopes of Mt Erebus. Captain Eden called soon after — he was not hopeful about the possibility of survivors.

The next few weeks saw a succession of mourners and sympathisers at the Collins home, including airline people, fellow pilots and other aircrew, and the crash investigators. It was during this time that Maria believes she began to feel the presence of two groups, the 'us' and the 'them'.

The 'us' were fellow pilots of Jim's who did not hold executive positions in the airline. They could offer nothing beyond sympathy; they too were shocked and perplexed about the cause of the crash.

The 'them' were executive pilots, a number of whom had also been close friends of Jim. Those are responsible for the administrative side of the airline's flight operations, although they fly frequently on scheduled operations.

Their administrative duties range from planning new services and maintaining training standards, to liaising with other departments such as the navigation section. As administrators, they have become part of the airline's heirarchical structure. A manager of flight operations is in line for the position of director of flight operations — who in turn is one of a small group from whom a new deputy manager or chief executive of the airline may be drawn.

The executive pilots were sympathetic and, to begin with, non-committal about the cause. They all said they were surprised the accident had happened to a pilot 'of Jim's calibre'.

Maria continues, 'As time went on I noticed oblique references creeping in. One pilot would say: "Of course, Jim was too low."

'Another would say: "It was so unlike Jim to contravene any regulations."

'I gained the impression they were trying to break it to me gently that Jim would be held to blame.

'Ron Chippindale (the Chief Inspector of Air Accidents) came to see me four days before Christmas. When I pressed him, he replied: "Well — Jim was inexplicably low." '

This confirmed Maria's growing fear, but it was the confidence of another pilot, a close and nameless friend of the family, who put fight into her. He confirmed that Collins was generally held to be blameworthy because of low flying, but added: 'They haven't told you about the errors made, have they? The company is going to come in for some severe criticism.'

Maria says: 'I didn't know at the time, but he was obviously talking about the navigation co-ordinates, and the failure to tell the crew about the alterations.'

Anne Cassin's experience was similar, but her knowledge of flying and navigation prompted immediate fears when news came through that the aircraft was overdue. Greg had left behind his briefing folder, and she had spent part of the day studying it.

Next day, one of the airline's captains came with his wife to offer condolences. He said he had been appointed liaison officer for the family. At his request, her brother-in-law, who was in the house, handed over her husband's DC10 manuals and a large volume of papers to do with flying. Anne Cassin was not present when the papers were handed over, and did not know that this had taken place. The briefing paper however remained with the Cassin family.

Much of the remaining energy Anne could summon in the next few months was aimed at regaining these documents which she believed went to the airline. But to this day, a number of them remain missing. They include passenger maps for the Antarctic flights, a page from the briefing in which her husband had recorded various numbers, and his DC10 flight training manual which, she claims, was his property.

Anne Cassin's was one of the strongest voices in calling later for a public inquiry. 'I welcome it and the very full way in which all aspects have been put before the commission because I want to ensure that other lives are not lost in the future, and that any mistakes or inadequacies that have been brought to the notice of the inquiry will go to prevent future tragedies.'

6. *On the Mountain*

IN WELLINGTON, THE Chief Inspector of Air Accidents, Ron Chippindale, began packing his bags when it became obvious that the missing aircraft's fuel reserves would be finished.

Early next morning he was in Christchurch being hurriedly outfitted with polar clothing by the Antarctic Division. With him were 35 others who boarded an RNZAF Hercules for the flight to McMurdo. They included 12 police under Inspector Bob Mitchell; New Zealand Press Association correspondent Paul Cavanagh, a small television team, and mountaineers.

Also among the first to fly down were the Minister of Science, Bill Birch; the airline's Flight Manager Technical, Captain Ian Gemmell; two air safety investigators, both Air New Zealand pilots but representing the New Zealand Airline Pilots' Association, First Officer Peter Rhodes and Captain Tony Foley; and the Director of the Antarctic Division of the D.S.I.R., Bob Thomson.

The numbers were swelled in the following days by further police, and by the arrival of American experts; Mr A.L. Crawford from the National Transportation Safety Board, Mr D.R. Grossi, a specialist in flight recorders from the same board, and by representatives from the aircraft maker, McDonnell Douglas, Mr L. Wright and Mr S. Lund.

At this stage, the cause of the accident seemed wholly inexplicable. The only reasonable guess seemed to be that it had suffered some form of technical failure, and this was already being suggested by the press. Mindful of the aircraft's unfortunate reputation and the public clamour, answers had to be found quickly. If technical failure was the cause, it had to be isolated and 40 airlines flying the DC10 had to be notified. So the top priority for the investigators was the recovery of the two 'black boxes', the cockpit voice recorder (CVR) and the flight data recorder (DFDR).

The mountaineers who first reached the accident had emphasized the hazardous nature of the site. Any retrieval operation would require a lot of 'roped together' work with experienced mountaineers always in attendance. They estimated that 50 to 70 bodies could be recovered, and that the team to camp on the mountain would have to be prepared to weather out several days under blizzard conditions. Because of the slope, a wooden helipad would have to be constructed to enable helicopters to put down.

The police and mountaineers assigned to the mountain experienced their first Antarctic blizzard within hours of arrival at Scott Base. It was not unfortunate — it gave them a valued day in which to plan and organise, and more importantly, to come to terms with the conditions they would have to live with over the next eight days.

An early grid pattern — the first step in the investigation of a crash of this magnitude — had been plotted from the air over the wreckage. A little over two days after the crash, a full working force (investigators, police and mountaineers) were air-lifted in from McMurdo, 28 miles south.

Both black boxes were discovered on 2 December. The CVR was among debris about two-thirds of the way along the wreckage trail from the impact point, and the DFDR was located under the ice nearby after a trench was cut. Captain Gemmell returned with the two vital instruments — actually bright orange in colour — to McMurdo that night, and next day they were on their way to the United States for processing.

The gruesome and dangerous job of body recovery, borne by the police and mountaineers, continued for several more days. The remains were placed inside plastic bags, tagged with the grid position they were located in, and then taken to the helipad to await lift-out by the US Navy helicopters.

It was a ghastly task made all the more horrendous by the attention of the voracious skua gulls, the scavengers of the Antarctic. But eventually all the bodies, except perhaps one or two encased deep within the mountain crevasses, were recovered. An exact count has never been possible because of the mutilation suffered by some.

However, over a number of months, 214 were positively identified at the Auckland Medical School mortuary where the remains were taken, and others were buried in a common grave at Auckland's Waikumete Cemetery.

In possibly the most difficult retrieval operation carried out in the aftermath of a major air accident, the authorities achieved one of the highest body identification rates — 83.4 per cent.

A listing in *The Bulletin*, the New Zealand Police magazine, summarised the methods used: fingerprints 39, dental records 115, X-ray records 11, visual 51, photographs on documents 5, documents with names 7, jewellery 80, clothing 60, uniforms 16, other property 12, operations, scars, tattoos, missing limbs, etc 23, other scientific comparisons 5, specific denture characteristics 3, blood groupings 3, hair comparisons 6.

Time on the mountain (and in the Auckland mortuary) left its scars on the young policemen. Long after their return, some were being counselled for psychological disturbances, sleepless nights, etc.

It was a trauma felt vividly in the following account written for this book by Captain Tony Foley, a professional pilot with Air New Zealand, who completed the University of Southern California's air accident investigation course in 1973 and is an accredited International Federation of Airline Pilots' Association air accident investigator.

He points out that while accident investigation might appear as a negative function for a pilot, the International Federation of Airline Pilots (IFALPA) considers it as extremely important because an informed and trained group of professional pilots are obviously in the best position to look after the interests of their colleagues, in particular, those who do not survive an accident.

It is of vital importance to the passengers and crews on an international basis that the true causes of a crash are established. They must not be masked by political or economic considerations.

> We arrived at Scott Base on 3 December, and were extremely pleased to learn that the flight data recorder and the cockpit voice recorder had been recovered. Ron Chippindale, who had preceded us by three days, assigned us our duties, depending on the expertise of the individual.
>
> For instance, the General Electric representative obviously was to concentrate on the engines. Peter Rhodes was assigned various duties, one of which was transcribing the tapes of transmissions between Flight 901 and Mac Centre. I was assigned meteorology and search and rescue.
>
> Our day would normally begin at 8 am. After breakfast we would go over the hill to McMurdo and spend the whole day at Naval Support Headquarters, getting back to Scott Base as late as midnight. At this time of the year it was daylight throughout the 24 hours.
>
> At Scott Base the comradeship and sense of loss was very apparent, and it affected everyone. The New Zealand flag flew at half mast all the time.
>
> At Naval Support Headquarters we were continually being confronted with photographs, some very graphic, that came back from the crash site. But the real horror and devastation that was awaiting us on Mt Erebus was yet to manifest

itself. One tended to become engrossed in the actual investigation and didn't really have time to think too much about the personal side of things. That was to change for me a few days later.

On Friday, 7 December, Peter Rhodes and I were advised that we were on standby to visit the crash site. However, our departure was delayed 24 hours because of bad weather on the mountain.

The delay highlighted a point a lot of people have overlooked, especially the media people who thought they had a divine right to be on the crash site. The logistics involved in actually getting people to the site by helicopter, and of sustaining them on the mountain before flying them out, was a problem which, unless you were there, you could never really appreciate.

The US Navy helicopter pilots, accompanied sometimes by RNZAF aircrew, flew virtually around the clock, with minimum rest periods.

Not only were they involved in the crash operations, but they also had their normal duties of supplying scientific field parties working in this part of Antarctica. These men were completely unselfish in their dedication to the job, and by the time it was over they must have been on their last legs. No praise is too high for them.

Peter and I arrived at the crash site around midnight, accompanied by two Lands and Survey men, Colin Fink and Peter Hall. The weather was completely clear, no cloud, unlimited visibility. We made our approach from the north over Lewis Bay, and even in these perfect conditions, approaching the landing pad was extremely difficult for the pilot. In fact we made three attempts and it wasn't until the people on the ground set off an orange smoke flare that the pilot could judge his height above the ground.

The lack of definition and depth perception made the operation extremely hazardous. We landed, disembarked quickly as the chopper barely hovered on the pad, and the outgoing members of the police party scrambled aboard to return to McMurdo.

After the noise of the machine receded into the distance the utter silence became deafening. We were only 28 miles from McMurdo, but for the isolation we felt we might have been on the moon.

Our first job, accompanied by NZ Face Rescue Squad

members (led by John Stanton), was to walk the length of the crash site, top to bottom, viewing the pieces of wreckage. The rescue squad people directed us around crevasses.

Viewing the total destruction and carnage, it was difficult to believe that this was all that was left of 250 tonnes of human technology I had seen some years earlier when visiting the McDonnell Douglas Long Beach plant where ZK-NZP was about to be rolled out for delivery to the airline.

Everywhere you looked you could not help but see, first, the once proud koro emblem on the tail fin, life jackets, coffee jugs, all manner of wreckage with the same motif and the name of the airline scattered everywhere.

It was at this point that it finally got through; that what I was looking at was *my* aeroplane, these were *my* pilots, colleagues, and *our* passengers who had met their untimely end.

For nearly two years prior to this point, the birth pains of Air New Zealand had involved us all with the merger problems with the old NAC and Air New Zealand 'Mark I', as we called it.

The bitterness and the feelings which I suppose are natural in the case of a merger of this nature had left their mark on all staff members of the new airline.

But for me the merger was complete at this point. I was no longer an NAC pilot. I was an Air New Zealand pilot — and a member of what I felt could still be a proud and successful airline.

I knew when I subsequently got home that a lot of people were feeling the same way. I only wished it could have been possible to bring down every staff member of the new airline, from top management down, and let them just stand back and look at the wreckage of one of the company's aeroplanes. I knew they would have felt as Peter Rhodes and I did.

Fortunately, our job at the site was made a little easier by the fact that the cockpit had broken off from the rest of the fuselage, and was thrown to one side thus escaping the intense fire which followed impact.

For 14 hours we worked among this wreckage, searching for the flight deck instruments, but every now and again coming across things like an old NAC cabin bag with the words 'NZ Women's Division Federated Farmers' Trip to

Australia 1964'. There was a small diary owned by a British tourist, a woman who was obviously very religious. The last pages included a prayer she had written, and a note to her family at home expressing how wonderful it was to be alive and on such a magnificent flight to the Antarctic.

We also came across a first officer's uniform jacket and two uniform caps which we placed beside the cockpit, to be left with the rest of the wreckage.

While searching the cockpit area we came across what subsequently turned out to be the last body recovered.

Up to this point we thought that all the bodies that were going to be found, had been recovered by the police party. These bodies were still on the crash site awaiting their return to McMurdo, and the RNZAF C130 flights to Auckland.

But Peter and I came across a small piece of the uniform of an Air New Zealand hostess. We brushed aside the snow to reveal the complete body of a hostess. She was lying face down in the snow, almost unmarked.

Earlier when we arrived at the crash site I had noticed a young fellow in a sleeping bag inside one of the tents, shivering as though suffering from exposure. Despite the warmth of the sleeping bag he was racked by uncontrollable bouts of shaking.

When we found the body, this photographer was called to come down and take this last photograph. He was not suffering from cold but from shock. After having spent several days photographing the victims' bodies as they were discovered, he just could not bring himself to take this last photograph. So we summoned the police people who photographed the body, and placed a numbered tag on her wrist.

We then excavated more of the snow and turned her over. Her face was immediately recognized by Peter. It was one of the hostesses who, only a week or two before, had been involved in taking 200 handicapped and underprivileged children to Disneyland, after two years of fund raising by cabin staff.

For me, this pitiful sight was all I could take. We removed her body, and placed her with the rest, then I told Peter I was going up to Base Camp to have a cup of coffee.

I am not ashamed to admit that as I walked the 100 yards to the camp, I cried the whole way. To this day, whenever I see that uniform, it brings back what was my worst memory of my visit to the Antarctic.

Back at Scott Base we learned that the base carpenter had begun the construction of a memorial cross. Everybody at the base, at some time, ventured down to the carpenter's shop and without saying a word, helped sand down or varnish the cross. They knew that they would be associated with a tangible memorial which was to be erected on a rock outcrop overlooking the site — there to remain in the sterile Antarctic atmosphere long after the pain of the disaster had disappeared.

On the flight south to McMurdo aboard a US Air Force C141 Starlifter, Captain Foley shot movie film of the mountain and crash site using an 8mm Minolta camera. The unique value of this film was not realised until 18 months later when the Royal Commission of Inquiry began sitting, and the film was screened before it.

Captain Foley describes the three passes the American aircraft made towards the 'black smudge' on the ice:

> On the first I noted in the view-finder a bank of fog at the bottom of the crash site at about the impact point. The fog was rolling in from Lewis Bay and *up* the northern slopes of Mt Erebus.
>
> On the second pass at a lower altitude, I noticed that the leading edge of the fog bank was now over halfway up the wreckage trail; and on the third pass the black smudge was completely obscured.
>
> The whole process had taken little over five minutes.

Captain Foley said it was not until I presented evidence at the inquiry on the possible effects of flying towards a mountain slope in whiteout conditions, under a cloud base that would seemingly displace the horizon upwards, that he recalled the movie footage.

> As Captain Vette described the possibility of a "ramp effect" — the presence of a fog bank coming in from Lewis Bay, up the crash slope obscuring the ice cliffs which would otherwise have alerted the crew — I knew that the film I had taken had, by pure chance, captured exactly what he told the inquiry might have happened.

In the outside world, rumour and speculation mounted quickly as the recovery operation progressed. It came from both the news media and individuals.

Bob Thomson, the forthright scientist who heads the New Zealand Antarctic Division, and who had flown a number of these trips in the same commentator role as Peter Mulgrew, announced within two days that the machine was on the wrong side of the mountain and 'certainly in the wrong place to come down to such a low altitude.' It was an obvious point that had to be said, but it aroused an angry reaction from both the Airline Pilots' Association and the airline's chief executive, Morrie Davis.

'It is quite untoward for anybody to make a judgment on the cause of the crash at this time,' Mr Davis said. He described Captain Collins as a man of massive integrity, 'probably one of the most outstanding pilots in the airline.'

'Anyone attempting to prejudge is either very brave or very stupid,' Davis said when he was asked to comment on a London *Sunday Times* report which claimed that the last three messages from the aircraft suggested that the crew misjudged their position by some 20 miles, confusing Mt Bird with Mt Erebus.

But speculation was rife within the first mystified week. New York safety experts were reported as saying that catastrophic winds swirling about the mountain were the most likely cause, creating turbulence so violent the crew lost control. Other reports blamed malfunctioning engines.

Clearly, some indication of the cause was necessary, particularly after some New Zealand news sources carried a story — subsequently discredited — which claimed that part of the tape from the CVR carried the exclamation from a crew member: 'Where the hell are we?'

McDonnell Douglas, anxious to see their machine cleared of blame as soon as possible, were first to rule out technical failure. Ray Towne, the company's popular public relations head, told long-distance telephone callers that initial CVR and FDR playbacks indicated the aircraft was at the right altitude and heading when it crashed. 'If there was an early indication of technical failure, we would have gone into action immediately.'

Snippets of information from the National Transportation Safety Board in Washington supported this. Gradually, speculation shifted to the navigation system, and then crew error.

On 11 December, the *Auckland Star* asked Mr Chippindale

if the way the navigation system was used aboard the DC10 had become a central feature of the investigation. The Chief Inspector confirmed this.

Asked to comment on remarks by DC10 pilots that the chances of technical failure in the navigation system would 'be so small as to be infinitesimal', Chippindale agreed that this was a reasonable assumption.

'We are looking at the way the system was used as well as the system itself — failure would virtually mean failure of all three (inertial systems).'

Loaded questions no doubt, and truthful answers. Without the benefit of hindsight and intellectual examination, the implication pointed to pilot error, and this came to be reflected in the media. In its dealings with correspondents Air New Zealand was now less inclined to dispel the notion of pilot error, and in time gave 'off the record' recognition to the theory.

The first indications that navigational error might have originated in the computerised systems outside the flight deck surfaced some six weeks later, again in the *Auckland Star,* and in its sister newspaper, the *8 O'Clock,* the latter containing a story written by Graeme Kennedy.

Meanwhile, the matching of the voice recorder tape with the flight pattern from the flight data recorder was progressing with urgency in America, largely overseen by the Auckland Inspector of Air Accidents, Milton Wylie.

Another consideration had entered the investigation. Almost every passenger who boarded Flight 901 carried a camera, mainly 35mm, and nearly all of them loaded with colour slide film.

Almost all of them were damaged during impact, but despite this they were collected and carefully unloaded. Some 900 frames were eventually processed, and found to carry an invaluable record of the progress of the flight. The film even provided a macabre record of the instant of impact — a shower of vapourising fuel droplets spattered on a window. The passenger who took it passed from life to death as his fingers pressed the button. (See Fig. 29 on page 73).

Some of the film was handed over to a young meteorologist, Mark Sinclair, who was on his second summer in Antarctica when the accident happened. Called back to Ross Island from the remote Dry Valleys area, he found himself immediately at the centre of a vital part of the investigation — his brief was virtually to transfix an instant of time in space; to build a conceptual, three-dimensional model of the weather over the area at the time of the crash. If possible, it had to be complete to the nearest wisp of cloud and trace of fog. He was given expert assistance by Antarctic Division head Bob Thomson. Most of the passenger photographs which were produced to the Royal Commission are printed as the numbered photos between pages 68 and 73.

Sinclair first obtained from the Americans a satellite photograph of the area taken on the day of the crash. This showed only a flat, two-dimensional map of a large stretch of cloud extending laterally across the edge of the continent. He then sought ultra-violet photographs taken by the same satellite.

By blowing up both photographs and then transposing the ultra-violet one over the ordinary picture, he was able, with the aid of balloon readings from the day, to assemble a three-dimensional picture of the multi-layered cloud bank, 8000 feet deep, hanging over the island.

As the job progressed amid the post-crash chaos at Scott Base, the first readings from the flight data recorder were coming in from America, showing the track, the figure-eight circuit, and the descent. Sinclair was then able to match the progress of the aircraft through the various clear spots among and between the cloud strata. Information from the voice recorder was now being radioed in from Seattle, and this too was matched with, and found to fit, the progress of the aircraft.

A clear scenario had now emerged, but still finer detail would add to the understanding of the final minute of the flight, and this was provided by the passenger photographs.

Using basic trigonometry and his knowledge of the universal 35mm camera lens, Sinclair was able to compute within a small margin of error where each camera button was pressed, the direction it was aimed at from the aircraft window, and the heading and bank of the aircraft at the time. Cloud shadow lying over the sea, the shape and size and position of the larger ice flows, the clarity of the atmosphere at various times throughout the descent, and the view from the flight deck were painstakingly added to complete the jigsaw.

If Flight 901 was the loneliest major air crash, its final minutes were surely the best documented. The flight data recorder, cockpit voice recorder, passenger colour slides and satellite photographs not only provided the information, they also acted as a double check on each other.

Yet for all this, they could not explain why the accident happened.

7. *Whiteout*

> *. . . in a whiteout condition a dark object is visible for many miles while a snow-covered object, even a mountain, next to the observer is invisible.*
>
> U.S. Navy Weather Research Facility

IN THE WEEKS immediately following the crash, the piecemeal release of little bits of information left me in a total quandary.

It was a relief to learn that the DC10 was not at fault, and that sabotage was not suspected. I am ashamed that I doubted the aircraft; it is a magnificent and unfairly maligned machine. In several thousands of hours flying it, I had developed a deep admiration as well as total confidence in it.

But what of the crew? I believed them to be of an equally high calibre.

Barring a mistake on their part, it seemed the answer could only lie in something which was totally unexpected and undetected by all members of the crew — one must emphasize all, including Peter Mulgrew. I was certain that Mulgrew would have been on the flight deck with them, just as he had been on my Antarctic flight when we approached Ross Island.

The dangers of loss of vision from whiteout appeared to have been covered in the airline's briefings. I was certain that the crew would have ensured they did not get anywhere near snow showers or blown snow.

I was approaching the problem knowing that this was a good crew who would have avoided any decision or action which could be called poor airmanship.

It was also obvious that the accident investigators would, naturally, have less regard for this assumption. They would be straining to find direct causes which, if by chance they reflected badly on the crew, would have to be accepted, and would also be a simple and uncomplicated solution to present to the public.

In effect, the official investigators were approaching the mystery along the conventional path, searching for mistakes and oversights from which they could deduce the cause of the accident.

My search was along an oblique path. I believed that Collins and his crew were acting totally in accordance with airline regulations. These, I began to realise, were formulated without regard to the unique conditions likely to be encountered in Antarctica.

However, fundamental allegations were emerging which seemed far better matched to the approach taken by the investigators.

Firstly, although the aircraft was a long way off the course programmed into the computers of aircraft on previous flights, the airline asserted it was on the correct track — lying over the top of Mt Erebus to McMurdo beyond.

Secondly, it was flying at 1500 feet when it hit the mountain. No pilot, except the most criminally foolhardy, would have taken the aircraft down to this height unless he was flying in clear, visual weather, and with the approval of the McMurdo controllers.

All preliminary indications were that the crew were not aware of their position when they decided to descend, and that during the latter stages of the flight, they were flying into conditions where visibility was very poor.

Even with the granted accuracy of the inertial navigation system, no pilot should descend in instrument metrological conditions beneath his minimum safe altitude until he is perfectly certain of his position. He can do this by checking with ground aids, or by visually identifying the terrain over which he is flying. Should he even look like straying into unclear weather at lower heights, he should fly out immediately — unless of course his position can be monitored exactly with ground radio aids or radar.

These facts were inescapable, and the feeling that I was manoeuvring myself into a position which went against all logic was growing . . . unless . . .

Was it possible that the crew of Flight 901 had positively identified what they believed was their position before descent — and were deceived?

And was it possible that they flew into a mountain in perfectly clear weather without ever seeing it?

The former, aided by careful analytical thought, could be possible. If by chance they had been misled about the track the aircraft would follow, there was a possibility — a very faint one — that the area they flew into bore a marked resemblance to another which they believed they were approaching.

Every pilot is aware that nature does throw up such coincidences, though rarely. The shape of a coastline, its cliffs and the sweep of land beyond it can bear a remarkable resemblence to another area just around the corner, especially when parts are blanketed off by cloud or low-lying fog.

It would need some checking out, with the aid of geographers familiar with the area, weather experts and maybe an artist to visualise my findings.

But the second theory seemed to belong to the most distant reaches of fantasy. Nothing is written into the flying manuals about an illusion so complete as a mountain directly ahead which cannot be seen in clear air.

However, I was surprised to learn that it was possible for even experienced ice pilots to lose the surface definition, or visual reference with the terrain, under what would normally be considered very good visual flying conditions.

For instance, a rolling area of ice cap normally shows up quite well in the sunlight because definition is gained from the shadows. If, however, a light layer of cloud diffuses the direct rays of the sun sufficiently, the shadows disappear, and with them disappears this crucial mental ability to define the surface and measure distance — even though one can see clearly for miles.

This went beyond the caution written into the Civil Aviation authority's operating specifications, and the airline's briefing team. They presumed no such conditions could exist without snow showers, or blown snow.

This, then, prompted my line of study which seemed to match events as they came to light, one by one.

Almost invariably, any operating hazard faced by a pilot announces itself with warnings which can often be acted on. They may be very subtle and muted, but are nevertheless there.

Could it be, though, that among the many and varied visual phenomena that are lumped together under the title of 'whiteout', there were conditions which existed without any hint or form of prior warning to alert the pilot?

And then, taking this hypothesis one step further, could these conditions be visually misleading? Could a pilot be looking at a mountain wall so visibly reduced and negated that he has an overwhelming impression of clear space, a flat, unobstructed vista extending safely ahead for miles?

I told Dr Barry Kirkwood, of Auckland University, about my concern. He agreed that there was a distinct possibility that this had happened, and that it probably would have affected all the crew to the same extent, and even an experienced Antarctic hand like Peter Mulgrew.

He added that the likelihood of this happening would have

been strengthened if there were other reinforcing stimuli. By this, he explained, he meant something probably aside from the immediate field of vision and the job of manually flying the aircraft. An ingrained belief, for instance, that they were not flying towards the mountain but over a clear track, would have fitted this theory.

Up to this point I had spoken with a number of experienced ice pilots, mainly Americans. Now Dr Kirkwood set me some solid reading of works by various specialists in visual perception: James J.Gibson, Helen Ross, Charles A.Baker, Savid Shinar, David Beaty, and others.

This served to convince me that the crew had flown into the base of the mountain in visual flight conditions, looking at the mountain and not seeing it.

It was a gradual but appalling revelation, especially when I considered the Antarctic flight I commanded some two years earlier.

I have been flying for 34 years, and like to think that I have never exposed my passengers to unnecessary danger. I know the same can be said for all my colleagues in the airline, and in other airlines, who have flown passengers around the sights of Antarctica — and below the height of its snow-covered mountains.

We enjoyed excellent viewing in what I took to be completely safe conditions which were exploited to the complete satisfaction of all aboard.

All of our flights would have descended below the altitude of the surrounding terrain, confident that because they had full visual conditions and plenty of space in which to manoeuvre, they were safe. This was the impression clearly conveyed at the airline's Antarctic briefings — along with the normal understanding that once the crew had reached their cloud break procedure height of 6000 feet, they could — if they maintained visual flight conditions — fly exactly as they were cleared to do by the McMurdo air traffic control. It was the same procedure as would be followed at any airport in the world.

My guess that the destination waypoint given Captain Collins and crew in their flight plan had been changed from that of earlier flights was soon confirmed beyond doubt — and relatively easily.

It will be recalled that Captain Simpson's flight of 14 November, just two weeks earlier, had caused this pilot some puzzlement. He had flown down the McMurdo Sound track and then turned eastwards to McMurdo Station. He was surprised at the distance between the navigational track and the station.

The reason for his surprise was coincidental to the scenario I was beginning to build, but it did result in his making a number of precise checks on the position of the track. No more convincing proof would have been possible that the final waypoint in his flight plan lay in the middle of the McMurdo Sound, and was approached along the sound, and not from over Ross Island.

Yet Simpson and his co-pilot had attended the same RCU briefing as Collins and Cassin, and absorbed the same simulator experience of descending into the McMurdo area. They too had been given the same sample flight plan for examination.

There was now one final piece missing from my hypothetical jigsaw puzzle. Was it possible that Captain Collins and crew had not been told about a late change made to their navigational track? (This was two months before the airline admitted that a late change had been made to the track.)

Dr Kirkwood's perception in sensing the presence of some 'outside stimuli' fitted my suspicion that the crew had not been told. If this were so, they undoubtedly flew south in full and complete confidence that they would pass well clear of Erebus.

Their senses and assumptions would have been attuned to the expectation of a clear route down McMurdo Sound. What pieces of terrain they were able to see through cloud and fog, they would have attempted to relate to this route.

Within the natural caution of airmen, they would have been conditioned to expect this. The term 'mental set' is often used by psychologists to explain this conditioning; in fact it is a normal part of our lives. Without it, our everyday actions would be abnormally complicated and time consuming.

It is an interesting exercise to put together the three points of my hypothesis and see how they relate.

Firstly, there was the probable mental set which conditioned the crew to believe they were flying on a safe track, used on all previous computer-generated programmes.

Secondly, there was a possibility that what they saw of the area they were flying into bore a sufficiently close resemblance to be mistaken for the area into which they thought they were flying.

And thirdly, there was the almost complete lack of knowledge by the crew, and the airline, of the true nature of the visual tricks which polar light could play on even the most experienced: the insidious lack of warning, an ability not only to destroy all perspective, but also to display mirage-like illusions which were totally misleading. I had now suspected, and ultimately concluded that these latter factors were the most significant ones in the crash.

Collins and crew (and in fact almost everyone who gave evidence at the inquiry except the scientists) believed that the only way you could collide with terrain was to fly into cloud and suffer reduced atmospheric transparency or visibility.

In assidiously avoiding this situation Collins was unknowingly setting up a 'Catch 22'.

In staying below and clear of cloud he was unknowingly rendering his visual perception more ineffective than had he flown into cloud. At least as he entered cloud he would know that his perception had failed and would immediately respond.There are regulations, and in fact basic rules intuitive to the professional pilot, which would have triggered off a caution, and then a decision on a new course of action which would have avoided the accident.

For instance, had Collins been a less meticulous pilot, and not carefully checked that the flight plan he examined at the RCU briefing showed a clear track down McMurdo Sound, he would never have let down, even in clear weather to proceed with the flight south.

Had the telltale glimpses of coastline and terrain not been as he expected, he would have been alerted to the fact that the aircraft was not where he expected it to be.

And had the crew been alerted to the visual traps inherent in polar conditions, they would never have flown in clear conditions beneath a light, but solid overcast.

But when these three factors existed in unison, there appeared to be nothing in the conduct of the crew for which they could be held blameworthy.

With the new-found knowledge I gained in my research of visual-perceptual psychology, nothing would induce me now to fly below the level of any white surface using visual means of collision avoidance.

Few, if any pilots, after studying the available research and talking with the experts on polar flying, would not agree that: 'there but for the Grace of God go I'.

In researching my theories, I had to get back to the fundamentals of what goes to make a pilot. He is human, subject to the same physical and mental stresses as anyone else — though sometimes airlines would happily have the community believe otherwise.

The question of 'command responsibility' was beginning to be bandied around. This tends to be an abstract idea, originally rooted in the days of sailing ships. It accepts the proposition that a command captain, having risen to a position of vested responsibility because of his skills and experience, must be capable of foreseeing possible hazards, and avoiding them.

As such, it can be wrongly used to vest in a captain a quality that approaches omnipotence and clairvoyance.

From my study of other accident reports, I was only too well aware that a failing of command responsibility could be a handy peg on which to hang an inconclusive investigation. All the signs now seemed to be pointing this way.

Much of the knowledge of the mental and physical qualities needed by a good pilot lie in the basic stock of psychologists, medical specialists and scientists around the world.

But I was surprised to find that while ten years had elapsed since man put astronauts on the moon (and meanwhile had sent space shots to the other side of the universe), little had been achieved in bringing together a cohesive picture of the commercial airline pilot.

There are, after all, thousands of these professionals flying the world's air routes, interfacing between their machines and a hostile environment. The airline pilot's responsibility is more awesome than that of an astronaut. The lives of perhaps 500 passengers rest with him.

The next chapter builds a profile of the airline pilot. In accepting the Erebus crew as human beings adapted by specialisation to an alien environment, I found there was a need to look beyond the superficial definitions held by airlines and aviation regulatory authorities. Conversely, the insights I gained brought a clearer understanding of the reasons behind this tragedy.

Also, it confirmed my belief that the time had come for the modern airline to look more deeply into its own understanding of what makes a pilot 'tick'.

The time has certainly come for a close look at the pilot training syllabi. These must surely be extended so that the pilot is aware of all the physical and psychological factors that go into that very delicate interface between man, machine and environment which we all so readily take for granted. Surely it must be even more important than the years of study we put into aerodynamics, thrust systems, hydraulics, electronics, etc.

And certainly the operations specifications procedure must be tightened to ensure that the deep fund of user experience stored at the International Civil Aviation Organisation headquarters is accessed in drafting new specifications, especially for new routes.

8. *Profile*

IF WE WANT to understand the task of an airline captain, we must first discover what demands his job makes upon him and then observe how he fulfils those demands.

He is the link between the machine and the environment; an environment, moreover, which is alien and often hostile. The machine operates at extremes of speed and energy, far beyond human capabilities and the environment may incorporate extremes of cold together with a rarefied atmosphere which would render unprotected human life impossible. Not only, therefore, does the pilot occupy this position as mediator between machine and environment, he is the weakest of the three components. Yet, without his meditation, the aircraft would not, could not, fly. Only he can interpret the environmental information and instruct and control the mighty forces governing the aircraft and prevent his, so much heavier than air, machine from obeying the dictates of gravity which seek to pluck it from the sky.

In making a profession of the art of flying, the pilot has learned to adapt the senses of vision, hearing, touch and balance to an extraordinarily high degree . . . perhaps one that he is hardly aware of himself.

Millions of years have passed during the evolution of *homo sapiens*, but for only the minutest fraction of this period has the species been involved in flight. Senses already highly adapted for moving upon the ground, for avoiding obstacles and for intercepting or throwing missiles, must be raised to an exponential degree of complexity to deal with events occurring at jet-speed in the upper reaches of the atmosphere enveloping our planet.

Most pilots are attracted to the job in the first place by a slight excess of the adventurous spirit. Ideally, they are not thrillseekers, but of necessity are able to cope with emergencies when they arise. They must also be fit in all respects, both mentally and physically. All their sensory capabilities must be in good working order and compulsory six-monthly medical checks will ensure that standards do not deteriorate below specified levels. The physiological and psychological attributes of the man will be, for critical periods of his professional life, "wired into" the mechanical and electronic attributes of the aircraft. This marriage of human and aeronautical engineering must be as near perfect as possible.

Although the range of our sensory receptors is immensely rich (Bruner noted our ability to perceive seven million shades of colour), perception of the world is not achieved by our sense organs alone. They provide the initial information, the data on which perceptual judgments are based, but these judgments depend upon additional information which is available from other senses and from the experience of the brain itself.

Thus, as James Gibson of Cornell University has claimed, there is a visual *world* which is distinct from the visual *field*. The visual field, for humans, is a constantly shifting pattern of light, 'an array', as Gibson calls it. It consists of all the light that reaches the back of the human eyeball, the retina. This retina is no cinema screen, no 'mirror' of reality. It contains up to 130 million photo-sensitive receptors, some wired singly, others connected in pairs and groups so that stimulation of the retinal receptors may be carried along the optic nerve, a fine cable carrying 800,000 strands, to the visual centres of the brain. The trouble is, that all of these interconnections and the optic nerve itself are in front of the retina, between the receptors and the world of light. The point of departure of the optic nerve for instance, is the cause of our characteristic blind spot causing us to lose vision for objects about 17 degrees below the horizontal in the temporal field. For us to see anything of the world, the light must travel through the detritus of dead cells in the humours of the eyeball and between the cells which connect up the retinal receptors. Because of head and eye movements the retinal image is constantly moving, because of the lens, the image is upside down as well as degraded and deficient due to the various obstructions. How is it then that what we normally *perceive* of the world is mostly crisp, complete, stable and the right way up?

The short answer is that no-one really knows except that, one way or another, the brain, over a lifetime of visual experience, learns to interpret the signals it gets from our eyes and integrates them into a useful and (relatively) accurate picture of the 'real world'. There is, therefore, no direct relationship between the world and what our eyes 'see' of it. As Cohen puts it, you have to conceive in order to be able to perceive, more simply, it is not a matter of 'seeing is believing', but of 'believing is seeing'.

Throughout this book I have noted the extreme reliance placed by pilots on textural cues in the visual array which offer clues as to the approximate size and distance of features in the landscape. Consider now, the human visual equipment upon which the pilot relies to supply this information. On the ground, travelling at 'human' speeds, most people avoid all but the most transitory illusions or mistakes about the real world. A fleeting notion of some bizarre construct in the world outside the front door is quickly corrected, we know it *cannot* be and our brain informs our eyes in effect, that the data they have provided is misleading and our perception is once more attuned to reality. The flying human is prone to many more illusions, ones that he is trained to recognise and to resist, to check against other information being received so that he may reset his datums. Human vestibular (balance) sensation and visual sensation may produce conflicting evidence as to which is 'the right way up'. The artificial horizon instrument on the instrument panel is there to settle the argument and the pilot is trained to believe, in this instance at least, in the reliability of the electronic instrument above his own senses.

I have stressed the importance of the visual input because, of all the wide array of human factors involved in making up that complex man/machine, the airline pilot, the eyes provide the most abundant source of information — about 80 per cent of the total in fact. Thus, a significant proportion of the general population are, from the outset, disqualified from the roll of possible pilots because of deficiencies in their visual equipment, colour-blindness, myopia, etc.

Cockpit instrumentation has been especially adapted to take advantage of our ability to discriminate colours. Information regarding "normal" states of affairs is usually shown in white or green on the instrument panel. Warning signals are carried in yellow, while red is reserved for important information signalling possible danger.

The pilot has to learn to maximise his visual efficiency both in the solar glare above the clouds and in the dim and difficult conditions encountered in night flying in poor weather. Constant exposure to bright light, or to any particular colour or wavelength of light tends to de-sensitise our ability to see. The pilot is trained to avoid exposing his sight to full illumination if he wishes to seek out some dim and distant object in the external night sky. He learns that, while the centre of the visual field is best for fine and accurate vision (due to the location of certain kinds of receptors in the central part of the retina, the fovea), the periphery of the visual field is best for detecting objects giving off very faint illumination as the rod receptors on the periphery of the retina seem to be maximally sensitive to such stimuli. Peripheral vision is also more sensitive to sudden movement of objects not in focus.

Scanning the landscape from high ground, one is often

unable to determine whether certain green vegetation is low scrub or more distant forest. Faced with such a dilemma we seek for other information in the array, especially objects of known size such as buildings or other, confidently assessable landmarks. Similarly the pilot draws on all his available sources of information to determine facts about his outlook. He will do this from charts, from experience on this and other routes and from a wealth of experience he has gained in learning to make sense of the visual information provided at altitude and at speed, which bears so little resemblance to the humdrum world beneath him.

Above all, he learns to understand the textural cues offered by the environment. When the eye picks up texture, it interprets variations in light and shading to determine both depth and distance. Very often such judgments are made from a pool of information rather than one single feature. Vertical planes (such as trees and cliffs) and horizontal planes (such as lakes) provide useful data to help settle questions left unresolved by gradually shelving, finely textured land especially if covered with sand or snow.

On earth, we rely very heavily on inter-positional cues to judge both size and distance. Not only are we surrounded by many objects of already known size, if we move, then closer objects appear to move relative to more distant ones. Walking along the street, we notice that lamp-posts, for instance, appear to move across our field of view, passing in front of the houses, offering incontrovertible evidence that they are closer to us than the houses and also giving us some idea of their relative size compared to the houses.

For the airline pilot such information is rarely available. Almost all of the objects in his environment are much further away from him than they are from one another and relative, inter-positional, movement is barely perceptible. The scale of things is altogether different up there, though the concept of interposition is still useful in the pilot's world when it involves the use of radar and radio beams and beacons. These *do* appear to change location relative to the aircraft and thus aid the pilot in 'seeing' his way through the sky.

Given the importance of good vision, it is not surprising it should be placed so highly on the physical requirements of the pilot. On the six-monthly examinations, vision is checked for any deterioration. As time goes on, some slight deterioration normally associated with ageing is considered acceptable, provided it can be demonstrated that experience allows a person to adapt progressively to that deterioration without endangering the safety of the operation.

Of the people who make initial application for pilot training, many fail because of searching medical tests and high educational demands. The military provides the most typical entry path for an airline pilot, and a typical air force course begins with an intake of about 50, possibly 10 to 20 per cent of the original applicants. Of this 50 about 30 eventually graduate, the others having slipped by the wayside. Of this 30, some 20 eventually find their way to the commercial airlines.

About 15 would gradually make it through to command, and only five out of the 15 might make it through to rank of captain of a large, international jet. Before retirement, about half of these jet captains may have lost their livelihood by failing renewal of their licences on medical grounds.

Now let us look at the high degree of specialisation required of these individuals to provide the best possible interface between machine and environment.

The cost of courses and training throughout an airline captain's working life runs to many hundreds of thousands of dollars. At a very early stage the trainees begin to learn about the results of errors of judgment, often in a dramatic and sometimes tragic way.

Those prone to gross mistakes tend to remove themselves at intervals from the programme, intentionally and unintentionally. For the other participants, the 'dicey do's' and 'near misses' have a quick effect in modifying behaviour patterns. Constant exposure to reports of accidents and near misses is a vital part of the learning process.

Aircraft are, of necessity, fragile when related to possible impact forces, and can do little to protect the vulnerable human frame from injury or death. So the consequences of an accident are usually extreme. This fact, at the back of the mind, tends to engender a cautious appraisal of situations — which is fine when time permits.

It takes time for a pilot to carry out the process of gathering information from his eyes or other senses, transmitting it to the brain for discrimination, and the neurons finally causing the muscles to manipulate the necessary controls. This process and the reaction time of the aircraft, or its controls, can add up to a dangerously lengthy period when coping with a sudden emergency.

The classic case is the setting of the lowest acceptable cloud base for an instrument approach on to a runway. By practice, with progressively lower bases, the response times can be shortened to quite low limits. Practice can bring it down to a limit which allows just enough reaction time for the pilot to recognise his position relative to the touch-down point, and either progressively arrest the rate of descent and speed to the point of touch-down, or quickly arrest the rate of descent and convert it into a safe climb, should he realise that landing conditions are unsatisfactory.

So it can be demonstrated that pilots, through this training, develop a short reaction time, especially in situations to which they are frequently exposed. While the machinery's response time remains constant, the pilot's response time has been reduced by using this reflex reaction. In other words, the time used for the brain to absorb the facts, discard irrelevant information, weigh up the conflicting information and determine the required action, is largely eliminated.

Similarly the pause before sending off the signals through the body's nervous system, and then turning off the action potential as the feedback from the various senses report back the progressive completion of the job, is also reduced.

'Conditioned reflex' training is so instilled in a pilot that he responds in an exact way when circumstances do not allow time for evaluation. He acts immediately on visual or other sensory signals. The shortest possible reaction time differs slightly for each individual but for all it should fall within the maximum time interval that can be planned for a particular operation involving a number of reactions.

The conditioned reflex is a learned process which, though stored firmly in the brain's long term memory bank, is actually carried out before the full analytical function of the brain can process the input. It follows that the stimulus signal must itself be absolutely reliable and unambiguous.

Much the same process seems to be involved in the skilled playing of a musical instrument. Long sequences of actions are 'programmed' through rehearsal and the brain (or aesthetic sensibility!) merely observes, monitors and adjusts a performance which is largely automated.

Obviously, very effective management can be achieved by stringing together a whole series of conditioned reflexes, interspersed by short analytical periods to allow for adaptation to changes to the original programme.

We have said that by specialisation and exposure, the overall man-machine response time can be reduced. It follows that the speed of analysis, although variable, is also reduced in situations which are repeated, where there are recurrent cues. Together with the development of a short-term memory capability, it is here where most of the years of training effort are applied.

Fig. 43

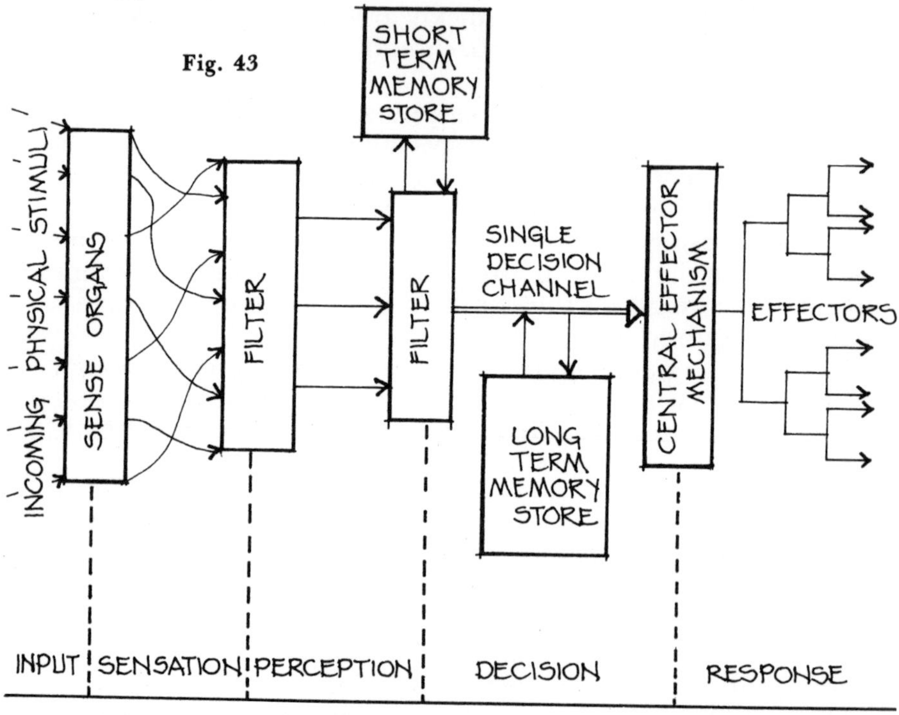

The long-term memory banks are probably called upon to store much less information than in many other professions, and most of what is stored is of specialised nature, relating to aerodynamics, regulations, meteorology and engineering, etc.

Since continual exposure to the simulator situations can reduce response time, perhaps further adaptations to this form of specialisation can be achieved on examining the neural system. For instance, those axons which, in the process of evolution, require the fastest and strongest transmissions of nerve signals, become sheathed in a white, fatty insulation material called myelin. Myelin prevents leakage of electrical action potential.

Could man in future adapt some technology to improve this transmission speed — which is at present about 90 mph? Unfortunately, there seems to exist an absolute limit on the rate at which the chemical processes occur.

Actually, the technology to do this is already available, but it would be difficult to apply. It is however the only remaining means of speeding up reaction time.

The G suit is a good example of how man's expertise has been used to overcome the unsuitability of the human frame to some aspects of aviation. In this case, in order to make an aircraft maneouvre rapidly, it is necessary to increase the lift force generated by the wings. When this is done, the pilot feels the increase in weight force, the G (gravity) effect, which drags down on the face and muscle tissue, and also drags down on the blood which is being pumped to the brain.

The result is that while the brain becomes starved of blood, it pools in the lower limbs and abdomen causing, progressively, loss of vision and finally unconsciousness.

The G suit merely uses the same forces to act on bladders located within the specially designed flying suit to squeeze the legs and abdomen. Blood is unable to pool in these areas and is forced upwards to the brain, thus avoiding the worst effects of accelerated gravity.

One area where specialised training is neglected is in the unnatural environment of instrument flying. Here, the inputs of the beautifully designed ear balance mechanism can lead to very strong sensory illusions.

If the head is moved fairly quickly to a new position, and held there, the fluid in the canal of the inner ear responds to natural inertia. As the canal moves, there is a corresponding surge in the fluid — as when a trough of water is moved and the water surges in the opposite direction to the trough's movement.

In the inner ear, little hair-like sensors on the wall of the canal bend with the moving fluid, transmitting the movement to the brain, and sensing when the movement stops. In effect, these sensory inputs reinforce what the eyes have also told the brain. However, if the head is now moved very slowly back to the original position, the fluid does not move against the canal walls in the same way because the movement is not forceful enough to overcome the friction — a trough of water very gently moved does not surge.

So the message does not get through to the brain that the

head has moved away from its last position. However, the eyes have detected the movement, and readjusted the datum for the brain.

There are many other minor inputs from body sensors which are not always accurate unless reinforced by the visual input. The alignment of weight or gravity through the lie of a muscle or the sag of an organ is an example.

The discriminatory power of the brain usually accepts the input from the eye, as paramount, over-ruling other sensory messages. It then resets all datums as dictated by the eye. This is known as 'visual capture' — the message from the eye is given first credence.

However, while the optic input is a fairly constant value, inputs from other body sensors are often affected by other considerations, for instance, acceleration and sound stimulations. The brain gives them varying amounts of recognition, depending on their strength.

Hence the vertigo effect of air or motion sickness. Here the sensors, responding to varying accelerations registered on the inner ear, return varying reports to the brain, and the brain has to continually act as judge and referee in resetting datums. Some people are more prone to suffer the discomfort of this reaction than others. It follows that the ability to keep one's visual reference — to keep seeing — must diminish or delay motion sickness.

If any example is needed, take two children travelling in a car along a twisting road. The one who watches the passing scenery with interest is less likely to become car-sick than the one who tries to read a book.

Training to understand and overcome these powerful sensations is particularly important, and badly neglected.

We now come to an illusion suffered occasionally by pilots who fly without visual reference, relying solely upon instruments.

This pilot's visual reference, in the normal sense, is taken away from him. So without the eye to act as a referee and continually reset datums, the information provided by the inner ear mechanism assumes a much greater importance. We get a situation where an acceleration is recorded and the brain resets its datum from this information. Then a 'sneaky' or unrecorded acceleration, as described a few paragraphs back, takes place. The body has now taken up a new position at variance to where the brain believes it to be. Now another large acceleration takes place and is reported back to the brain — which then calls up an entirely wrong response for the situation. The resulting position then leads on to an even more confusing input of signals to the brain.

The pilot finding himself in such a situation can only resolve his problem (and reset his datum) by reference to instruments, which are much less likely to supply wrong information.

But instruments require skilled interpretation, and do not provide the same instantaneous datum-resetting capability until the pilot has acquired a lot of exposure to these sensory illusions.

It is not uncommon to see a pilot in his early learning phase flying his aircraft by reference to his instruments, nice and level, but leaning across in his seat in a compromise with the sensory illusion — a response called the 'leans'.

There are other illusionary sensory inputs which, if a pilot did not know about them, would be very dangerous. One of them is the auto-kinetic effect of light at night. Here the eye muscles get tired of looking at a light source, say the light of another aircraft he may be flying alongside or in formation with, and relax into a new position. The light then appears to shift position. Because the muscle has not reported back that it has decided to take a rest, the brain assumes that the position of the light has changed. The consequence can be disastrous.

In aviation, as in everyday life, there is the predictable behaviour pattern of an individual in response to familiar and recurrent stimuli.

A motorist travelling the well-worn route between office and home does not bother to read the road signs, and he moves automatically into the correct lane for his turn-off. It has all been carefully checked before, and the outcome is predictable.

Under certain circumstances, particularly if a pattern has been reinforced by being proved reliable over many repetitions, a range of expectations is built up in the mind so that a mental 'set' is formed. Events will then be perceived as fitting into the 'set' and confirming our expectations. Naturally, conflicting information which threatens the validity of the set will still be received, but the strength of the set can be measured by the amount of conflict or 'noise' that is tolerated before the set is 'unlocked', that is to say, queried, examined and then amended or abandoned.

All humans rely on both automatic behaviour (to perform such mundane but nonetheless highly complex actions such as walking) and set (which absolves the mind from the need to process all information about the environment to an exhaustive degree). Perhaps the pilot has greater need than most for such reliance. Certain frequently recurring episodes in his life (approach and landing, for instance) involve such a vast amount of sensory input that some form of pre-filtering is essential so that his attention can be focussed on new and important information (wind-shear for example) while the rest of his actions follow a well-worn route and form.

Mental sets are not entirely reliable. We have all experienced the feeling of dread that our car has been stolen after a minute or two of fruitless search in a featureless carpark. How relieved we are to find that our set was in error — we have been looking in the wrong carpark having left the building by a different door. Two things have probably occurred. Firstly, our attention has been drawn to the search because of anxiety over the car and this has led to our being more aware of our surroundings and thus more likely to realise that we are in the wrong carpark. Consider a second example. We are again in the wrong carpark, but there is a car of the expected make and colour in approximately the 'right' location. How long will we spend trying to get the key into the lock, before we notice that the belongings on the back seat are not ours and that our car has a radio aerial anyway? Having discovered our error, we slink away, embarrassed lest anyone has seen us making such a silly mistake, or worse, seen us trying to 'steal their car'!

Fortunately, the environment at ground level is usually rich in clues, so that errors like that occur only rarely in our sets. At ten, twenty or thirty thousand feet, however, the array is impoverished in the extreme, which is one reason why the airline pilot has recourse to charts and landmarks, to radio beacons and radar tracking and why he uses the AINS system (albeit with two back-up platforms) to reinforce his set. Indeed the AINS provides a major set. Just as important is the loop procedure in the cockpit whereby each member of the crew monitors the others' reactions. It would be a unique circumstance indeed, which saw four trained, adult flyers experience the same set error and maintain it for any time. It is certainly true that the airline pilot's proficiency depends on his ability to discriminate between various conflicting stimuli. Often the strongest stimulus, the one most reliable under normal conditions, is the one which must be discarded.

Safety depends upon continual on-the-job pilot training, and an important part of this is simulation under safe conditions of all situations, plus the lesson of frequent checking. Generally, the requirement is for simulators and in-flight checking at not more than three-month intervals.

This way a pilot learns the shortcomings of the human as an interface between environment and machine. He learns to reset his datum when necessary, even when under strong and conflicting stimuli.

The important thing is that he knows when such a stimulus may be in error. He must know never to reset the datum until satisfied with the integrity of the resetting stimuli.

Quite often, knowing is not enough and special training is required to be able to reset the datum.

The human pilot is bathed most of the time in a wash of sensory stimuli. Because he has only one pipeline through which this information can be processed, information must be held by the short-term memory while other information, which has been given higher priority by the decision-making component of the mind, is processed.

A fascinating point is that this master discriminator, which apparently controls the direct output of the brain by an understanding of the brain's task and the rate at which it can work, appears to have no work-rate limit itself. It is almost as if a mastermind of unlimited rate and capacity is filtering, sorting, encoding and setting priorities from a vast mass of sensory inputs — at an un-human speed. It reduces them to a rate it knows the limited-capacity human brain can handle.

The fact that we are positively limited to a rate which has to be greatly reduced by this much faster, upper discriminatory part of the mind, certainly suggests that as individuals, we are merely a subordinate and inferior computer drawing on the supply services of some ethereal master computer service, just as our metabolic and circadian rhythms are.

If information is held too long in the short-term memory bank it decays and is lost, unless the stimulus reasserts itself. The design of the short-term memory is such that, ideally, it should hold just long enough for the priorities ahead to be cleared. But when more information than it can cope with keeps crowding in, inevitably some is lost before it can be acted on.

This is to be expected, considering the thousands of trivial inputs it deals with in the course of a day — a scratch to an itch on the face, a prompt to look left or right, etc. Some items will obviously find their way from the short-term to the more limited-capacity, long-term memory bank — for instance, home and office telephone numbers, a recommendation on where to go for your annual holiday, etc.

The pilot is called on to respond very quickly to some stimuli which flood in beyond the normal capacity designed for the short-term memory bank. By training, I believe he learns much more effective coding or 'thinking' strategies. Above all, I believe he bypasses the problems associated with short-term memory by storing in the long-term memory, characteristic patterns of important stimuli.

This way, he can more quickly process numerous pieces of vital information from the short-term memory by feeding little bursts of stimuli, one after the other, into the part of the mind which is constantly evaluating and analysing the flight.

It follows a cyclic pattern. Little pieces of stimuli are very rapidly processed, then before these stimuli have had a chance to fade, they are worked over again.

The concept can be likened to the way an international telephone link sends several different conversations down the one line at the same time.

By taking a small sample (the minutest fraction of a second) of each conversation, and running a continued cycle of all conversations, a complete record can be transmitted. Obviously a straight receiver at the other end would only get an indecipherable noise. But when sets of receivers are used, and each is synchronised to listen only at the exact time to its piece of conversation, the individual voices come through clearly — even though only a fraction of the voice sound has been transmitted. The time lost between each break is so small it cannot be detected by the human ear — which, as the eye does, tends to fill in the gaps by joining up the edges.

This ability is of greatest advantage to the pilot when, after a prolonged period of receiving non-contradictory inputs (when everything is running smoothly), he is suddenly thrust into a situation where there are few cues, and these are impoverished. In such situations, this developed characteristic is put to maximum use by rapidly sampling the available cues for datum resetting and decision-making.

My old IFALPA colleague, Captain Peter Bressey sums up the situation well when he says: 'The claim for sufficient time in which to consider any major business decision is signalled by the use of cliches.

'The logic of this situation is apparent and it is unnecessary to labour the point beyond — "I must have time to think about this" . . . "I'll ring you back" . . . "Consult my partners" . . . "Sleep on it and let you know in the morning."

'It would therefore be regarded as a grotesque aberration if a decision on the disposition of equipment worth many millions of dollars and the lives of several hundred people — including the executive from whom the decision is requested — were to be insisted on in the space of three to five seconds.'

Yet this is the decision time-frame for an airline pilot making an instrument approach to land in marginal weather conditions. In the most graphic and basic terms, he is sliding down an electronic bannister in the sky with some 300 tonnes of aircraft strapped to his posterior, while the ground is coming up to meet him at an approximate speed of 230 feet a second.

He knows that if he sees anything at all before he reaches his decision height, he will have less than five seconds in which to make his operational decision; namely to carry on with the landing, or to overshoot. And he knows — in the face of the stark alternative — that the decision *must* be the right one.

He cannot slow the aircraft down to gain time for thought; he cannot put off the decision. He cannot even take time out to consult his co-pilot. He has less than five seconds to form a decision from fast sampling of priority cues held in the short-term memory; and to act on them.

In the most critical cases of aborted take-off, the time limit to recognize and act is down to less than one second, though usually there are much better cues.

Control goes hand-in-hand with prediction. It is often observed that an action or event precedes certain other actions or events. Therefore, if we use this knowledge to predict the occurrence of a phenomenon indicated by certain existing conditions, we can prevent or control the occurrence by changing these pre-existing conditions. Obviously, we can also make it occur by creating the right conditions.

Confidence is a vital requirement in the make-up of an airline captain. It is difficult to define exactly, but could be described as his own satisfaction that he is able to depend on himself to respond correctly to the situation which confronts him. As William Blake said: 'If the sun and moon should doubt, they'd immediately go out.'

At its ideal level, confidence exactly matches the ability of the individual at that stage of his development. Over-confidence can lead to an unwanted predicament, as indeed can under-confidence. So it behoves the individual to try to judge his own ability and limitations honestly, and to attempt only those tasks within these limits.

Most successful pilots do just that, and have their judgments reinforced by the progressively happy outcome of events. Naturally, this carefully matched equation of confidence and ability must work within the limits of effectiveness and thrift.

It would be fair to say that the airline captain, over many years and constant exposure to the various pressures, has arrived at a suitable level of confidence, a fact the airline industry depends on.

Tight economics and safety can come into conflict, yet nothing threatens an airline's balance sheet so much as lack of safety. The airline that values itself also values the exposure it gives its pilots, and that includes exposure to simulator exercises as well as flying training.

The good airline pilot has to have the confidence to handle difficulties — for example, the Hong Kong harbour approach or the colony's instrument guidance system approach to Runway 13; a Wellington landing in boisterous conditions, and a low cloud base and low visibility approach to Los Angeles.

However, accident statistics all too often show that this necessary level of confidence can lead to an accident when some previously unexperienced phenomonen occurs. Unfortunately, unless it can be detected early enough by someone else in the crew, or by designing a device to prevent or detect that particular phenomenon, we will remain vulnerable.

The essential ingredient that is frequently tested in a pilot's exposure to simulator training, is how accurately his confidence matches his ability in the particular exercise he is carrying out. It obviously varies from individual to individual, and with it there must be associated a cost factor in terms of cost-effectiveness to the company. But that confidence level has to be preserved intact, and it is vital to the company that it should be.

So the pilot therefore believes that should there be any change required to his standard operating procedures for a particular new route, then this will be encompassed in his training, and his datums can be appropriately reset.

He believes that for every area to which he is exposed, he is trained to meet the shortcomings associated with it. This, of course, is where shortcomings — if there are any in the operating specifications — become apparent.

It is important in this context to understand that Captain Collins and crew would confidently approach the visual sightseeing flight to Antarctica in the knowledge that it would be a relatively simple task to fly up the 40-mile-wide, clear sea ice of McMurdo Sound at 1500 feet provided that they observed the warnings about remaining clear of cloud and blown snow and retaining 20 kilometres visibility.

The following section concentrates upon the harbour circling approach and landing at Hong Kong's Kai Tak airport. It provides a good example of a pilot and his cockpit crew operating at peak efficiency, exploiting to the utmost their training, the aircraft's instruments and, most significantly, their response to visual stimuli.

I use this example of the confident application of interfacing techniques during an extremely difficult airport approach to indicate that a straight flight path in good visibility up the wide expanse of McMurdo Sound at 1500 feet, monitored by the area navigation system, would seem to a crew, ignorant of the traps of polar vision, to be not only a simple task, but free from any possibility of complication so long as the current weather conditions remained constant.

The habour circuit at Hong Kong, which I describe below, has now been discontinued at Kai Tak with the advent of wide-bodied jets. However the IGS system which now takes you down through 600 feet still requires a manual visual approach on the final stages of the gully flight, and demands good interfacing between pilot and machine.

Should this type of manoeuvring be attempted over most cities of the world, the crew would find themselves in prison. At Hong Kong, it was the only way the airport could be used in certain weather conditions.

When the wind was blowing off the Pacific towards the China land mass, the 100-tonne jet had to be swung and tightly manoeuvred for eight miles through the narrow sea channel which divides the mass of mainland Kowloon and the nearly vertical walls of Hong Kong Island.

This was precision flying of a sort rarely demanded of big jets anywhere. At one early point, the Lei Yue Mun Gap, the channel constricted to a throat less than half-a-mile wide, and the shore lights whipped past the wingtips like glinted moisture drops in a tunnel.

At night, and when the cloud base was down to 1500 feet or lower, it was akin to moving at high speed through a convoluted tunnel — only without the benefit of rails.

Here, surely was the most exacting interface between man and machine; the point where a pilot became the organic cortex with nerve ends grafted on to a mass of metal moving along at 150 knots — a speed only just sufficient to give control, but uncomfortably fast for the environment or rugged coastline, ships' masts and high rise buildings.

Every featherweight of pressure applied to the control yoke and rudder had to be precise, every change of flap setting and throttle applied to the second. It was a time when all that counted was the intuition, feel and the reflexive recall of a good pilot.

On let down from above 30,000 feet, the aircraft was vectored on a radio beam towards Checker Board Hill, an escarpment lit by sodium lights hard by the Kowloon runway. This beam brought the aircraft down in a long, easy descent to position it over the Tathong Channel, the south-eastern gateway to Hong Kong. The intention was to get the aircraft below cloud base before it entered the Lei Yue Gap — into the 'tunnel'.

Should the landing be from the south-east, the runway was nicely positioned two miles ahead, a few degrees turn to the right and a final alignment just before touch-down.

But more often the landing was from the opposite end, Runway 13, from inland Kowloon.

When Runway 13 was in use, the two pilots had to quickly establish visual references to orientate themselves as they broke cloud above the narrow gap. Then manual skill took over. They were 'eyeballing it'.

The tight twists of Victoria Harbour and the turn-around over Kowloon were navigated in a way more familiar to navigators.

A series of coloured lights signalled the way, and the aircraft was lined up on these one at a time. As each light slipped beneath the nose of the machine, the aircraft was swung on course for the next, a mile-and-a-half, maybe two, ahead — 30 to 40 seconds flying time.

These variously coloured lights were so critical for pilots that neon flashing lights were not allowed in the colony lest they confused or delayed recognition of a marker for a moment.

But memorizing the positions and colours of these lights was only half the pilot load. They also had to memorize the different overshoot procedure for each should they inadvertently find themselves a fraction off the right heading and unable to make the exactly timed swing from one light to the next.

It was memorizing for instant recall, a precise compass heading and climb-out rate — no time to ponder when flying through a canyon at a mile every 20 seconds.

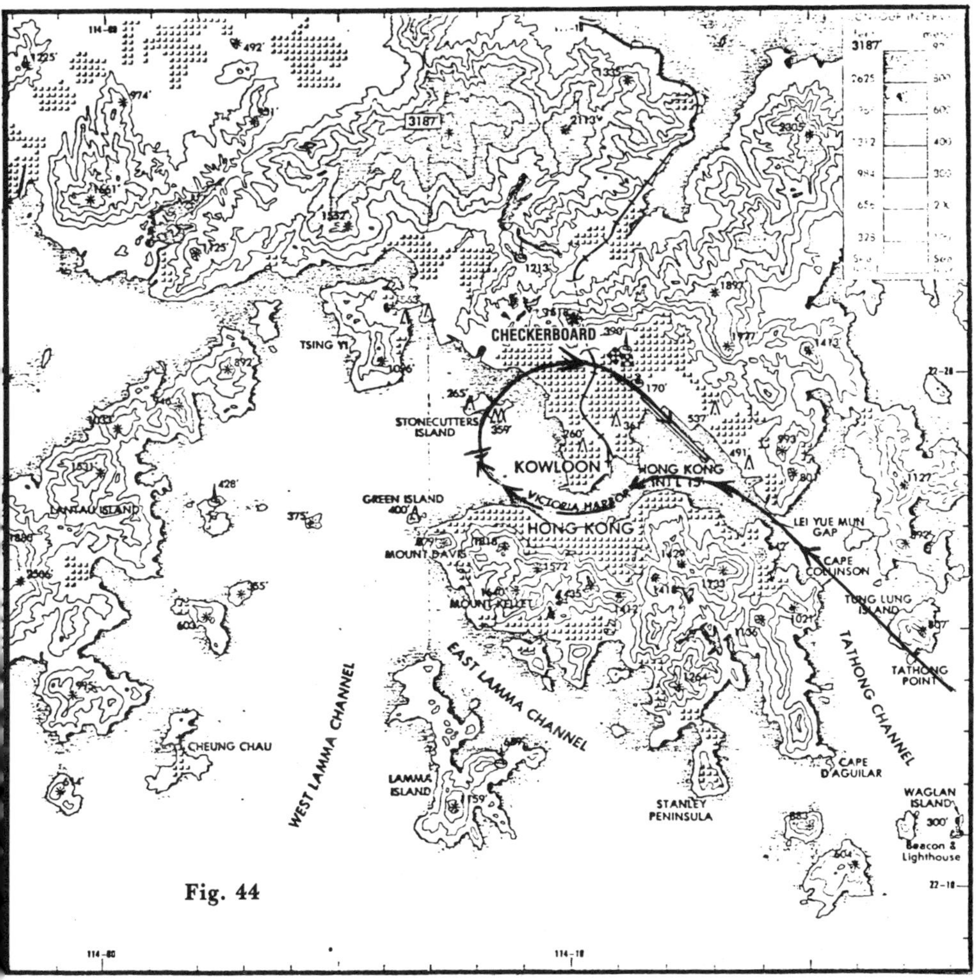

Fig. 44

And while every habitable yard of Hong Kong and Kowloon resembles a tightly packed glow worm cave at night, there was no safety in the black voids. Peaks, as abrupt as stalagmites, rear up within two miles of Runway 13's threshold, and there are unlit, high-rise buildings.

Swinging low through the basin of light reflected between harbour and cloud, these granite teeth usually lay beyond the visual rim, swarthed in misty cloud.

Despite the captain's warning to passengers what to expect, words over a PA system could never adequately prepare them for the experience. Some momentarily thought the captain had taken leave of his senses as they peered transfixedly out of the windows — *upwards* at apartment building balconies.

As the aircraft swung more tightly into the approach, they noticed what appeared to be a wall of buildings converging on the nose of the aircraft, now bouncing and shaking only 50 per cent above stalling speed. For a few, this brief glance was enough. They stared trancelike at the carpet, lips moving in urgent communication with the powers that be.

For others, the only concession allowed this unusual approach were white knuckles squeezing the arm-rest molecules into denser composition.

But for most, it was spellbound unreality; a fairyland of lights on all sides, racing, ever-changing.

On the flight deck, the luxury of this fixity of viewing — to focus, relate, absorb — was only for the pilot in the jump seat, the man observing the approach. For the captain at the controls, it was only the interpretation of a peripheral blur that enabled him to accurately position the machine.

First the entry through the Gap; a careful left-hand turn, assessing and regulating his angle of bank; applying just enough back pressure on the control column to get the right radius of turn without losing height — but careful, a fraction too much back pressure and you're back in the cloud without visibility.

A gradual right-hand turn, adjusting for wind drift as the lights of the North Point, Causeway and Wanchai slip past on the left. Skirting the thrusting bulk of West Point on Hong Kong Island, and rolling the machine more steeply to bring it around inside Green Island, on heading for Stone Cutters.

The aircraft should be coming around on the 50 degree arc (200 to 250 degrees true) of the flashing red directional lights on Stone Cutters Island.

Ease off the back pressure fractionally to stay below that convex cloud ahead, while scanning to confirm heading, height and airspeed, and that all the landing checks, with the exception of final flap, are completed.

You listen briefly to the Airport Terminal Information Service to confirm landing conditions, joining the others in the crew at the same time in trying to pick out the lighted array of aerial masts atop Stonecutters. When they show up, easing a further adjustment to put them over to the left. As the shoreline of Kowloon comes around, power settings are adjusted to set the aircraft descent at 500 feet for the approach.

The directional flashing white lights of Cheungshwawan and Yau Yat Chuen are buried deep among the apartment buildings of Kowloon. Now the crew are scanning systematically for the flashing red lights of the Checker Board Hill. It is sighted, and the power and speed are adjusted again to bring the aircraft around to the right of the checker board towards the threshold lights. The machine has now come around almost the full 180 degrees, and is committed to the carefully regulated swing through the gully.

The pilot has to try not to cut the corner to the runway, yet avoid closing in too near the rising apartment buildings, or the cliff face of the Checker Board. His senses are pitched taut, his judgments turning over, his cue sampling rate at its fastest.

While the passengers are subjected to a close inspection of the drying garments poled out of apartment windows, the pilot is unconsciously using the same cues — but using the rate at which they blurr past in his peripheral vision — to negotiate the final swing to the right through the canyon.

His fixity of focus is now centered on a point about one third of the way along the runway. If his trajectory is good, it will remain relatively stationary, though closing. The effect is not unlike watching the lights of an approaching car at night on a long stretch of straight road. They seem unmoving, but then at the last moment flash away at a tangent. But a car driver is not coping with wind shear, drift and varying sink rates. . . .

The co-pilot will be calling speeds, descent rates, altitude — and will have set up the overshoot beacons, headings and heights should the aircraft not key into the narrow slot for landing, or (unusually) a sudden wind change occur.

A final thought. It could be nicely on trajectory, but a

fraction too much speed and it will overrun the far sea wall. Too slow, and there will be a stall. You will help take in the washing. But you reach the right point to initiate your flare. Power levers are snapped into reverse as the gear touches. A quick glance to see the nose is aligned with the runway centre line, and another check of the distance remaining.

The nose gear touches down. Full reverse power selected, and braking as required, while the aircraft slows to 80 knots.

The long taxi through the massed parking places to the docking position gives time for the pulse to slacken.

You've done it many times before, but there is always the sense of pride in having your skills and judgments tested to the limits.

9. *Epilogue*

IN THE FINAL analysis, two questions remain. Was Collins entitled, on the briefing given to him, to descend beneath 16,000 feet? Secondly, if so, what are the implications of low level antarctic flights conducted under visual flight rules?

Fig. 45. A view of a glacier from a transparency distributed to interested parties by Vern Mitchell of Air New Zealand's public affairs department in August 1979. It formed one of a set of three transparencies. An extract from publicity material which accompanied the transparency follows. Observe that the airline says the flights are bound for McMurdo Sound and they make a low level circuit of Mount Erebus. Both statements contradict the case which the airline subsequently maintained.

"Air New Zealand has settled into a pattern of four such trips each November, all bound from Auckland to historic McMurdo Sound, base for the Scott and Shackleton expeditions and present-day site of the year-round American and New Zealand bases.

> The DC10 flights don't land, but return to New Zealand after about three hours of Antarctic viewing from various altitudes in the course of an 11-hour flight.
>
> They make their Antarctic landfall at Cape Hallett, at the entrance to the Ross Sea on the northern Antarctic coastline. Then they fly for nearly an hour down the Ross Sea, with the Victoria Land mountains on the right, to reach McMurdo Sound.
>
> At that stage they are at relatively low altitude and they spend about 45 minutes 'touring' the McMurdo area — winging over Scott's Hut on Ross Island and on over the Great Ice Shelf, to turn not far short of the spot where Captain Scott and his remaining companions died in 1911 after their tragic march back from the Pole.
>
> The flights make a low-level circuit of steaming Mount Erebus, sweep over the existing McMurdo bases, the airfield and the New Zealand dog lines, cross to the mainland to explore the Dry Valleys and cruise above the massive glaciers which descend from the Victoria Land mountains.
>
> When they turn for home they begin a slow climb which gives plenty of viewing of bays, ice falls, icebergs and mountains before they are back at cruising altitude at Cape Hallett.''

The Air New Zealand case presented to the Royal Commission, which the airline maintained to the end, was that pilots were forbidden to descend beneath what I refer to as 16,000 feet, but which is more precisely stated as flight level 160, until they were overhead a 20-mile sector around McMurdo airfield identified by radio beacons, first the NDB, then from 8 November 1979, the TACAN.

They were then only permitted, according to the airline's case, to descend under radar guidance to 6000 feet, while still remaining within the 20-mile sector, provided they were in visual meteorological conditions, they had visibility of 20 kilometres or more, and that there were no snow showers in the area. Flying was then permitted at 10,000 feet VMC off the Victoria Land coast to Cape Hallett. The Civil Aviation Division advanced a similar interpretation.

If the case put forward by the airline and CAD is accepted, these parties suggested that the cause of the crash rested with the crew who failed to obey the height instructions which the airline gave them. It would also follow that re-routing the flight path to run over Mt Erebus without telling the crew was irrelevant, since the crew should have remained at 16,000 feet until they had confirmed their location by means of the radio beacon. At that point they would have discovered that the position of the aircraft as they had plotted it, was some 25 to 27 miles different from their true position.

Furthermore, if the airline case is correct, it would follow that the belief of Mulgrew and the crew that they had obtained a positive visual fix on the entrance to McMurdo Sound is rendered irrelevant as a contributing cause of the crash, since they breached airline instructions in descending and making that fix.

Finally, the airline case means that the airline's not briefing the crew on whiteout in general and on sector whiteout in particular, and the actual existence of sector whiteout which created the visual illusion of causing Erebus to appear to be McMurdo Sound, are both irrelevant, again because the crew breached instructions to fly above Erebus.

The logical consequence of the airline case is that nothing which was done or omitted to be done by the airline had any part in the crash; the sole cause of which was the breach of duty by the crew in failing to fly at the height which the airline instructed them.

This alleged breach of duty by the aircrew was only one of thirteen pilot errors that the airline and CAD asserted were committed by the crew and which led to the crash. Justice Mahon found that none of them had been established to his satisfaction.

Ron Chippindale, in his press release dated 18 January 1982 says:

"Had it been flown by a crew of automatons the aircraft would have flown over Mt Erebus, turned and returned safely.

"In the event the crew decided to descend the aircraft in an area on the opposite side of Mt Erebus to that approved for any descents and they must be responsible for this decision. Had the aircraft not descended it would not have collided with Mt Erebus let alone the coastal slopes of Ross Island at less than 1500 feet above sea level seven miles short of the summit of Mt Erebus."

I turn now to consider whether or not Jim Collins and his crew were justified in descending beneath 16,000 feet before they overheaded the radio beacon at McMurdo Station. I do not wish to become involved in any way in the controversy over the bona fides of those who maintain that the aircrew knowingly breached height restrictions given to them, and that all those in authority were ignorant of the universal practice of descents before reaching McMurdo Station. Nevertheless, if any reader remains unconvinced by my reasons for maintaining that Collins was justified in descending when and where he did, then I suggest that note 4 on page 335, researched and written by an independent observer, be read. My intention is merely to demonstrate that pursuant to his briefing and in the climate of knowledge at the time, Collins was justified in believing that the airline sanctioned, if not encouraged, flying beneath 16,000 feet before reaching the McMurdo radio beacon.

The knowledge, or lack of it, held by those in authority is not relevant to the justification Collins had for his belief that he possessed his airline's authority to descend when he did.

From Collins' viewpoint, what situation faced him? He was sent by the airline to conduct a visual let-down for a sightseeing trip of Antarctica, the tacit purpose of which was to provide passengers with value for money. The major area of scenic interest was McMurdo. If reported weather conditions at McMurdo had been too poor for sightseeing at the time he reached the Cape Hallett waypoint, his instructions were not to fly on to McMurdo but instead to divert to the South Magnetic Pole, otherwise he was to proceed to McMurdo.

The weather reports which he in fact received justified him in proceeding to McMurdo. The existence of cloud cover then became apparent. All Antarctic flights prior to 8 November were despatched on instrument flight rules, in the sense that if crews encountered cloud cover in the general McMurdo area then they were instructed to carry out a cloud break, using the NDB, in order to descend through the cloud so as to conduct sightseeing beneath the clouds.

However Collins had been despatched with only a visual descent available, so that the only let-down procedure available to him was visually in clear air and of course in accordance with visual flight rules under the regulations but with visibility increased from 8 km to 20 km. There was a space clear of clouds of many square miles located almost upon his navigation track and his options were either to fly away without providing his passengers with McMurdo sightseeing, or to descend. He elected, quite sensibly to make his descent over the clear sea ice in visual conditions at the entrance to McMurdo Sound.

Visual flight rules require a pilot above 3000 feet, or 1000 feet above terrain, whichever is the higher, to maintain a distance from cloud horizontally of one mile and vertically 1000 feet; whereas at or below those heights the pilot need only remain clear of clouds and in sight of the ground or water. Visibility for Antarctic flights was increased to 20 km.

The airline's briefing document of 23 October 1979 contained a section headed 'Descent to 6000 feet for cloud penetration at McMurdo' which says:

> **Descent to 6000 ft for cloud penetration at McMurdo.**
> Permission has been given to descend to 6000 ft QNH in VMC conditions, or using the approved NDB procedure in IMC conditions, provided that:
> a) cloud base reported to be 7000 ft or better.
> b) visibility reported to be 20 kms or better.
> c) ASR is available and used to monitor flight below Flight Level 160.

d) No snow showers in area.
Pilots will have received a comprehensive briefing and completed a simulator detail involving a let-down and climb-out procedure, particular emphasis is being placed in the use of grid navigation procedures.
Flight in the McMurdo area below FL 160 will be restricted to an arc corresponding to a bearing of 120 °G through 360 °G to 270 °G from NDB within 20 kms to keep well clear of the Mt Erebus region.

I interpret that section, and most pilots (except for some of the executive pilots responsible for administration: see note headed *The Chippindale Report* on page 309) interpreted that section to relate only to cloud penetration also known as cloud break procedure. This is a procedure employed to break through a layer of cloud in order to reach clear air beneath it at which stage visual flight rules may be resumed.

While descending through cloud in the McMurdo area, Mt Erebus will be obscured by that cloud. Consequently, to avoid risk of collision with it, the section concluded with the instruction that flight in the McMurdo area beneath 16,000 feet is restricted to a defined arc "from the NDB within 20 nautical miles to keep well clear of the Mt Erebus region". The phrase "McMurdo area" read in context means within the area of McMurdo Station, with reference to a cloud break based on the radio beacon at the station.

The logic behind such a procedure becomes clear when one remembers that the navigation track originally planned by the airline ran direct from Cape Hallett to Williams Field, directly over Mt Erebus. Should the whole region be cloud covered, then in order to enable sightseeing, a let-down under instrument flight rules was provided for. The flight would follow the navigation track at 16,000 feet to avoid Mt Erebus which was directly on the track, then with its precise position located by the NDB it would make the cloud break under instrument rules. Under such conditions a height limitation of 16,000 feet until above the NDB makes sense.

This commonsense meaning of 'McMurdo area' being the area around McMurdo Station, means the airline had no justification for claiming, as they did after the accident, that Collins had breached his briefing instructions when he descended well outside the McMurdo area. He believed he was descending at the entrance to McMurdo Sound.

To present their case, it was necessary for the airline to fasten onto the ambiguity inherent in the phrase McMurdo area and assert that it meant not only the area around McMurdo Station, but McMurdo Sound, Lewis Bay, and in fact the entire McMurdo route over the Ross Sea from Cape Hallett on.

If weather conditions on the McMurdo route as reported to Collins had been considered by him to justify his action, Collins had full authority of the airline to turn west at Cape Hallett and head towards the South Magnetic Pole descending to such a height as he felt appropriate provided he did not breach VMC rules. The briefing suggested that he fly at 10,000 feet to view terrain of 9000 feet. Such a flight would be in polar conditions and subject to whiteout (see map on page 40).

At this point most readers may have concluded that the entire airline case against the crew self-destructs. Any who still believe the aircrew breached their briefing should read on as I spell out the issues in detail.

The McMurdo authorities were in fact not allowed to accept traffic directly to McMurdo Station over the top of Mt Erebus. All other traffic handled by McMurdo centre was planned to fly down McMurdo Sound on instruments at around 6000 or less to the Byrd reporting point where, under guidance, the aircraft would fly in to Williams Field, avoiding Mt Erebus, and so obviating the need to maintain 16,000 feet.

When the Air New Zealand navigation track was re-routed in September 1978 to approximate the military track down McMurdo Sound, there was no need to maintain a 16,000 feet limitation even for an instrument flight rules approach. It is not for me to say whether this re-routing was done by accident or design; the fact remains that it was re-routed down McMurdo Sound, Collins knew it was routed down McMurdo Sound, and his knowledge would contribute to his interpretation of his briefing instructions, leading him to place a commonsense meaning on them.

Whether or not those in authority knew that the route ran down McMurdo Sound is not within my province; nevertheless I suggest that those who have criticised Collins so harshly should, before continuing with their criticism, carefully consider the 54 mistakes as to flight path made by the airline, documented in note 3 on page 329.

Collins has been impliedly criticised for flying beneath cloud. I believe his briefing justified his belief that operation below cloud was sanctioned. I do not think that he or any other pilot could be expected to interpret his briefing to mean that he was permitted to fly beneath cloud only in one restricted area under ground control. Such an interpretation would be logical only if the airline knew that cloud cover in the Antarctic created whiteout, but all concerned agree that whiteout was unknown to the airline, which was why it never explained whiteout to its pilots. In fact, their understanding of whiteout is covered by their remarks in the briefing sheet, to the effect that avoidance of blown snow is, in effect, avoidance of whiteout. They obviously believed that whiteout was a reduction in atmospheric transparency.

To me, the meaning of the extract previously quoted from the briefing is clear.

When the whole McMurdo area is covered by cloud, but the cloud base is 7000 feet or better, and visibility beneath that cloud is reported to be 20 kilometres or better, and there are no snow-showers in the area, then, in order to enable sightseeing beneath cloud, pilots may let-down through the cloud, using the approved procedure.

Beneath cloud, under those conditions they would be operating VMC to avoid terrain, and could see any obstruction, including Mt Erebus, with the usual visibility requirement of eight kilometres increased for safety reasons to 20 kilometres.

The briefing went on to say:

> **Fuel management departing McMurdo northbound to Christchurch**
> Flight Operations Despatch will provide the crew with a fuel figure to cover the depressurised fuel requirement from McMurdo to NZCH.
> It is mandatory that the aircraft departs McMurdo northbound when the fuel on board reduces to this depressurised fuel figure.
> Provided that VMC conditions can be maintained it is permissible to climb the aircraft to 10,000 ft flying off the coast to Cape Hallett for sightseeing purposes. This will not jeopardise the depressurised fuel position.
> If VMC cannot be maintained FL 160 is the minimum safe altitude.
> From Cape Hallett the aircraft must climb to en route cruising flight level.

Terrain up the Victoria Land coast ranges as high as Mt Murchison at 11,487 feet, while Coulman Island itself is 7500 feet.

Yet there is no suggestion in the briefing that terrain clearance of 6000 feet must be maintained above these features. The clear implication is that the pilots will be operating only in VMC and thus will see and be able to avoid any obstructions.

The airline's case was that if a crew chose to fly down the flat expanse of McMurdo Sound in VMC, then their briefing required them to do so at no less than 16,000 feet, giving them terrain clearance of 16,000 feet and to overhead the radio beacon at McMurdo Station so as to verify their position and ensure their navigation system was accurate.

Next, according to the airline's case, the crew should descend in a tight circuit, not easy to fly, but only under radar control, to a height of 6000 feet. After that they were permitted to fly back up the Victoria Land coast at 10,000 feet regardless of the terrain height.

The next section in the briefing read:

Sightseeing S.M.P./Ninnis Glacier
This route covers many glaciers and mountains on the coast. No let-down procedure is available. Study of contours and spot heights suggests 10,000 as a suitable viewing altitude.

Thus the alternative procedure open to a crew was not to fly to the McMurdo area at all, which meant not updating the accuracy of the navigation system, but instead flying at 10,000 feet to the South Magnetic Pole and Ninnis Glacier.

Such a procedure would, of course, have appeared perfectly safe given the knowledge of the time, since flight would be VMC and the airline was unaware of the visual illusions of polar flight which could lead to pilots failing to see obstructions.

Confirmation that the airline took for granted that VMC flights would be made with no extraordinary height limitations is given by Brian Hewitt's memorandum of 25 November 1976 suggesting the South Magnetic Pole route and recommending that for flying over the Ninnis and Mertz glaciers crews should descend for 30 minutes to 9000 feet being approximately 1000 feet above terrain.

To me as a pilot, the section in the 'briefing' headed **Fuel management departing McMurdo northbound to Christchurch** means exactly what the heading implies. A major technical constraint upon flying to Antarctica is to retain sufficient fuel reserves for a safe flight back to Christchurch.

The lower the altitude the DC10 is flown at, the greater the rate of fuel consumption. Thus low level flight and fuel consumption must be carefully monitored. When the briefing states that flying off the coast at 10,000 feet will not jeopardise the depressurised fuel position but that from Cape Hallett the aircraft must climb to en route cruising flight level, I read that as telling me that for fuel conservation purposes only I must climb to 10,000 feet flying up McMurdo Sound. I do not believe I am being instructed to maintain terrain clearance of 10,000 feet over McMurdo Sound when flying VMC. Instead I am being told that I cannot fly back at the low level permitted under VMC rules, or I would jeopardise my fuel reserves. I am instructed that if I am not in VMC then 16,000 feet is the minimum safe altitude. No terrain clearance is given for the S.M.P. route implying that normal VMC rules apply there, and further confirming that the McMurdo reference to 6000 feet refers to cloud break not to terrain clearance when in VMC.

The next stage in the airline case was the internal memo dated 8

November 1979 from Captain Ross Johnson to Antarctic crews which says:

McMURDO NDB NOT AVAILABLE

Delete all reference in briefing dated 23/10/79.

Note that the only let-down procedure available is VMC below FL160 to 6000 ft as follows:-

1. Vis. 20 km plus.
2. No snow shower in area.
3. Avoid MT EREBUS area by operating in an arc from 120° Grid through 360G to 270G from McMurdo Field, within 20 km of TACAN CH29.
4. Descent to be co-ordinated with local radar control as they may have other traffic in the area.

To me as a pilot, this means that the non-directional beacon referred to for a cloud break procedure on instruments is no longer available for that purpose. Because the DC10 does not possess the equipment to use the TACAN to give bearings as were obtainable from the NDB no instrument let-down could be achieved. Reading the earlier briefing together with this memo I would still take the reference to 6000 feet to relate to cloud penetration. Undoubtedly some ambiguity is becoming apparent, but I believe the memo is merely amending the briefing by deleting reference to the approved NDB procedure in IMC conditions, and is saying make your own way visually to 6000 feet in the McMurdo area. Co-ordinate with McMurdo for traffic clearances and final radar vectoring around the McMurdo base.

Roughly the same interpretation is offered by Captain Simpson who said he believed the 8 November letter meant that cloud break procedure no longer applied and descent had to be visual. It was still his impression that if one was still visual at 6000 feet, then descent below this could still be co-ordinated with the local controller. He considered a pilot would be restricted to the arc referred to in point three if he were in the area around McMurdo field which was clear of cloud but Ross Island and Erebus were completely obscured by cloud. To remain within that arc would keep him clear of the high ground that he could no longer see.

Captain Wilson, who briefed Collins, believed that the cloud penetration section of the briefing represented conditions laid down by the McMurdo authorities. Consequently he believed those authorities retained the ability to give consent to descend lower than 6000 feet.

He also believed that the arc to the south of Ross Island was purely and simply for the purposes of ensuring terrain clearance while the aircraft was making an IMC descent. If VMC conditions prevailed, crews were at liberty to fly in areas outside that sector, maintaining their own terrain avoidance visually, but not below 6000 feet, unless cleared by the McMurdo authorities to fly lower.

The question of what reasonable pilots would have believed their oral briefing meant, can be judged by inference only from the evidence of those pilots who attended the same briefing as Collins and Cassin, but who survived. Those three others were Captain Simpson, and First Officers Irvine and Gabriel.

First Officer Irvine recalls that the 6000 feet altitude was discussed, someone mentioned other flights had gone below 6000 feet, and it was said that due to the speed of the aircraft and the small passenger windows, it would defeat the purpose of the flight to descend too low as the passengers would get only a brief view of features as they went past. Following the discussion he did not regard 6000 feet as a minimum, and thought the crew would obtain from the local air traffic controller both visual flight rules clearance and altitude clearance.

Although First Officer Gabriel does not remember the discussion of flight below 6000 feet he says this may have occurred during the informal discussion at the end of the briefing. Since this officer did not participate in the discussion with Irvine and since Captain Simpson makes no reference to it, one may infer that Collins and Cassin were participants, probably with Wilson, the briefing officer, since he had discussed the topic at several earlier briefings. Captain Simpson's opinion is noted above.

It was my understanding that the significance of 6000 feet in the briefing was that McMurdo radar, when in its ordinary mode of tracking and bringing flights into Williams Field could operate no higher than 6000 feet. Although it could be switched to another mode for higher altitudes, in that mode it could no longer carry out its primary task of traffic surveillance from 6000 feet down.

My understanding of "descent to be co-ordinated with local radar control as they may have other traffic in the area" is that it means literally what it says. That is, the instruction has no relation to using radar as a let-down procedure from 16,000 feet (which it was not capable of doing while handling other traffic on approach) in order to locate the aircraft geographically during a visual descent. Hence our requirement initially for an NDB approach down to 6000 feet. Instead it means that the pilot must co-ordinate his descent for the safety purpose of ensuring separation from other traffic. This Collins did, when he received approval to descend from McMurdo Air Traffic Control.

I do not believe that Collins could have been expected to believe that the references to heights meant what the airline after the accident said they meant.

Why should a pilot maintain a terrain clearance of 16,000 feet contrary to normal visual flight rules, down McMurdo Sound, when he need maintain only 6000 feet in IMC conditions within the vicinity of high ground around McMurdo Station?

It is ridiculous to suggest that the briefing required pilots to maintain 16,000 feet in full visual conditions over McMurdo Sound, and only require by regulations 6000 feet in the McMurdo area.

The more incongruous factor is a requirement of 6000 feet clearance in VMC conditions over the flat terrain of the McMurdo Base area, but only require 2000 feet whether VMC or IMC over the top of an active volcano.

Why 16,000 feet southbound down McMurdo Sound but 10,000 feet northbound over the Ross Sea alongside terrain over 11,000 feet on Victoria Land and possibly across Coulman Island at 7500 feet?

Why 16,000 feet down McMurdo Sound when he was permitted to fly visually at 10,000 feet on the South Magnetic Pole route containing high terrain of 9000 feet and over?

Why 16,000 feet down McMurdo Sound until he had verified his geographical location by overheading the TACAN, whereas he could fly the SMP route visually without overheading the TACAN at normal visual flight rules' heights?

Anyway how was he expected to overhead the TACAN? To overhead the NDB is straightforward, because the onboard indicator needle points directly at the radio beacon. So it is a simple matter for the pilot to fly directly towards the beacon or even fly to intercept a particular heading and then turn towards the beacon on that heading. But Collins had been incorrectly told the NDB was not available. The TACAN beams no directions receivable by a DC10, only the distance to run, decreasing when approaching the beacon, increasing when flying further from it.

Crews on the Antarctic flights were permitted to deviate to the right or left of the navigation track. That is, they were permitted to fly true south along the Victoria Land coast. If the airline's case is correct, it follows that crews were permitted to fly south at 16,000 feet, in VMC, but provided they descended in the arc around McMurdo Station, they could fly back north on the same route at 10,000 feet.

Should a crew not have been able to descend in the particular arc, say because it alone in the entire McMurdo area was cloud covered, and the NDB was withdrawn, then the crew, if the airline's interpretation is correct, was not permitted to descend to less than 16,000 feet in VMC when flying back up the Victoria Land coast.

Opposing my views, executive pilots such as Ian Gemmell maintain that the briefing specifically prohibited crews from descending beneath 16,000 feet even under VMC anywhere in the McMurdo area except within the arc around McMurdo Station, and even then only to 6000 feet. The logic of this interpretation is unclear to me.

Supposing the visual illusions of polar flight do not exist (and the fact is that the airline was unaware of them), then under VMC crews can see any obstruction and avoid it. Why then prohibit flight beneath 16,000 feet?

Why increase terrain clearance over McMurdo Sound and the Ross Sea, still under VMC, to 16,000 feet southbound but 10,000 feet northbound? Some surrounding terrain is higher than 10,000 feet, so that if that terrain cannot be seen and avoided visually, then risk of collision exists.

Increasing terrain avoidance to 6000 feet in the McMurdo area, while leaving it unaltered from normal on the alternative SMP route is contrary to common sense and the inconsistency of the airline's interpretation is one factor causing pilots to place the commonsense interpretation upon it that it related to cloud penetration.

One explanation of the otherwise illogical 6000 limit is to be found in the evidence of a witness propounding the interpretation which opposes my views, namely Captain Ian Gemmell.

He testified that the company decided that within the boundaries stipulated in its descent instruction, safe flight could also be carried out within a 20 nautical miles radius but not below 6000 feet. This allowed for a 10 nautical miles buffer zone in all directions should the boundaries be exceeded inadvertently during turning manoeuvres provided the 6000 feet minimum altitude was maintained.

If there is cloud cover, then in order to descend for sightseeing in the McMurdo area, crews must make a cloud penetration under instrument flight rules using the NDB at McMurdo Station. But a descent in this area involves descending close to the high ground of White Island and Black Island. To safeguard against an aircraft inadvertently exceeding those boundaries during its turning manoeuvres, the airline may have created a 10 nautical miles buffer zone by providing that the turns should be carried out at not less than 6000 feet. Possibly reference to that height during turns under VMC was added as a further precaution, maybe because the problems of descent manoeuvres had not been examined in depth at the stage of initial planning.

It is irrelevant whether Captain Gemmell is right or wrong. Had Collins believed the rationale for referring in the briefing to 6000 feet was to provide a safety zone to avoid collision with the Black and White Islands while circuiting the McMurdo Station area, such a belief could only reinforce his understanding that references to 6000 feet bore no relevance when flying over the Ross Sea or McMurdo Sound. On the other hand, I think it is more likely Collins thought 6000 feet related to radar limitations.

We understood the requirement was that you should retain 6000 feet until intercepted by the radar, for radar monitoring and separation from other traffic. Certainly not as a separation factor from terrain. It seemed incongruous to have to maintain 6000 feet clearance over this flat area of McMurdo Base and yet only have to clear Mt Erebus by 2000 feet under the regulations. The true issue is not what was actually in the minds of those airline officers who prepared the briefing materials. Collins has been harshly criticised by Captain Gemmell, among others, but all that is relevant is what an airline pilot in Collins' position, in the context of general public knowledge of low level flights from 1977, without any criticism from the airline or CAD, would reasonably believe his instructions meant.

Justice Mahon said that in his opinion all Antarctic flights from and including 18 October 1977 involved a let-down in the McMurdo area to altitudes considerably less than 6000 feet, and that in the main the flights down McMurdo Sound and across the Ross Ice Shelf to the south of Mt Erebus were conducted at altitudes ranging from 1500 feet to 3000 feet.

I believe Collins was fully justified in concluding, as I and most non-executive pilots concluded, that when in Antarctica, he was entitled to fly anywhere in accordance with visual flight rules with an increased visibility requirement. Except possibly should he descend within the specified arc around McMurdo Sound.

Captain Gemmell testified that he considered that Collins' proceeding to McMurdo beneath 16,000 feet was an 'approach', and that using the AINS for an approach was specifically prohibited under the limitations section of the Flight Crew Operating Manual — an FAA limitation which has been endorsed by the Civil Aviation Division.

I agree with Justice Mahon who said of this evidence that the limitation cited related to a descent for landing purposes. The prohibition is against fixing an exact point such as a landing field.

There is no similarity between that and merely using the AINS to take the aircraft in VMC to the 40-mile-wide opening of McMurdo Sound. The AINS must not be used in IMC beneath MSA.

Captain Gemmell cannot be correct, or flying the South Magnetic Pole route at 10,000 feet would be an 'approach'. How can he suggest that visual flight is permitted, but that visual flight in conjunction with using the AINS is prohibited? Is he implying that most Antarctic pilots from 18 October 1977 onwards breached Operations Manual instructions? His criticism seems inconsistent with his testimony that in conjunction with the DME function of the TACAN, the AINS could be readily used to keep the aircraft within the specified arc around McMurdo Station.

If my argument is accepted, that Collins was entitled in terms of his briefing to descend as he did beneath 16,000 feet when flying in Antarctica, the issue now arises as to what are the implications of visual flight in polar conditions.

Is it safe to operate in the Antarctic in VMC below 6000 feet? It is possible for illusions to occur at any altitude below the top of textureless terrain, therefore VMC conditions do not provide adequate protection in low texture terrain at any altitude below adjacent terrain, unless the crews have received considerable specialised training in polar flying. But even this training will not enable them to avoid the pitfalls of whiteout conditions. It merely gives them a much greater chance of recognising the areas and conditions under which they may occur.

The operations specification and Antarctic briefing allowed for a cloud break procedure down to 6000 feet. There are four types of aberration or illusion which could occur when flying at 6000 feet above ground level while flying below a cloud layer.

Firstly where overcast cuts the light down then certain white surfaces with low edge definition would disappear — quite insidiously since it leaves other prominent edges and texture showing, thus deluding the pilot into believing that he is secure within his visual meteorological conditions. In other words, he could be flying visually towards a defined landmark and crash into a poorly edged interceding white obstruction.

Lack of depth perception can occur in reportedly very good meteorological conditions where there is very little texture contrast with the white conditions or no other objects of known proportions in view, such as trees, houses, vehicles etc., to give the comparison on heights.

The above, but where known objects are in view but their placement in distance relative to the other object, can in fact create an extremely serious distance illusion. For example, a large basalt face on a white edgeless (visually) white mountain may appear to be a larger face of a mountain at a much greater distance. This could be further reinforced by lack of depth perception causing a false estimate of the distance to the known object.

Finally, total lack of texture cues could lead to the eyes going to a 10 foot focus, inducing an illusion of clear space extending ahead into the distance, known as the Gansfeld effect.

Were VMC rules infringed by flight TE 901? I am certain that Jim Collins would not intentionally infringe these conditions and I feel sure that he and all the persons in the cockpit must have been satisfied that they had achieved these conditions and were maintaining them to the satisfaction of all of them in their run into McMurdo. They were visual at about 10,000 feet and McMurdo had a cloud base of about 3000 feet and 40 miles visibility, which is well above VMC.

It was well known, and Collins would have known, how McMurdo reported visibility. The officer looked out the window and if he clearly saw the Victoria Land coast on the other side of McMurdo Sound, some 40 miles away, then he reported visibility was 40 miles, which really meant that under Antarctic conditions true visibility would be closer to 200 miles. If he saw the Victoria Land coast but visibility was impaired, then he reported a nominal visibility, such as 38 or 39 miles, meaning that although visibility is at least 40 miles, some loss in atmospheric transparency is present.

Therefore, I can only assume that they must have crashed in visual conditions. This contention is also supported by the passenger photographs and by the meteorological study of the clouds at the time. Knowing Peter Mulgrew and the crew concerned I have not the slightest doubt that if they had approached anywhere near cloud then the crew would have responded immediately and comments on the voice recorder would have clearly indicated the situation from the operating crew loop members.

We in Air New Zealand all believed that, as provided for in the regulations, you were either in VMC where, provided you maintained the required number of kilometres of atmospheric transparency or visibility, you would be able to see, recognise and avoid any obstruction or else you would be aware (because the eyes would report it) that the transparency, and therefore visibility, was being eroded with, say, dust or rain, or snow showers. In each case, when the transparency decreased to a certain number of kilometres, you would then be in Instrument Meteorological Conditions (IMC) where a totally different set of regulations would apply, because you would no longer be able to rely on your eyes to provide you with obstacle avoidance.

These rules are ingrained into all professional pilots who have survived for many years in an environment of balancing safety and economics, and they are appropriate to almost all conditions of flight where there is adequate visual texture. But such rules rely upon the ability of the eye to report any erosion of transparency or visibility. My investigation into whiteout revealed that in certain polar conditions the Catch 22 is that there are numerous visual illusions; all of which will remain undetected by the human eye. The significant point is that the eye is unable to report its misfunction in providing obstruction avoidance. Therefore attempts to conduct visual flight at any altitude below any white surface must be totally inappropriate — particularly below any overcast. Some people failed, even after exposure to the documentation on whiteout, to grasp the vital fact that the eye will not report to you that, although you have plenty of visibility, *you are effectively in IMC* while you have no way of knowing it.

Ron Chippindale failed to grasp this vital point, see his report Para 3.37 p. 53.

> 3.37 Probable cause: The probable cause of this accident was the decision of the captain to continue the flight at low level toward an area of poor surface and horizon definition when the crew was not certain of their position and the subsequent inability to detect the rising terrain which intercepted the aircraft's flight plan.

This paragraph clearly presupposes that the pilot would be able to detect this sector whiteout phenomenon, but this supposition is contrary to all the scientific evidence.

So we have the two groups; the first group including Chippindale subscribes to the theory which believes that, without fail, when pilots untrained in polar flights are in sector whiteout those pilots are negligent if they fail to detect the existence of that sector whiteout. The other group, to which I belong, does not believe this. In any case, had the Air New Zealand operations specification for the Antarctic contained the simple warning that visual separation from white surface should not be attempted, particularly below an overcast sky at any altitude, the accident would not have occurred. Furthermore the remainder of the Antarctic crews, including myself, would not have been exposed to this unknown danger.

In my opinion the protection the public should have been entitled to, would have been the preparation of the Operations Specification by the CAD. If they had adequately researched the route with the

ICAO they would quickly have discovered the dangers of attempting any form of visual sightseeing below an overcast sky in Antarctic conditions. They certainly then would not have condoned a descent to 6000 feet under overcast to continue visually in an area of the most prolific visual illusory conditions. The two positions are apparent throughout the evidence which was given before the Royal Commission and subsequently. Those who comprehended the phenomenon of whiteout believed that Collins, together with his crew and the very experienced Antarctic explorer Mulgrew, were totally unaware of the simultaneous misfunctioning of their visual perception in clear conditions. Whereas those who failed to comprehend whiteout stuck to the position that the crew must either have been in cloud or else they must have been aware that their perception was misfunctioning.

What is not clearly spelt out in the available literature is the problem of 'sector whiteout' that is, where only a narrow sector of the total horizon is in whiteout. The aircrew which is untrained in polar flight can be testing the visibility over a very wide arc and proving to themselves, by the distance to known visual texture, that their atmospheric transparency puts them safely within the visual rules. However, if the sector in unrecognised whiteout happens to be in front of them, then they can become the unknowing victims of this 'malevolent trick of the polar light'. Clearly this is what happened to Peter Mulgrew, Jim Collins and crew. We should be looking at what needs to be done to ensure that the vital lessons that could be learned from this tragedy are not lost.

If it is possible to fly into an ice face which is in the visual field, but which is not *perceived* by a full operating crew and an experienced Antarctic explorer, when the obstacle was straight ahead in their field of view for a considerable time and with them apparently not being in any sort of stress-filled environment, then a strong argument could be developed that, without highly specialised aircrew training in recognition and avoidance of this phenomenon, no civil aircraft should be permitted below the height of surrounding terrain except under suitable instrument approach guidance. In other words polar flights should be treated as restricted visibility or cloud flying.

When I flew visually in the Antarctic I believed that there was no problem. Prior to my research on the Antarctic crash, I would have scoffed at this requirement. I am now firmly convinced that under certain lighting conditions an aircrew could fly into terrain, even with that terrain in the field of view and with plenty of time to take avoiding action. Therefore descent below the top of Mt Erebus, or other polar terrain, even in clear conditions, is hazardous. It appears to me that those of us who conducted the Antarctic flights may unwittingly have exposed ourselves, our passengers and crew, to a similar danger.

Appendices

Briefs of Evidence to and Memorandum by Royal Commission.

Appendix A: Captain Vette

Appendix B: Captain Cooper

Appendix C: Professor Day

Appendix D: J. H. Wilson

Appendix E: First Officer Rhodes

Appendix F: Memorandum of the Royal Commissioner as to Evidence and Information Obtained Overseas

Appendix A

Evidence presented to the Royal Commission by Alwyn Gordon Vette.

Qualifications and Experience

I joined TEAL (now Air New Zealand) as an engineering apprentice in 1948. At the same time I commenced flying training at Auckland Aero Club. I subsequently joined the Air Training Corp and completed my pilot training with CMT and subsequent reserve commitments. In 1953 I graduated as an aircraft maintenance engineer and I also completed an advanced aero-dynamics course at Auckland University. I was awarded the RNZAF trophy and the Geoffrey Roberts prize for top engineering apprentice. After graduation I applied for a short service commission in the RNZAF and was accepted. I served as a flying instructor and progressed through the various categories and finally gained a Service A1 and Civil A instructors category. These are the highest instructor categories available to military and civil instructors. (See Civil Aviation Regulation 249) I was also an accredited Instrument Rating Examiner. Whilst in the Air Force I passed my ALTP examinations and most of the subjects for a flight navigator's licence. In 1958 I rejoined Air New Zealand as a co-pilot on DC-6 aircraft. In 1960 I gained my command on DC-6 type aircraft and in 1964 I was appointed to training and check captain DC-3's and Electra's. In 1965 I was appointed to training and check captain DC-8. After having served a posting in Los Angeles I returned to New Zealand and became Senior Training Captain. In 1972 I was appointed to training and check captain DC-10 and in 1976 I was appointed to Senior Route Captain North America. In 1979 I was appointed to Line Superintendent North America. I am currently Flight Instructor for DC-10 type aircraft and check pilots at all levels of seniority. I have also kept current my instructor's qualifications for all types of aircraft. I train pilots from *ab initio* to top professional. I served as President of the New Zealand Airline Pilots Association for two terms and have represented the Association at world and regional conferences.

Briefing for Antarctic Flight

I captained an aircraft to the Antarctic on the 15th of November 1977. My briefing for this flight was conducted on the 13th of October 1977 by Captain A. Lawson. My crew consisted of Captain A. Potts as a Second Captain, First Officer Clapshaw. In addition there were two flight engineers. This was the first flight to the Antarctic for all the flight crew. On our flight to the Antarctic we carried a commentator, Mr Peter Mulgrew. Both Captain Potts and First Officer Clapshaw were briefed for the flight besides myself.

The briefing for this flight consisted of an audio-visual presentation and a talk through of written briefing notes together with a flight in the simulator. During the talk through of the written notes conducted by Captain Lawson at the stage when the NDB let-down procedure for cloud penetration was being discussed, I asked Captain Lawson what we would do once we had reached 6000 feet following a cloud penetration and were at that stage in VMC conditions. Captain Lawson answered to the effect that we were to maintain VMC. I understood his reply to infer no prohibition against descent below 6000 feet. In fact, if one is to maintain VMC below a cloud base of 7000 feet it is often preferable to descend further rather than flying close to the cloud base. Furthermore, it would be incongruous to consider 6000 feet to be a minimum altitude over almost flat terrain in VMC conditions when we were authorised to overfly the most threatening terrain (namely Mt Erebus) at approximately 2000 feet (subject to accurate QNH) and authorised to overfly other terrain at 2000 feet on other sectors of the route. I took the 6000 feet altitude as an intermediate level due to possible traffic in the McMurdo area, and subject to clearance by the local air traffic controller I considered I was at liberty to descend further so long as VMC conditions could be maintained.

I have seen the audio-visual presentation as demonstrated to the Commission as having been given to Captain Collins and can say that it is not the same as was given on my briefing. I recall the slides as being different. I am not certain of other aspects of the audio-visual briefing. The audio-visual briefing was for the purposes of familiarisation and not for the purposes of navigation or operational procedures. The change to grid navigation and the let-down procedure was conducted in the simulator. At the conclusion of the briefing I did not believe that my descent was limited to 6000 feet. I regarded the descent procedure only as a cloud break procedure to get to 6000 feet with cloud at 7000 feet or above. I understood the procedure to be to enable me to get to 6000 feet and proceed from that level in VMC conditions as cleared by McMurdo. The cloud base was in my view unusually high and I considered that this was a courtesy to U.S. authorities who had other traffic in the area. Had an absolute prohibition of flight below 6000 feet been intended I would have expected the terms *flight below 6000 feet prohibited* to have been used as in normal aviation terminology together with an explanation. As it appeared only to apply in the McMurdo area and not elsewhere I assumed it to be an intermediate approach altitude from which McMurdo could be clear to suit your mutual convenience. I knew from my briefing that the McMurdo co-ordinates on the flight plan were the co-ordinates of the NDB and that the flight plan route went from Cape Hallett to the NDB. I did not expect to have to use the flight plan route however unless IMC conditions prevailed.

I did not regard it as any requirement to overhead McMurdo NDB unless I needed to carry out the approved cloud break procedure. I did not carry out this procedure and did not overhead the NDB as conditions were excellent. I felt free to descend and proceed for sightseeing purposes at my discretion so long as I was not infringing VMC regulations and co-ordinated my flight by MAC Centre.

My flight was monitored by radar only after I had completed my visual let-down procedure. The only vector instruction I received was in relation to another aircraft and this was to position my aircraft behind and to the right of that aircraft on its final approach to view the landing.

My flight proceeded to the Balleny Islands. At this point I checked the accuracy of the AINS by means of a line of constant bearing. I decided that the aircraft was tracking within two to three miles of its correct track. It was difficult to determine when my aircraft was directly overhead Balleny Islands but I estimated that it was within five miles of the recorded distance. I did not conduct such a specific check of the AINS at Cape Hallett but found the aircraft to be on track and did not see any sign that the system was operating less accurately than normally.

The accuracy of the AINS system has to some extent been challenged. My experience in operation with this system has shown it to be extremely accurate. The system operates with three stabilised gyros and if any one becomes inoperative a warning light in the cockpit appears. I also understand that the computer system associated with the AINS has the ability to discard any information that is obviously inconsistent. At the conclusion of each flight after the aircraft is positioned at the ramp a check of the accuracy of the AINS is made. Each ramp in, for instance, Los Angeles has its own co-ordinates available on the aircraft. A reading is taken from each platform of the ISDU and compared with the set of tolerances on the flight plan. If any platform is outside the tolerances maintenance action is called for in engineering log. In my eight years experience on the DC-10 I only recall having to make such an entry on three or four occasions and then only in respect of one single platform.

I am a qualified navigator and have kept my navigator's licence in force. I regularly make a flight on DC-8 type aircraft in order to do this. I regard the AINS navigation system as more accurate than an astronomical fix or a three-bearing radio fix. I would however also have in mind that there could be an error of 1.2 miles per hour or greater if any part of the system was indicated as defective. In estimating any error in the system I would have regard not to the whole flight time but to the time from the last point at which I had been able to make an accurate check of the system. In the case of my flight this would have been at the Balleny Islands.

My flight went from Balleny Island to Cape Hallett and from this point I departed from the planned flight route and followed the Victoria Land coast. Because of favourable weather conditions I then crossed McMurdo Sound to pass to the east of Mt Erebus. This decision was made after consultation with Mr Mulgrew as to the best areas in view in the conditions pertaining.

I departed my cruising altitude at approximately 30 miles before Mt Erebus and descended to Flight Level 160. I maintained this flight level until I reached Mt Erebus. Mt Erebus was clearly in view for 150 miles or more. I did not regard myself as limited to this altitude at this time and flew at this level with the intention of passing close to the summit of Mt Erebus to give the passengers a view. At this time the volcano on Mt Erebus was very active. I would not have overflown the plume of the volcano because of the discomfort through the airconditioning system and because of the possible damage to various types of filters. At the time of my flight I estimated that the plume of Mt Erebus rose to 4000 or 5000 feet above the mountain. After coming abeam or slightly past Mt Erebus I turned to the left and flew to the east of Mt Terror. I then circled to the right towards Scott Base. At Mr Mulgrew's suggestion we visited the site of Scott's hut and Shackleton's hut and then the area where Scott had died.

When over the ice shelf, I was cleared by McMurdo to descend further to about 3000 feet. I was then further cleared by the Air Traffic Controller to circuit height of 1500 feet and vectored to follow a Starlifter on final landing approach. We were positioned to this aircraft's right rear and observed its landing. I was later asked to do another low level run over the airfield so that photographs could be taken from the ground. After I had completed my sightseeing I climbed out along the Victoria Land coast and reached an altitude of 10,000 feet.

In completing sightseeing in this manner I did not regard myself as breaching any regulations or company operating instructions. I had as passengers on the aircraft, Mr J. C. Brizendine, the President of the Douglas Aircraft Company and Mr A. R. Watson who I think then held the title of assistant or deputy manager. I always attempt to operate my aircraft as safely as possible while endeavouring to achieve the task. I would certainly not have flown the aircraft in any dangerous manner with Mr Brizendine on board and nor would I have breached any operational instructions with Mr Watson present. Both Mr Watson and Mr Brizendine are ex-pilots. I spoke to them after the flight and both complimented me on the conduct of the flight. After the flight Mr Brizendine wrote an article which was published in an aviation magazine (Exhibit 56). I know that a copy of this article was received by Air New Zealand as the company sent me a copy. During the whole of the flight my flight deck was opened to passengers and at various times both Mr Brizendine and Mr Watson were present on the flight deck.

The weather conditions for my flight were uncertain. The forecast did not indicate good viewing conditions in the McMurdo area and my first thought was that I would have to do the alternative route to the South Pole. The forecast received at about the time we reached Balleny Islands indicated an improvement and I then decided to go to the McMurdo area although the forecast was still not particularly good. I discussed with Mr Mulgrew what alternative viewing might be available if we were not able to visit Mt Erebus and the Scott Base area.

It is standard company procedure to carry out a pre-descent briefing to inform all crew of the pilot's intentions. I carried out this procedure with both my second pilot and my first officer present. I would be very surprised if Captain Collins had not had all his flight crew at his pre-descent briefing. The purpose of the briefing is to inform all crew members how the Captain intends to handle the task of descent with the information available and if any uncertainty exists to discuss alternatives.

Mr Mulgrew played a significant part in my planning. Mr Mulgrew had a knowledge of the area and I did not have to refer to any of my topographical charts to fix the aircraft's position. Although not a navigator or trained observer Mr Mulgrew was perfectly at home in the air and demonstrated that he was very competent in identifying topographical features. I had no hesitation in using his expertise to assist me but his views were available for assistance and not for determining the aircraft's route.

With my knowledge of Mr Mulgrew's ability and knowledge of the area I am confident that if Beaufort Island had been perceived from the flight deck as has been contended, it would have been positively identified by him. There is no record on the cockpit voice recorder of Beaufort Island having been either perceived as an island or identified as such and I can only conclude that this landmark was never available to Mr Mulgrew for identification. I would be happy to fly with Mr Mulgrew as an aircraft observer even though he is not so trained. I believe that he would be of greater assistance to a captain in identifying landmarks than another pilot who had been there previously.

With the knowledge I had at that time I would not have been concerned flying my aircraft at 1500 feet with a reported visibility of 40 miles. It is standard practice at all airports throughout the world when approaching a circuit area visually to approach at 1500 feet. Since this disaster, my knowledge of Antarctic conditions has been greatly increased and I am now aware of the significance of horizon or surface definition and of the dangers of flying in what are regarded as normal VMC conditions. Conditions when there is less than 'good surface' or horizon definition in all directions are dangerous for VMC flying and such dangers are not appreciated by those who have not flown in Antarctica. Even in reported 'good conditions' there are also dangers.

Post-Accident Study

After this accident was reported and before any accident report was published, I could not understand how Captain Collins could fly towards a mountain for more than 25 miles and collide with it in reported 40-miles visibility without taking some avoiding action. This suggested to me that he must have been misled in some way.

I have had a close association with Captain Collins and feel competent to speak as to his capabilities. I first instructed him in the Air Force in about 1950. I believe I cleared him for his first night solo flight. I flew with him when he was in the Air Force and he was a student when I was an instructor in the flying training school. I subsequently checked him in my capacity as a training and check captain on DC-6 aircraft, Electras, DC-8s and later DC-10s. It is my view that Captain Collins was an extremely competent and conscientious officer. He was well aware of operational procedures and was meticulous in complying with such procedures. In all my association with Captain Collins I have no knowledge of any adverse comment having been made about his ability as a pilot but I am aware of many commendations. I have heard Captain Collins briefing his crew at various times and I have no hesitation in saying that he does this very well in a clear and unambiguous manner.

Captain Collins was a person who was well prepared to delegate responsibility in accordance with the 'failsafe' concept of crew training. I have had little knowledge of First Officer Cassin and do not feel qualified to speak as to his ability. The people I have spoken to who knew him held him in high regard.

I knew First Officer Lucas very well. He was a very positive and sometimes outspoken person. I feel confident from my knowledge of him that even although he may not have been on the flight deck he would have not remained silent if he had observed anything with which he was unhappy. This was his first flight to the Antarctic and I feel certain that he would have been observing the scenery from some vantage point. The low level of flight would have been apparent to him and if visibility conditions had deteriorated to a dangerously low level I am certain that he would have made his concern known.

I have made reference to a 'failsafe' crew training system. This is a system of training used by Air New Zealand in the training of both pilots and engineers to ensure that the actions of all persons on the flight deck are monitored and co-ordinated. It is the responsibility of the engineers to monitor not only their engineering instruments but also some flight instruments. The flying of a DC-10 involves all three crew members, the captain, the co-pilot and the engineer. All personnel are trained to draw attention to any irregularities and departures from operational instructions or matters which could affect the safety of the aircraft. Any matters which are drawn to the attention of a crew member must be responded to, failing which, the matter is repeated. This is referred to as a challenge procedure and apart from ensuring correct flying techniques it is designed to detect any incapacity on the part of a crew member.

I knew Engineer Brooks particularly well. He was an instructor in the 'failsafe' crew training methods. I had high regard for him and for his positive and outspoken manner. In this regard he was very similar to First Officer Lucas. With my knowledge of Engineer Brooks I can not accept that he was uneasy before he uttered the words at 0049:24: 'I don't like this'. Knowing the man I believe that this comment showed no great unease. It may have been the first sign that he was becoming uneasy but not that he was concerned about the safety of the flight at that stage. Engineer Brooks' comments had he been concerned about the safety of the flight, I am certain, would have been much more positive and direct.

With the 'failsafe' or 'crew cooperative' training the threshold of demuring is very much lower than in the everyday situation in which non-aviation persons are involved. If I can cite an example, a passenger in a motor car may well feel uneasy because the driver is travelling too close behind another vehicle. This uneasiness would not normally be expressed but a similar level of uneasiness in an aircraft would be expressed and expected. Various personalities would use different expressions and tones but would certainly get their message through.

With my knowledge of Captain Collins and his crew I sought to pursue my studies further to find an answer as to how this accident occurred. During the course of my studies I became aware of the part that 'psychological set' or 'preconditioned state' has played in aircraft accidents. I now believe that this 'set' or 'state' is recognized as having been a contributing factor to many aircraft accidents. I have studied many authorities on what are now generally regarded as 'human factors' contributing to air accidents. Before the accident investigation report was released and before hearing Captain Kippenberger's evidence, I had formed the view that one of the contributing causes of this accident fell in the field of 'human factors'. Since reading the report and hearing the evidence my views are very much confirmed.

I believe that 'psychological set' or a 'conditioned state' contributed to this accident in two areas. In the first place, I believe that Captain Collins was led to expect that the route his aircraft would follow was different from that which it was in fact programmed to do. Secondly, from my past experience in textured conditions (i.e. in conditions of contrast); I believe he was conditioned to interpret visual clues according to experience. Because of inexperience in white surface flying, he was unable to appreciate visual deceptions that may have been present.

Dealing with visual deceptions or illusions I refer to the AROWA publication, a portion of which has already been produced as exhibit 103. I now produce the entire AROWA publication and draw attention to the visual and optical deceptions which pose a danger in Antarctic flying which are referred to on pages 43 (Whiteout) and following and page 51 and following. Exhibit (103). I have prepared a paper on the subject of visual and optical deception and human factors which is produced as appendix A to my evidence. Until I commenced my investigations I was not aware of these matters and I believe that other pilots would similarly be lacking in knowledge including the people responsible for structuring the requirements of the route specification and those responsible for implementing the associated training.

I suggest that Captain Collins was led to expect that his route would pass up the McMurdo Sound clear of high ground for the following reasons and I group these reasons into two categories. Firstly, positive factors which are established and secondly, possibilities which can not be disregarded.

For positive factors indicating a track up McMurdo Sound I refer to the following: the RNC chart, the passenger map, Annex H referred to in the Accident Investigations Report, Annex "I" in the Accident Investigation Report and the slide appearing in the audio visual presentation which refers to Mt Erebus ahead but showing the high ground depicted in the slide to the left of the track.

For other possible factors confirming a track up McMurdo Sound I list the following. The reliance to be placed on any of these items may vary from individual to individual but I believe that they are all worthy of consideration.

(1) The knowledge of the existence of the Byrd Reporting Point for radar pick up and as the arrival point for the area.
(2) A reference to an atlas. In particular the atlas produced in evidence would appear to indicate a logical route both for safety and sightseeing purposes up McMurdo Sound over the sea ice.
(3) As a trained navigator and prudent pilot, Captain Collins could well be expected to have mentally noted the co-ordinates given at the route qualification briefing and to have plotted approximately such co-ordinates on any other typographical material he had available, such as the atlas he was known to have possessed and taken to the briefing.
(4) In any communication Captain Collins had with MacCentre concerning sightseeing, the existence of Mt Erebus would be known to the American ATC and the non reference to Mt Erebus as a possible en route danger would have suggested

that the route was clear of the mountain.

(5) The fact that the flight was virtually without ground navigational aids meant that it was VMC. An approach in VMC conditions could be expected to give the longest possible opportunity for orientation.

(6) One cannot discount the possibility that Captain Collins had discussed this flight with other persons who have approached along what is generally referred to as the military route.

For reasons I will later give, I believe that once a route up McMurdo Sound had been established in Captain Collins' mind, it would require some positive stimulus to alert him to the fact that the route, rather than being up McMurdo Sound, was to overfly Ross Island and Mt Erebus.

I believe that the fact that the flight was continued at low level to a distance known to be 26 miles out clearly indicates that Captain Collins considered he was flying over sea ice and not approaching high ground. I believe that this view can be substantiated by superimposing the let down pattern of the aircraft over the 'incorrect' route and relating the cockpit voice recorder comments to a flight along this route. I produce Annex 'B' showing a transposed flight path. My intention in this presentation is to show that the visual sightings made by the crew and recorded on the CVR are consistent with sightings that would have been expected for an approach up the McMurdo Sound. That is, none of the sightings made by the crew were sufficiently at variance with what they would have expected to see.

I draw attention to the radio message timed at 0042:01 where First Officer Cassin informs Mac Centre: 'We'd like to let down on a grid of 180 and proceed visually to McMurdo.' This message indicates an intention to vacate 10,000 feet on grid 180. On grid 180 over sea ice, minimum safe altitude in VMC conditions is 1000 feet. Having vacated 10,000 feet and visually inspected an area below the aircraft and having satisfied themselves that VMC conditions could be maintained at a lower level, the crew turned the aircraft back to proceed visually to McMurdo as agreed with the air traffic controller. In turning the aircraft back toward McMurdo, they were tracking over the ground that had just been surveyed from a higher altitude and had been found to be clear. The ground below had been identified as nearly flat sea ice.

Following this conversation there is a further conversation at 0042:15. The aircraft is instructed to do four things;

(1) To maintain VMC
(2) To keep McMurdo advised of the aircraft's altitude
(3) To approach McMurdo
(4) To report 10 DME.

Apart from the fourth item I believe the instructions were complied with. The statements made by the Inspector of Air Accidents on this matter have seriously misled the public. In order to comply with the Inspector's requirements, the aircraft would have had to cover approximately 22,000 miles on a heading of grid 180 until it reached a position of 10 DME McMurdo.

At this point two further factors should be made clear. Firstly Captain Collins has been criticised for not positively identifying his position prior to descent. If this proposition is correct then no aircraft should descend over the sea where there are no positive visual reference points. As a matter of practice, the safety of an aircraft's position can be established either by positive or negative means. The criticism only allows for a positive fix by sighting recognisable terrain but I believe that it is an acceptable procedure in VMC to establish the aircraft's position by visual reference alone, given there is adequate visual range. In this regard I draw attention to the fact that at 0031:20 Captain Collins says 'We have contact with the ground'. This comment is made before the first descent orbit. Furthermore on at least six subsequent occasions after that time, reference is made by the crew to being VMC. If Captain Collins advised that he was VMC then there is no doubt in my mind *that in his judgment good VMC existed.* Secondly, airmen are trained from *ab initio* training that when tracking normally they read their expected position on a map and then identify visually the terrain features on the ground. This is a standard procedure and is what the crew of the DC10 appear to have been doing. It is only when doubt exists as to the aircraft's position that the 'lost procedure' is adopted. This is the reverse procedure whereby the airman looks at the ground to identify significant terrain features and then endeavours to locate those features on his map. As I have endeavoured to demonstrate, the CVR suggests that the crew were getting confirmatory sightings and were indeed seeing what they expected to see.

Visual Perception

I turn now to the field of visual perception. This is a topic that will be dealt with in depth later by an expert witness, but an understanding of the problem is necessary at this stage to enable me to explain what I consider the reasons for this accident. I have extensively investigated this area and I believe that visual deception occurs at two levels. My views on this topic are elaborated in my prepared Appendix 'A'.

I believe from the evidence already cited that there can be conditions when the human eye is incapable of determining what is ahead. This occurs because of a complete lack of texture for reference points (i.e. contrasts). When the eye has no reference points whatsoever it tends to fix on a point approximately one to two metres away and to lose all distance and depth perception. This is a true visual deception.

A second type of deception occurs when although the eye is capable of determining what is ahead, the image seen is not recognised because of a lack of experience in interpreting the image. A reported 40 mile visibility ahead would condition a recipient of this information to believe that what was being seen could be an object up to 40 miles distant from him. The object could in fact be a similar but smaller object in a very much closer proximity. For example a line of black cliffs 100 feet high would subtend the same retinal angle as a line of cliffs 300 feet high and three times the distance away. Without knowing the height of the object seen or having reference points between the object and eye, the distance of the object cannot be determined. The expected visibility may cause the distance to be grossly over estimated. The cliffs with the ice above could well be 300 feet high at twenty miles from the viewer or 100 feet high at seven miles. A texture trained pilot used to trusting reported visibility calculated down to metres, and used to relying on his own judgment, would be seriously disadvantaged in believing that a reported visibility (i.e. atmospheric transparency) would give the same protection in white surface flying. New Zealand Airline pilots are conditioned to fly in a textured environment. That is, they are accustomed to seeing horizon definition and contrast. When thrown into an environment where these conditions do not exist, they are untrained to interpret what is seen since the human mind tends to interpret what is seen in light of past experience.

It is highly likely in my opinion that the cloud base which can be clearly seen on photographs 9 and 10 would have also occurred ahead of the aircraft and would have at the point where it met the snow and ice slope of Ross Island created the illusion of a horizon ahead of the aircraft, especially to a crew expecting to see an horizon ahead of them for a considerable distance.

What the eye sees is not necessarily what the mind perceives. In some circumstances the mind will impose a horizon where it believes or is conditioned to expect the horizon to be. This may give an entirely false impression of distance to the horizon. I believe this to be another instance of a psychologically induced state. The merging of cloud and sea fog or cloud and snow on the slopes of Mt Erebus could well occur over a considerable arc. In these circumstances the mind would impose an horizon where it expected one to be. There would in fact be no horizon. The false perception of an horizon would give an impression of distance where no distance existed.

The authorities to which I have referred all very clearly indicate that to arouse a person from a preconditioned state it is necessary for there to be some positive and unusual stimulus which is inconsistent with their expectations. Where more than one person is so preconditioned, the stimulus required would be greater. I can only suggest that had Captain Collins been informed that the co-ordinates of his flight had been changed on the day of his flight he would have been aroused to check the route he anticipated was being flown. Similarly had the phenomenon of visual illusions, depth and distance perception, been known to him his threshold of doubt would have responded accordingly. First Officer Rhodes, in his evidence, has referred to the practice of the U.S. pilots being specially trained for VMC flying in the Antarctic. My research has led me to realise that such training is absolutely necessary before flight is undertaken in Antarctic conditions.

The DC10 aircraft is in my view a very safe aircraft under VMC conditions. Not only can it fly a programmed track and move from a programmed to a designated track, but also it can maintain a safe altitude and a set speed with minimal internal cockpit monitoring. The flight crew are accordingly free to carry out their prime duty in VMC flying, which is to monitor the aircraft's progress from visual clues outside the cockpit.

Bearing in mind Captain Collins' training and experience as a texture trained pilot I do not subscribe to the view that the decision to descend early, with the obvious intention of descending in VMC conditions and maintaining VMC conditions to McMurdo with no navigational aids available, was incorrect. If Captain Collins had been adequately warned of either the possible visual difficulties and the invalidity of VMC and VFR in Antarctic conditions, or the change to the planned flight track he would have reacted differently and this disaster would not have occurred.

I would go further and say that from my recent study of the visual phenomena in white surface conditions and my discussions with people experienced in white surface flying, including helicopter pilots, attempts to operate below the height of any white surfaced high ground, particularly in overcast conditions, are much more dangerous than I had previously thought. The more so for texture trained pilots who are warned only to avoid blown snow or cloud (reduced atmospheric transparency) and who would not realise that the confident application of texture flying techniques could lead to being trapped without warning, by losing the very protection expected in conditions of atmospheric transparency (i.e. the distance in which a pilot would expect to see, recognise, and have sufficient time to avoid any obstruction lying in the path of the aircraft). This danger is so insidious and undetectable that being with another person who has previously been to the Antarctic may not help very much because he too can be subjected to the same illusions.

I consider that apart from a detailed knowledge of all the information I have submitted, there are many other requirements which would be needed to qualify a 'texture trained' pilot for flight below the height of white surfaces. The most vital requirement would be the understanding that whereas previously he was used to working with visibilities down to feet or metres, one hundred miles of visibility in Antarctic overcast conditions doesn't necessarily guarantee obstacle avoidance. By way of comparison with the stated conditions in the McMurdo area of 40 miles visibility and a cloud base of 2000 feet, it is worthwhile noting the minimum conditions required for a visual circuit approach to various fields. For example, at Manila, minimum visibility is 4000 metres with a cloud base of 1075 feet. For Pago Pago, there is a minimum visibility requirement of 3½ miles with a cloud base of 1030 feet. For Las Vegas airport, there is a minimum visibility requirement of 3½ miles with a cloud base of 3180 feet, but as Las Vegas is 2171 feet above sea level, the minimum cloud conditions are little over 1000 feet above terrain. I believe a comparison of these airport minimums with the reported visibility and cloud base at McMurdo as related to the crew and indeed the actual conditions encountered by the aircraft during its descent orbit and approach to Ross Island, demonstrates that both the reported McMurdo conditions and the actual conditions encountered by the aircraft were many times more favourable than the visual airport circling circuit minima that the crew would have previously experienced.

Captain Cooper in his evidence expressed the view that some comments recorded on CVR may have come from the galley area. If any of these comments are taken as indicating unfavourable visual conditions it should be borne in mind that visibility from this area would be of a window frame nature with probably only the sky or cloud areas visible to a person standing at the rear of the flight deck and the galley area. Such a view would not give a true indication of the visual condition and should not be regarded as an accurate assessment of whether or not VMC existed.

Summary

Hypothesis: In my opinion, the crew of TE 901 by virtue of the content and manner of their briefing were led to believe that they would approach the McMurdo area up McMurdo Sound. In fact until the co-ordinates of 'McMurdo' were changed on the evening prior to the flight, that is precisely where they would have gone. In the case of Captain Collins there will be evidence to show that following his briefing on the evening prior to his flight of 28 November 1979, he had obtained maps of the McMurdo area which he was working with in preparation for his flight. Having known Captain Collins, I have no doubt that he would have plotted the co-ordinates of the McMurdo Sound way point from the flight plan shown to him at the briefing and he would have noted the track to that way point relative to the radar pick up point, and the distance to run to the McMurdo way point from the Byrd reporting point. In normal practice it is unusual to check the flight plan co-ordinates of a briefed position against a topographical map. I believe that Captain Collins having been briefed and having done his 'homework' would have considered the possibility of the co-ordinates of a critical waypoint being significantly changed without notification as remote a possibility as the digits changing themselves after having been entered and checked off the flight plan.

The crew's expectations and actions during the flight were totally appropriate for a flight up the McMurdo Sound towards the Byrd reporting point, being the radar pick-up point.

Notwithstanding the above, the crew would have under normal texture conditions an additional protection in that even if their navigation equipment was in error they would expect to be able to see, recognise, and avoid any unexpected obstruction in VMC conditions, especially where visibility was reported at 40 miles. In fact the onboard photographs taken by passengers show an even greater visibility range.

With the expectation of a route up McMurdo Sound and the belief that good visibility would provide more than adequate protection to achieve terrain avoidance, the crew of the DC10 in letting down in VMC and intentionally maintaining the nav track were availing themselves of a double protection; a route over sea ice, and flight in conditions where obstructions would be readily visually detected.

The expectation of where they were flying and their understanding of where they thought they were, would have been reinforced by both the knowledge and approval given to their intended actions by the ATC McMurdo, and the topographical sightings which they made whilst visually sweeping the area prior to descent. As I endeavoured to demonstrate, the visual cues encountered while flying the nav track towards Ross Island were consistent with what they were expecting to see, and would have in fact seen if flying up McMurdo Sound. It is of note that Mr Mulgrew was also satisfied that the sightings made of features were consistent with flight up McMurdo Sound and he of course had a good knowledge of the area.

We now know that for a considerable period the aircraft was flown straight toward Ross Island. As I have stated in my evidence the reason the high ground was not identified (although obviously in view for 25 miles) was due to —

(i) The highly likely existence of the fog ramp which would have covered the Lewis Bay ice cliffs and created the illusion of an apparent continuation of the flat sea ice beneath the aircraft to a distant (and false) horizon.

(ii) The existence of the regular cloud base of the stratus cloud (at a height of 2000–2500 feet) above the aircraft and cutting across the snow and ice slope of Mt Erebus ahead would have created the illusion of a far distant horizon.

(iii) The combination of the sea fog, regular ice slope, and cloud base would have in addition to the distant horizon effect given the impression of an obstruction free ice shelf extending into the distance ahead of the aircraft.

(iv) The sighting of terrain on each side of this apparent ice shelf presented an image consistent, if not almost visually identical with, that expected by the crew if flying up McMurdo Sound.

This illusion must have been a powerful one to simultaneously deceive at least five people. Perhaps it would be worth noting here that visibility (atmospheric transparency) on either side of the aircraft was very good. This would have reassured the viewer that they had a similar visibility range ahead of them. In my opinion there is absolutely no evidence to support the contention that the aircraft was at any stage flying through cloud and I do not consider it at all possible that Captain Collins would have continued to fly towards an area that he recognised to be one of poor surface or horizon definition or an area of poor visibility. In my opinion Captain Collins clearly thought he could in fact see well ahead of the aircraft across the ice shelf and this view must have been supported by the other crew members.

The crew's demeanour as described by Captain Cooper indicates that they were confidently applying their texture trained skills. The fact that no one on the flight deck apparently saw the high ground ahead at any time also supports this thesis.

The unexpected presence of Beaufort Island and the existence of large dark holes and leads in the sea ice when viewed from altitude could have led to the crew failing to recognise the image of Beaufort Island as an island. I believe that when the island was ahead of the aircraft at the position shown as 'p' on Appendix 3 of Mr Chippendale's report (CVR time 43:27), the crew would (as they are trained) be looking towards the direction into which the aircraft was then turning i.e. towards McMurdo (Ross Island). These two factors must have contributed to the failure to perceive Beaufort Island. Although Beaufort Island is at this stage highly visible in the photographs taken by passengers on the right hand side of the aircraft, this does not necessarily suggest that the crew would have obtained as clear a sighting. The crew at that stage were looking in a different direction, namely ahead and to the left of the aircraft where they thought the threatening terrain of Ross Island would lie.

Having proceeded toward the McMurdo way point on nav track over the sea ice in VMC, Captain Collins was apparently motivated to climb out not as a result of his having recognised poor visual conditions but by the failure of the aircraft to establish VHF communication and consequent radar contact at a time he thought he was closing on a position abeam the Byrd reporting point. Captain Collins says, 'We are 26 miles north, we'll have to climb out of this'.

This decision to climb out was made 20 seconds prior to impact and prior to the GPWS being activated, with the crew still believing they had at least 30 miles unobstructed visibility ahead to climb out over the Ross ice shelf.

Unfortunately this crew were operating in accordance with their training for a textured visual environment, endeavouring to maintain VMC when the environment in which they were flying was such that VFR provided no guarantee of protection. I see little difference in this situation than if the crew had been told that the engines were serviceable when in fact they were not. I do not see that any degree of culpability or blameworthiness can be attached to this crew.

An Underlying Cause: In addition to the many proximate causes, each of a greater or lesser degree, which have been referred to in evidence, I consider that an underlying cause of this accident is the failure of the system set up to transfer information. The information of white surface flying conditions was readily available and presumably would have been known to the International Aviation Authority ICAO. The transfer of information of ICAO may have been inadequate to prepare the New Zealand Civil Aviation Department to write the operation specification and to adequately monitor the Air New Zealand proposal for Antarctic flights. It may well be that this information was available and not sought by CAD. Obviously there was a wealth of experience in New Zealand with the RNZAF and Deep Freeze operations. Unless our system draws on all available expertise, then the possibility of further accidents arising from ignorance will exist.

Paper by Captain A. G. Vette on visual perceptions and human factors relative to DC10 crash Mt Erebus.

Definition of Texture — Texture is the causitive factor of a light array reaching the visual receptors in such a manner as to cause the visual perception to interpret an array of such things as colour, shapes, edges, verticals, horizontals and facets of such magnitude as to provide a supply of information to make computation and judgments of relative sizes, distances and depth perceptions. That is, a contrast structure or irregularity on which the eye is able to focus.

Whiteout — Unfortunately this term means different things to different people depending on their particular experience with white surfaces. To most skiers, it is the term used to express the total loss of visual perception in a blizzard or blown snow conditions and generally leads to the 'Gansfeld Effect' referred to by Cohen, 1957, where the eyes failing to find a focal reference point or texture, take up their natural focal length of approximately 4-10 feet. Although you may believe you are looking in to the far distance, you are in fact focussing on a position in very close proximity. Many people have experienced the situation described above and in the absence of further education would be firmly convinced that they understood the phenomenon of 'whiteout'. Most 'Texture' trained pilots would come into this category and would believe that they were reasonably safe because the whiteout condition would require a reduction in atmosphere transparency (which they would quickly detect) caused by cloud, fog or blown snow. The fact that this is the general understanding accepted by ourselves and the D.C.A. is indicated by our briefing material. This situation is however warned against by the Flight Safety Foundation Aviation in a safety letter dated May 1979, entitled 'Learn from the mistakes of others; you'll not live long enough to make them all yourself'.

An extract reads: 'While it's white outside the pilot can become literally clueless, why else would a pilot on a perfectly clear day fly into the surface of a lake? It happens every year. Snow on the ground obscures detail, hides contrasts, and destroys shadows with its reflectivity. Whether you realise it or not, that final approach over featureless fields calls for a lot more visual concentration than in the summertime. Little wonder that we have accidents when the eye fails to see reality. Failing to see reality when you think you do is a dangerously insidious illusion and it is called "whiteout". Whiteout then is an environmental condition in which the pilot reasonably expects to have all the visual clues needed for depth perception.

The classic case is of the pilot flying over forrested terrain at low altitude on a day when a milky sky presents the same colour and light intensity as the snow covered terrain. He flies into an area which lacks surface features — usually a frozen lake. Confident that he has miles of visibility (and he has), he continues to fly heads up peering at what he thinks is the horizon (which it isn't). Unless he gets on the dials quickly he'll prang.

What isn't whiteout, is running into a snow shower or becoming engulfed in your snow swirl when you land or takeoff in a helicopter, these are visibility problems. Let's remain alert for "whiteout" as well as misusing the term to describe something which could otherwise just as easily take place in fog in the summertime.' Ex 230.

I bitterly regret not having discovered this article until I had almost finished my research into the visual phenomena of Antarctic flying. It is even more regrettable that my colleagues, whom I hold in very high esteem, from the D.C.A., Air New Zealand and the Inspector of Accidents had apparently not read this article. If they had, they would not be puzzling as to why such an excellent (but texture trained) crew should have acted in such an uncharacteristic manner and, inexplicably, in total collusion (at least five of them). I am also sure that they would not have allowed an intermediate let-down procedure below M.S.A. to below a 7000 foot cloud base where the most prolific visual illusions would abound. This is a trap for any confident "texture trained pilot" attempting to operate visually in VMC conditions even with approximately 20 km visibility.

Before I list the factors which I believe should have been covered, I must clearly state that I am speaking with 20/20 hindsight, and as a senior instructor with this company I must share some of the responsibility for not having researched white surface flying prior to going to Antarctica. With hindsight I am now aware of how the situation could have occurred and how I could return from having operated the Antarctic area at normal circuit height without any apparent difficulty.

Whiteout

I am certain that I would have been much better prepared had I been given a copy of the A.R.O.W.A. document on Antarctic weather analysis and forecasting which I have already in part tabled in Exhibit No. 228.

This document is the most useful of the M.E.T. documents I have discovered so far and would give a texture trained pilot a much better chance of anticipating possible problems. It deals with situations leading to: strong winds, sudden changes, low level jet streams, katabatic winds, blowing snow, fog, ice fog, sea smoke, whiteout.

Whiteout

The whiteout is a condition which occurs when the sky is overcast and the ground is completely covered with snow. In this situation, even though the sun is above the horizon greyish-white snow blends with grey clouds to form a uniform background lacking shadows, detail and perspective. The horizon disappears and only dark objects are discernible. It has been succinctly stated that in a whiteout condition, a dark object is visible for many miles while a snow-covered object, even a mountain, next to the observer is invisible. In the absence of colour contrast, an object becomes visible only when the ratio of the difference in brightness between the object and its background exceeds a certain threshold valve. Over snow-free terrain, with overcast skies this ratio is approximately three to one. That is the zenith sky is about three times brighter than the horizon sky. This state prevails since snow-free terrain absorbs a substantial amount of the solar radiation, providing a differential in brightness between the overhead sky and the ground, which is minimal for visual acuity. However, a snow-covered surface acts as a diffuse reflector. Instead of absorbing the incident light rays, the snow reflects nearly the entire amount of light energy. This reflected energy (albedo) is directed into the atmosphere where it strikes the cloud layer, and due to the high reflective capacity of the clouds is once again re-directed to the snow surface to repeat the cycle. As a result, the trapped incident light rays travel many times between the rebounding surfaces of snow and clouds. With little or no absorption, the degree of brightness between ground, horizon sky and zenith sky become equal and the threshold contrast is zero. Therefore whiteout can be anticipated whenever no variation in brightness occurs as an observer scans the horizontal sky and the zenith sky. This is the diffuse brightness referred to in all descriptions of this vexatious problem.

Whiteout Forecasting

The snow prerequisite for whiteout is a constant feature which will prevail over the Antarctic continent, particularly over areas which are devoid of projecting mountain ranges. Sky conditions comparable to whiteout are those in which clouds of uniform colour cover the entire sky. The clouds may vary in height and in vertical development. Thus a sky made overcast by cirrus clouds (15,000 to 20,000 feet) or alto stratus clouds (8000 to 15,000 feet) can prove as effective as a low overcast (500 to 7000 feet) in producing this diffusion in which loss of horizontal and vertical visibility is experienced.

The fundamental process for the production of extensive clouds in the Antarctic is the cyclonic storm or low pressure area. Such storms or depressions are usually accompanied by cloud decks of uniform texture which are projected 500 to 800 miles in advance of the approaching storm. Accordingly, any storm travelling along the periphery of the Antarctic continent has the potential of creating a whiteout cloud condition for a radius of approximately 500 miles over the interior. This then underlines the close relationship between the forecast of cloud cover and the possibility of whiteout.

A stratus cloud overcast can be formed by non-frontal or non-cyclonic conditions. The occurrence of this type of cloud depends upon the characteristics of the air mass, its trajectory and other thermodynamic considerations. These factors are also capable of predetermination by forecast.

In view of the limited observations made of whiteout phenomena, its forecast must of necessity be based upon empirical rules. However, it is feasible to provide forecasts of the contributory meteorological processes and such other factors as topography, and the density and crystalline structure of the snow. A forecast of conditions in which whiteout may occur can be introduced as an essential element of the forecast.

It has been noted that the occurrence of the whiteout is seasonal in nature, appearing usually in late winter and early spring. This bears out the contention that the reflective capability of snow is directly proportional to its density. In winter and early spring snow density is at a maximum. This, however, does not preclude its occurrence at any time if the prerequisites of terrain and weather conditions are present.

Whiteout will not occur in the vicinity of water areas or expanses of snow-free ice. It is considered that this phenomenon presents its greatest hazard to aircraft making landings on uncleared snow areas on the continent. Terminal conditions under whiteout conditions will be acceptable for operations if base structures are of a sufficiently dark colour so as to be discernible and electronic landing aids are utilised. The extent of the snow-free terrain at McMurdo Sound precludes the occurrence of a total whiteout condition.

Whiteout has been encountered in flight, but according to pilot reports, it is inconsequential since the flight can be safely conducted on instruments. Visual flight in whiteout conditions may expose the pilot to the possibility of vertigo. However, experience has shown that in most whiteout conditions encountered in flight, there has never been a segment of the horizon which is not clear and distinct to the pilot particularly if the flight altitude is above 8000 feet.

Recent publications mention whiteout as being produced during precipitation and blowing snow. It is considered that in these cases hydrometeors suspended in the atmosphere constitute an obstruction to vision and amply forewarn the traveller that reduced visibility is immiment. In a whiteout condition where hydrometeors are not present, forward visibility is apparently reduced even though the atmosphere is actually transparent.

Since it is the inability to distinguish snow-covered objects in one's path which characterises the dangerous aspect of whiteout, this allusion should be reserved to identify a unique atmospheric condition rather than one experienced anywhere where hydrometeors are suspended in the air.

Antarctic Optical Phenomena

Ice blink, water sky, mirage (upright, inverted or looming), fatamorgana, prismatic refraction, iridescent clouds, glory or brokenbow.

The above touches on the main thrust of the document. However, there are additional factors which have been revealed to me by questioning Antarctic pilots. One such factor is the phenomena of suspended super-cooled ice crystals aligned in such a manner in one instant as not to interfere with the atmospheric transparency suddenly falling into such alignment as to cause a sudden reduction in actual visibility by the scattering effect of the prisms on the light.

This just about covers my research on meteorology and its optical effects. I would now like to expand it with some study in the field of visual perceptual psychology — in the area that may relate to the accident.

When I learned that Captain Collins had flown for at least twenty-five miles in Nav Mode and in VMC conditions straight into Mt Erebus, I decided to rule out the possibility of drunkeness, drugs, uncharacteristic behaviour, suicidal tendancies. Because of the degree of collusion necessary, the fail-safe crew concept ruled out incapacitation. That left me with only one possibility and that was some form of unexpected and powerful visual illusion, shared in common by the operating crew and I suspect, by Mr Mulgrew. I discussed this possibility with my two sons, Karl and Mark, both science graduates. They agreed with me and put me in touch with Dr Barry Kirkwood of Auckland University. He considered this to be highly probable and gave me an extensive list of authors, in the field to research such as Gibson, Pearce, Davidoff, Cohen, Nelson, Bartley, Brown, Herrnstein and David Shinar etc. I also studied *Pilot Error* edited by Ronald Hurst and *Human Factors* by David Beatty. I also held some discussion with Professor Day of Monash University.

Because I make absolutely no claim to any degree of expertise, I will state my observations using a broad brush principle. Hopefully it will find support from experts and be expanded by them for the Commission.

I claim that preconditioning can and does induce a 'psychological set' and that if this is reinforced or at least not disqualified by other tests it can lead to an extremely high level of expectation.

This can be illustrated verbally and visually —
Interpretation of the Visual World — page 177 Gibson
Structuring of Light by Artifice — page 247 Gibson
Human Factors — page 107 David Beatty

Gibson refers to the visual field as a constantly shifting light pattern and Brunner to the ability of the human being to perceive seven million shades of colour. Gibson also refers to the visual world as distinct from this visual field. In the formation of a visual world, sensory data from 'other sources' are used to correct the visual field (Reset Datums). These 'other sources' are the conceptual framework. Seeing is a synthetic process incorporating our conceptual assumptions and expectations. Edward Hall points out that we are less actively aware of seeing than we are of talking. It is difficult to grasp that talking and understanding are synthetic processes overlapping and incorporating an intricate network of varied responses. Much more difficult is the idea that "seeing" is an active process. The variables that enter into seeing are enormous.

Cohen wrote that the world we see is far from an exact image of the physical world and that since perception is highly variable and often erroneous, we can only perceive what we can conceive. Cohen also wrote that we tend to see only what can be incorporated into our established frame of reference and tend to reject anything not fitting.

Pearce wrote that the condition of reality exists as an ever current sum total of our representations and responses. Whatever we see is what reality is for us. Professor Day wrote 'perception refers to the meaningful representation in experience of internal conditions and of the external environment'. As a result of this usually veridical representation, the individual is able to adapt to, survive in, and cope efficiently with the environment, consequent on stimulation of the photic receptors of the retinas by light within a specified range of wave lengths. The pattern of light impinging on the retinas is called visual stimulus. Its effectiveness depends on wavelength composition, intensity, contrast, movement and various other physical properties. The term visual perception encompasses such activities as visual search, vigilance, detection, discrimination, identification and recognition. It is to be noted that in man, about 80 per cent of responses are visually determined and guided. It is critically important to understand that the final perceptual experience initiated by the stimulus is determined by a complex interplay of processes at various levels in the peripheral and central nervous system. At the most fundamental level stimulus energy at the retina is transduced, (changed from light energy to electrical impulses in the neural fibres), encoded (the information in the stimulus array — colour, movement, shape, size — is preserved in the pattern of impulses), and transmitted to the visual areas of the cortex of the brain. At a higher level, the incoming information is decoded and interpreted as a meaningful representation of external events. Interpretation involves those 'higher mental processes' collectively called cognition. In particular it involves memory and 'mental set' or perceptual expectancy. A simple example will serve to make these processes clearer. In the case of an airport control-tower operator, the visual stimulus is an aircraft silhouetted against the background of the sky. This stimulus, specifiable in terms of its luminous intensity, luminous contrast, visual angle (size), and motion, initiates transduction, encoding and transmission processes. Its identification as a particular aircraft is dependent on these basic processes as well as on memory for prior information about it, leading in turn to expectancies about its position relative to the tower and time at which it is most likely to be visible.

Failures in perception can occur at the point of stimulation itself, at the level of neural processing, and at the level of cognition. If the stimulus is too low in contrast, so small that it is below the threshold for detection, masked by visual "noise" or has graded rather than sharp contours, it may not trigger the basic processes in the retina visual pathway. If the visual system is not intact or is malfunctioning, the stimulus at the retina may not be transduced, encoded or transmitted. If there is insufficient or distorted memory or false expectancies, the transmitted information may not be fully interpreted or, worse, misinterpreted. It is to be stressed that all these processes are equally involved in perception and failures at any level can result in inadequate representation or mis-perception is effectively a loss of perception.

One of the interesting characteristics of the human eye that I discovered in S. Howard Bartley's *Principles of Perception*, 1969, is that where the human eye is presented with a gradual change in luminosity or light shading, such as may occur where a greyish-white cloud base merges with a slightly differently shaded mass of fog or snow, the eye has a neural predisposition to put in a boundary line or horizon which does not exist. This is called "Mach" effect.

One further and very significant fact is that a mind or minds with a strong 'psychological set' can artificially induce a visual image particularly where it tends to reinforce a less distinct situation or reinforce what could well be an artificial horizon existing by illusion, e.g. a displaced cloud baseline cutting a concealed slope on Erebus, Bird or Terror.

Gibson, Davidoff and others refer to the establishment of the size of an object by relating the angle the object subtends at the eye to the angle subtended by another known object at the same distance. Similarly distance or depth perception is established by relating the number of identifiable units of texture intervening between objects of known size.

With all of the above we can now set about establishing an hypothesis for the cause of the Erebus crash which I am satisfied will leave the previously demonstrated high level of integrity of the crew of Flight 901 intact, they being no more responsible than had a bomb been detonated on board their magnificent DC-10.

The first leg of my hypothesis requires that the crew believed the Nav was taking them up the clear waters of McMurdo Sound. The evidence so far leaves little doubt but more evidence is going to be produced to support this aspect. In any case I have no doubt whatsoever that Captain Collins, his crew and Peter Mulgrew believed this to be the case and had this 'psychological set' reinforced when they visually swept the area with turns in each direction as they descended visually further reinforced by the accuracy of Nav at Balleny Island. Their set was further reinforced by mis-identification of edges of Ross Island — mistook Tennyson Cliffs — Cape Royd Cliffs — Cape Bird for Cape Bernacchi and the fact that they had, as expected, textured cliffs on each side at the correct retinal angle. In actual fact, approximately one-third the height but also at approximately one-third of the distance away.

They had just traversed an area of exceptionally good flight conditions with visibility of approximately 100 miles and expected this to gradually decrease to a 2000 feet cloud base and visibility of 40 miles as they proceeded towards the radar pick-up point and line of sight to McMurdo. The 'set' was further reinforced by McMurdo accepting their visual descent.

The 'set' was further reinforced by the fact that it was being shared by each of the members of the 'fail-safe' crew without any demurring until Captain Collins gave the first indication that the level of conflicting inputs — within forty miles of McMurdo but no visual sighting of any structures at base on the apparent visual plateau, approaching the twenty-three mile point abeam the radar pick-up point and now within line of visibility without getting VHF or radar contact — had built up to a stage where he wasn't satisfied. The fact that his decision to overshoot was overtaken by an unexpected series of events is now a matter of record.

The second leg of my hypothesis should be capable of standing on its own but is of course strongly reinforced if the first leg is accepted.

I contend that as they turned back visually for the radar pick-up point, or any other point, they believed that in arriving at approximately 2000 feet they had met their undertaking with the controlling authority (McMurdo) and that it was only a matter of remaining VMC until the radar picked them up or they had a visual sighting of the McMurdo base area. Then they would get a clearance to run across the field at circuit height 1500 feet as already indicated by McMurdo and as flown by previous aircraft under the protection of the radar monitoring separation from other aircraft and helicopters.

They would be expecting to see a long white surface, something like a mosaic of white with black and dappled patches, as already experienced over the Sound, running into a solid white plateau at the head of the Sound where it would meet the cloud base to form a horizon at least forty miles away.

They would expect to see some definition of coastline on each side with a visibility of forty miles, but in any case, and most importantly, as 'texture' pilots they would expect to see and recognise any obstruction with plenty of time to respond as long as they remained clear of precipitation and retained the protection of at least 20 km visibility.

I believe their expectations were thoroughly met almost up to the point of impact by the expected texture on each side being apparent proof that they were retaining their visibility. The large sector from the end of the textured cliffs on Cape Tennyson to the start of the textured cliffs on Cape Bird was either in whiteout or more probably obscured by the fog ramps demonstrated in Prof. Anakawa's experiment using the model of Ross Island and also observed as a fact by the rescue team under identical conditions and clearly shown in Capt. Foley's film taken over the site. This ramp would form the expected transition in the middle distance of broken sea ice to snow-covered land. The cloud layer cutting the lower snow of fog-covered slopes of Mt Bird, Erebus and Terror would form the expected horizon in the far distance. The elevated artificial horizon is clearly shown on Mt Bird in one of the passengers photographs and also in the movie films taken at the same time. The elevation of this false horizon would give the same effect as a far distant horizon almost up to impact.

I further contend that if the area was insidiously in 'whiteout' (that is sector whiteout) then unless an artificial horizon was induced by cognitive means over 'whiteout' or Gansfeld effect or by Mach effect the crew would have responded quite differently, and this would have been shown in no uncertain terms by the cockpit voice recorder.

I am also satisfied that absolutely no regulations were infringed by this crew until they cut the coastline shortly before impact.

Section 2 : Visual Flight Rules

2.1 Introduction

Regulation 96 of the Civil Aviation Regulations 1953 prescribes:

"96 Meteorologicals Requirements for VFR Flights — The Director may from time to time prescribe the requirements (including requirements relating to the distance from cloud and visibility) to be met by pilots in command when flying aircraft under visual flight rules."

"52. Carriage of co-pilots. A co-pilot licensed pursuant to these regulations shall be carried in an aircraft:

(a) when required by the certificate of airworthiness or the associated flight manual or other approved document.

(b) when the Director considers that the safety of the operation requires the carriage of a co-pilot."

2.2 Object

This section prescribes the requirements to be complied with by pilots in command when flying aircraft under visual flight rules.

2.3 Visual Flight Rule Table

2.3.1 VFR flights shall be conducted so that the aircraft is flown in conditions of visibility and distance from clouds equal to or greater than those specified in the following table except as provided in paragraphs 2.5 and 2.6 for operations in control zones.

	Within Controlled Airspace	Outside Controlled Airspace	
		Above	At or Below
		3000 ft AMSL or 1000 ft above (1) terrain whichever is higher	
Flight Visibility	8 km	8 km	5000m (2)(3)(4)
Distance from cloud (a) Horizontal (b) Vertical	1 mile 1000 ft (5)	1 mile 1000 ft (6)	Clear of clouds and in sight of the ground or water

(1) Unless a higher plane of division is prescribed by the director as applicable in defined areas.

(2) Except that helicopters may operate with a flight visibility below 5000 m, clear of clouds and in sight of the ground or water, if manoeuvred at a speed that will give adequate opportunity to observe other traffic or any obstructions in time to avoid collision.

(3) Except that aircraft engaged in agricultural operations may operate with a flight visibility of not less than 1500 m, clear of clouds and in sight of the ground or water.

(4) Except that dual low flying flight training in designated low flying areas is approved with a flight visibility of not less than 1500 m.

(5) Except that gliders may fly to withins 500 ft below cloud up to 11,000 ft AMSL within the Rotorua control area (visual exempted).

(6) Except that gliders may fly to within 500 ft below cloud up to 11,000 ft AMSL.

2.4 VFR Flights by night

VFR flights shall not be operated by night except in accordance with the provisions of regulation 41 of the Civil Aviation Regulations 1953 and except as provided for in paragraphs 2.5 and 2.6 hereunder.

C.V.R.

Position References

Note: Map reading from map to terrain. Lost procedure would require reading from terrain to map.

Time
34.21 *"There's Wilson? F1*
43.27 *There you go there's some land ahead c5 P.*

45.36	*Taylor. Right*
46.39	*Erebus left 20-25 miles etc. c5 (c3 and c4) z* ....
47.02	*That's the edge c5 AA*
48.10	*I reckon Byrds through here. Ross Island there.* *Erebus should be here. c5 HH* ....
48.55	....
48.59	....
49.08	*That looks like the edge of Ross Island there c5 HH* ....
49.24	....
49.30	*We are 26 miles north. I'll have to climb out of this c1 JJ*
49.33	*It's clear to the right etc. c2 KK* .... *No negative c1*
49.35	*You can see Ross Island right (meaning, am I right?) Fine.* *c5 KK* ....

Significant inputs from "fail safe" crew members in the loop and Mr Mulgrew

(Mainly looking left towards area of expected threatening terrain (Ross Island) passengers mainly looking right towards previous viewing area (Victoria Land Coast)
Confirmatory tone in voices

Confirmatory tone

48.55	What's um — have we got them on tower? c1 By tone — this is the first demure
48.59	No — I'll try again c2 By tone — First confirmatory demure
49.08	Confirmatory tone for expected position
49.24	I don't like this c3 First verbal demure (completion of loop)
49.30	This is the Captain responding to his own and fail-safe crew inputs and acknowledging that they were getting close to the radar pickup point — Pt Byrd without satisfactory results (visual sighting of base buildings or radar contact) Indication of intention to overshoot straight ahead possibly as previously briefed from abeam Pt Byrd Still no dramatic tone indicating Ross Island safely passing on the left and under observation

Appendix B: Evidence of Captain Cooper

I am an airline pilot residing at . . . Auckland. Prior to joining Air New Zealand (Teal) in January 1965 and after graduating from an R.N.Z.A.F. Wings Course in 1957 I was employed firstly by NAC and secondly by SPANZ consecutively. In January 1968 I was promoted to Captain and have flown as Pilot in Command on Lockheed L 188Cs aircraft, DC-3's, DC-8's and DC-10's. I have in excess of 13,000 hours flying time including over 2,000 hours as Pilot in Command of DC-10's. I have also held a Flight Navigator's Licence and operated as Navigator on DC-8 aircraft. However this licence lapsed in 1973.

At the time of the accident I held the position of Vice Chairman Overseas Branch Council New Zealand Airline Pilots Association. On 4th December 1979 in my capacity as an ALPA representative, I was invited to and subsequently became a member of the team formed to transcribe the CVR at the National Transportation Safety Boards Laboratories in Washington. Other members of the team were Captain B. J. Wyatt, Fleet Captain DC-10's and D. C. Olliff, Chief Flight Engineer. We were directly responsible at all times to Air Accident Inspector, Mr M. Wylie, whom we accompanied first to Seattle for the removal of the DFDR and CVR tapes from their protective containers at the manufacturers (Sundstrand Incorporated) premises and thence on to Washington.

The Audio Laboratory of the NTSB was under the supervision of Colonel P. Turner USAF (retired) who then held a current FAA Pilot's Licence. For successful transcribing of a CVR tape Colonel Turner maintained that the three fundamental requirements necessary if possible were —

1. Familiarisation with the person (persons) whose voice (voices) had been recorded.
2. Familiarisation of the aircraft type.
3. Familiarisation of the flight deck procedure and terminology used.

The importance of these requirements became evident at an early stage of the transcribing when various interested American experts who had occasion to sit in from time to time found that they had the utmost difficulty without exception in understanding what was said by the New Zealand crew and it was not uncommon for a phrase or sentence to be completely misinterpreted by them. For this reason I believe the American transcript may be more accurate in the difficult areas than that produced in the United Kingdom.

Colonel Turner also recommended certain guidelines to follow and amongst these that I recall were —

a. Only transcribe word by word those that we unanimously agreed were spoken.
b. Do not credit any words to any specific person unless there was

unanimous agreement as to who that person was.

c. Be cautious of making editorial comments.

d. Be cautious of interpreting what might have been said.

e. While transcribing the tape do not attempt to analyse it at the same time.

During the transcribing, which took five days to complete, several consecutive tapes were made of the original, each one being an improvement on its successor. This quality was improved by the utilisation of various electronic filters for inhibiting extraneous noises evident on the original tape and in achieving this we were fortunate in having the cooperation of the FBI who made available for one morning the facilities of their Audio Laboratory.

The CVR in the DC-10 records from four separate channels. Three of these RDO's, 1, 2 and 3 are very clear and present little problem in transcribing. The fourth channel is from the cockpit area microphone (CAM) and when the source of information is from the CAM, when interpreting and analysing the transcript many factors must be considered. Some of the obvious are —

1. The direction and distance of the voice source from the CAM.
2. The strength of the voice source.
3. The articulation of the voice source.
4. Whether the voice source is a question, an answer, or taking part in a general conversation.
5. Whether the voice source is relating to an operational consideration or only a topic of general nature.
6. Whether there is more than one conversation taking place at a time and who is involved in the relevant conversations.
7. Not taking comments out of context.

The use of voice recorders is a sensitive area with airline pilots and I produce as Annex 1 [*omitted from this reprint*] a copy of the current conditions under which New Zealand airline pilots are prepared to allow cockpit voice recorders to be used. This is in line with International Federation of Airline Pilots Association Policy. Some Airline Pilots Associations have more stringent restrictions. It is a matter of concern that cockpit voice recorders are used only for accident investigation purposes by properly qualified persons and are not used to monitor performance. I fully endorse the views expressed by Mr Tench as contained on page 5 or Exhibit 221.

The confidentiality of remarks made on a flight deck is a matter of concern. Security as to the contents of the tape we transcribed was strictly observed and to my knowledge, at the end of each day's transcription, Mr Wylie would retain in safe-keeping the only written copy of the transcript in existence together with the CVR tape and all copies. After all parties were agreed that the maximum had been extracted from the tapes the transcript was retained by Mr Wylie and my understanding was that on the return of Mr Wylie to New Zealand, the transcript would be typed, sighted, accepted and signed by all members of the team that took part. I subsequently learnt that in fact Mr Chippendale, the Chief Accident Investigator, took the tape to the United Kingdom for further analysis. I have compared the analysis obtained in America with the analysis in the United Kingdom and agree that they are substantially the same except for one or two areas on which I will comment later. I am also satisfied that there has been no material matter omitted and that the transcripts as presented to the Commission of Enquiry are a full disclosure of events which took place on the flight deck.

Mr M. A. Burgess of the Office of Flight Operations of the Federal Aviation Administration was present at several transcribing sessions. I understand that he is an expert in CVR transcription and analysis. He expressed himself as very satisfied with our efforts and complimented the New Zealand team on their dedication and skills in transcribing and said that ours was an exceptional effort.

The CVR tape does not have a time base and it was necessary to establish our own base while transcribing. I understand that the time base was more accurately established by reference to the DFDR. I do not disagree with the recorded times as maintained in the final accident report.

As previously stated, great care is needed in interpreting the tape so that only what is actually said is recorded and difficult passages are not distorted to record what one expects to hear. On the other hand what was understood to be said should be logically consistent with known flight deck procedures and to illustrate this point I quote the remark attributed to First Officer Cassin in the Official Accident Report, page 74 at time 0034:21 — "Damn I had whole series of er figures, visual figures holding TACAN er during the descent". This statement makes no sense whatsoever and in my opinion was never made by First Officer Cassin. It will be noted that it was not recorded in the American Transcript.

It is also important to understand the three crew "fail-safe" concept utilised in flying an Air New Zealand DC-10. One of the basic fundamentals of this philosophy is that it is the inherent responsibility of every crew member, if he be unsure, unhappy or whatever, to question the pilot in command as to the nature of his concern. Indeed, it would not be going too far to say that, if a pilot in command were to create an atmosphere whereby one of his crew members would be hesitant to comment on any action, then he would be failing in his duty as pilot in command. I have no doubt that if any crew member on Captain Collin's flight had any concern about the progress they would not have hesitated to advise Captain Collins of that concern. Some remarks in the tape transcript are suggested as having been directed to the Captain and to have indicated concern on the part of the person who made the remark. Any such remark would be in the nature of a challenge in the "fail-safe" concept, and would be responded to. When the transcript shows that there has been no response to a remark, the remark must be regarded as conversational and not a challenge nor critical or important as to the operation of the aircraft and indeed it must be a matter of doubt as to whether Captain Collins was aware of the remark having been made at all.

To illustrate this point I present Appendix 2. This annex records all remarks by Captain Collins and First Officer Cassin between the critical period from 0046:14 to 0049:24 and shows that there was no acknowledgement of the remarks attributed to the Engineers and suggested as evidencing concern on their part. If they had been made as a challenge and heard by the Captain they would have responded to. If they were made as a challenge and not heard and responded to they would have been repeated. I do not believe that Captain Collins or First Officer Cassin took any remarks by the Engineers heard by them as a criticism, challenge or sign of uneasiness.

Criticism has been made of Captain Collins in allowing passengers on the flight deck in the final stages of descent. I do not believe that such a criticism is justified. There appeared to me to be a significant reduction in background noise from about 0037:54 and from this time I formed the impression that passenger access to the flight deck had been restricted. If, as I believe, the flight deck door was open at all times, then it is my opinion that from about this time nearly all extraneous conversations emanated from the front galley. The distance from the CAM to the centre of this galley being only approximately 10 feet.

Also it has been suggested that passengers on the flight deck had been of a distractive nature to the operating crew. I do not accept this either and I present Appendix 3 in support of my contention. This appendix lists the occasions when the flight crew made remarks to persons other than fellow crew members.

Criticism has also been made of the amount of time that First Officer Cassin spent on the radio communications. Once again I do not believe that this criticism is justified. Apart from speaking into his microphone and listening through his earphones there was very little else that he was required to do in this respect. By reference to Appendix 4 and the diagramatic radio presentation I believe it can be illustrated that radio communication should have in no way inhibited his ability to maintain an adequate visual lookout.

In the context of distractions two further factors should be considered —

1. The period of flight in question did not require normal approach and landing checklist procedures to be carried out and monitored and confirmed. This is a precise procedure requiring concentration within the cockpit.
2. First Officer Cassin was not required to make any radio transmissions from 0045:00 until 0049:10, a critical period of the

flight.

Therefore I believe that there is no evidence to show that First Officer Cassin was distracted from and did not keep a satisfactory lookout.

I now wish to comment on three specific remarks which I think have been taken in isolation and I believe have had placed on them undue emphasis at this Enquiry. The first is that at 0048:05 attributed to Engineer Officer Moloney. Extensive effort was put into endeavouring to transcribe this statement accurately as it was felt at the time that it could have been very significant. I believe the perusal of the handwritten American Transcript Exhibit No. 225 indicates this clearly. The comment was made quietly and the articulation was poor. It will be noted that the comment appears to have been made in the form of a question to which there has been no obvious reply from any member of the flight crew. I urge the Commission to be cautious in attaching any significance ot this comment. The second is that 0047:55 from an unidentified source "Bit thick here eh Burt". As this was transcribed as unintelligible in the American version and as no member of the technical crew and especially Engineer Officer Moloney were known by the name of Burt, I do not accept that this comment was made on the flight deck. I could however accept that the comment may have been made not on the flight deck but from the galley area. Thirdly — the comments between 0046:39 and 0047:02 commencing with the question asked by Engineer Officer Brooks as to the position of Erebus. This conversation took place between Engineer Officer Moloney and the commentator, Mr Mulgrew. Because these two gentlemen had very similar voices, it was not possible to positively establish who spoke which comment. However, I am satisfied that no other person took part in that conversation. The significant thing is that neither of the pilots were involved in the conversation, and that remarks were not addressed to the pilots. The conversation appears to end with a positive statement and not a note of doubt.

At time 0044:47 First Officer Cassin advised MAC Centre that the cloud base was 10,000 feet at 50 miles north and followed by at 0045:00 that they were at 6,000 descending to 2,000 VMC. This was acknowledged by MAC Centre. I believe that from this point many remarks not of concern but of assurance appear on the transcript before the final decision to climb out was made and I point to these as follows —

1. At 0045:18 Engineer Brooks says, "Yes we will be going down to 5,000 feet and then to McMurdo we're going down to 2". No criticism of the decision can be inferred from this remark. Note: I believe the American transcript reports the comment more accurately but the substance is the same.
2. At 0045:26 and 0045:33 a conversation between Captain Collins and Mr Mulgrew indicates that they are satisfied that they

can change their route if necessary.

3. At 0046:19 a comment on the public address system refers to going down in altitude and "it won't be long before we get quite a good view". This comment would probably have been heard by most parties on the flight deck and there is no indication of disagreement.
4. At 0046:39 there is the discussion about Erebus being on the left and at 0046:48 Engineer Officer Brooks appears to be satisfied about the position of Erebus.
5. At 0047:06 both Captain Collins and First Officer Cassin appear to respond to a comment on the aircraft arriving at 2,000 feet. The acceptance of this altitude is not queried.
6. At 0047:43 both Captain Collins and First Officer Cassin with the best forward vision concur in the decision to descent to 1,500 feet. Although First Officer Cassin comments "Probably see further anyway", he follows this up with a comment at 0047:49. "It's not too bad". This is before commencement of descent from 2,000 feet to 1,500 feet.

The remarks at 0048:05 which are attributed to Mr Mulgrew should be read as one sentence — "I reckon Byrd's through here and Ross Island there, Erebus should be here". This assessment appears to have been concurred with by both Captain Collins and Engineer Officer Moloney.

From my personal knowledge of the parties and my involvement in the transcript I believe that the first positive indication of anxiety or concern appears at the time of 0048:55 when Captain Collins asks "Have we got them on the tower?" This is before the time to comply with the instruction to report 10 DME and is just 35 seconds before the decision to climb out was made. Therefore, I believe that isolated remarks identified by various witnesses at this Enquiry as indicating concern do not in my opinion, when viewed against the above remarks, indicate that any real concern was shown by any parties on the flight deck until 55 seconds before impact.

I do not profess to be an expert on whiteout or human factors but believe that I now have a reasonable understanding of these matters. I believe that the conversations recorded after 0049:33 show that visual clues were being interpreted as indicating that the track ahead of the aircraft was clear of obstacles. I certainly do not believe that they are capable of indicating that the aircraft was not flying in VMC conditions and I believe that photographs produced confirm this.

Photograph 10 produced by Mr Thompson is I believe of great importance. This photograph was identified as having been taken at approximately the same time as the decision to climb out was made. This is the one and only photograph which indicates any lowering of the cloud base in the direction of the aircraft path and would indicate that the decision to climb out was made at about the same time as this becomes apparent. Photograph 10 and photograph 11 taken together show that at this time visibility to left and right was adequate for normal VMC operation. These photographs showing land to the left and right would reinforce any belief that the aircraft was tracking up the Sound. The movie film Exhibit 171 further substantiates these remarks.

I also believe that the words used in the record at 0049:30 are particularly relevant. The text of the recording shows that the aircraft was flying Nav. track and that Captain Collins was obviously confident as to where this track would take him. That the decision to climb out is prefixed by the words "We are 26 miles north" I believe clearly indicates that this was a significant factor in Captain Collins decision to climb out. Whether he related it to the position of the Byrd reporting point or his anticipated way-point would be purely speculative but nevertheless the remark cannot be overlooked as indicating a decision to climb out at some pre-determined point. In fact, if the observed weather conditions had been the reason for the climb out, one could have expected a more positive statement and a more positive reaction, that is, some remark of concern relating to the weather. I appreciate that this interpretation is not conclusive but believe that it should be considered. This view can be further confirmed when it is noted that Captain Collins decided not to turn right when advised by the First Officer that it was clear to do so. I believe his intention was to climb out straight ahead indicating that, even at this late stage of the flight, he believed his position to be McMurdo Sound. Once the decision to climb out was made, if the aircraft was in the Sound, to climb out to the true south or south-east could be accepted as the most prudent action.

There are three matters relating to the AINS that require clarification.

1. The reading of latitude and longitude from the AINS is no more or less accurate that the position given from the track and distance display. If there is an error in the system it will appear in both displays. So long as a pilot is satisfied with the accuracy of his "from" way-point and his "to" way-point, there is no cause to prefer the latitude and longitude reading to the track and distance reading. Captain Collins would not have any cause to doubt either the "from" or "to" way-points.
2. The AINS system is specifically designed to avoid the necessity of plotting latitude and longitude, and facilities for this are not provided on the aircraft. Charts for plotting and plotting equipment are not supplied. The provision of topographical maps and charts on the Antarctic flights would be solely for sight-seeing purposes.
3. There is distinction between a checking of the systems accuracy and a manual update. Checking and monitoring system accuracy is generally an ongoing exercise. This is usually accom-

plished by noting bearings and/or distances from adjacent ground navigational aids, or prominent geographical features when often the airborne radar can be utilised. It is not necessary to overhead a specific geographical position to check the system accuracy. Conversely, in order to carry out a manual update, it is necessary to overhead a specific position for which the geographical coordinates are known. A manual update in practice would only ever be carried out if for some reason it had been established that there was a significant error in the system accuracy, for it must be realised that to manually update the system by overheading an NDB or DME station at cruising flight level, would build in a probability system error of not less than 5 NMS. To put this into its proper perspective, I can say that apart from training I have never found it necessary to carry out a manual update on a DC-10. The premise that it was prudent to plan the Antarctic flights direct to the McMurdo NDB/TACAN in order that a check on this AINS system accuracy could be made is not valid.

It is probable that Captain Collins would have established the accuracy and integrity of the system (and would be fully aware of the system's published limitations) to his satisfaction on one or more occasions south of the Balleny Islands. In saying this, I can state, as an experienced DC-10 Pilot, it is not impossible to establish the accuracy of one's position in relation to the area NAV readout to within say two nautical miles at altitude given the right geographical features. If by the time that he had arrived over Cape Hallett Captain Collins was satisfied with the accuracy of the AINS equipment he would, I believe, have no question to doubt its future accuracy on that flight. If the accuracy of the equipment had been established by Cape Hallett, then any additional error between Cape Hallett and McMurdo Sound would be, in my view, insignificant.

Captain R. T. Johnson in his brief made certain qualified comments on the use of AINS equipment (paragraphs 5.19, 5.20) which I and I believe all other Air New Zealand DC-10 pilots would in general endorse. However, I believe Captain Collins could be justified in his reliance on this equipment as in service it has proved extremely accurate. Errors of the magnitude suggested by Captain Kippenberger have never to my knowledge been experienced. As an indication of the accuracy of this equipment I can say from my own experience that it is not uncommon on the Auckland–Honolulu route for the northbound aircraft at 35,000 feet to have its radio altimeter triggered by the southbound aircraft at 33,000 feet on a reciprocal track passing directly underneath in the middle of the Pacific Ocean.

Criticism has been made that First Officer Lucas was not present on the flight deck. This criticism assumes that because he was off the flight deck he was taking no interest in the aircraft's progress.

Realistically I believe that it must be appreciated that as this was his first flight to Antarctica he would probably have been as interested as anyone in the scenery. I cannot credit that if he had in any way been worried about the weather conditions and the height at which the aircraft was then flying he would not have returned to the flight deck and made his misgivings known; remembering that it can be assumed that he would have heard all the PA announcements making the Captain's intentions known. I might add that First Officer Lucas had a forceful nature and did not hesitate in making his views known in respect to aircraft operations.

In his evidence Mr Amies placed great emphasis on the need for pilots to accept the information contained in a computer flight plan with caution on the basis that the transmitting of plans through the various communications networks world-wide allows errors to creep in in the form of transposed lines or jumbled figures and letters. While this is correct, I believe that Mr Amies has rather overstated the matter. The types of errors that he suggests are readily obvious and are indeed quite common. It merely requires, if need be, the calling up of a replacement plan. The possibility of receiving an apparent perfectly transmitted flight plan with only one figure in error I believe strictly speaking would not happen. If I am wrong on this assumption, then I for one have been living under an illusion. I understand that one of the prime concepts of use of computers is to remove human errors. I believe that a pilot has the right to expect that the base information contained in a computerised flight plan is factual and correct unless otherwise advised. By this I mean in practical terms that if I were to operate from A to B on a certain date and had satisfied myself during that flight that the base information contained in the flight plan was accurate, I would expect that information contained to be equally accurate say one week later unless notified of any change. Some years ago I attended a load sheet refresher course about the time Air New Zealand was changing over to computerised load sheets. By signing the load sheet prior to a flight the pilot in command is accepting the responsibility that the aircraft has been correctly and properly loaded to his satisfaction. While attending this refresher course I recall specifically asking the instructor what check that I, as a pilot in command, could carry out to satisfy myself that the contents contained in the load sheet were in fact correct. His reply was that apart from checking the aircraft registration, date, crew complement number, fuel and zero fuel weight — none. The computer would always be correct. I believe that I am entitled to expect the same form of accuracy in computerised flight plans.

The essential function of any briefing is for the pilot being briefed to be made aware of all the relevant factors that are at variance with the normal operating envelope. If this were not so, there would be little point in holding them. If at a briefing for one reason or another, reference to a significant factor is omitted by implication in the form of a double negative, it would be logical for a pilot attending that briefing to never contemplate the existence of that significant factor. Unless at an Antarctic briefing the pilot being briefed was left in no doubt whatsoever that Mt Erebus lay directly on the flight planned track, then for it do so could well be the last thing that he would expect. I believe, with the benefit of hindsight, that this is what has occurred at many of the Antarctic briefings. It is my belief that what is presented at a briefing must be —

(i) Factual

(ii) Adequate

Provided these two criteria are met, I would see little difficulty in a briefing such as the Antarctic one being held some time prior to the flight. Provided the take home information was adequate, certain advantages would result if this was the case, in that it would allow for the pilot to digest in his own time the relevant factors involved and to expand or follow up in his own way any lines or avenues that he may wish and presumably that was the Company's intention. However, having seen the material supplied to pilots on the Antarctic flights, I feel obliged to express some criticism. I do not regard the supply of a photocopy of a portion of a topographical chart as operationally acceptable, particularly after three seasons of operation. Such a document is difficult to read and has no contrast. To my knowledge copies of MS 135 (Exhibit 4) which were admirably suited for the Antarctic briefing were readily available from the Land and Survey Department at a cost of approximately $1.50 retail. The photocopy of the track and distance diagram in my opinion is unacceptable as an operational document for navigation and is no substitute for an RNC chart. Its quality of reproduction would make it dangerous for such purpose. Whilst an RNC chart does not show topographical features it does show the coastal profile. When compared to the excellent material including the specially produced maps of high quality prepared for the passengers, one is bound to conclude that perhaps the order of economic priorities in planning these flights was allowed to become a little distorted.

Hypothesis

I am a DC-10 Pilot of some experience. Although I have not flown to the Antarctic I have attended the Enquiry almost full-time and I have had the opportunity to hear and study the evidence presented to date. On this basis I wish to make comment on the decisions as they were made by Captain Collins during the final 30 minutes of the flight. In doing so I have made the following assumptions —

(i) He believed that his track was from Cape Hallett to 77.53S 164.48E and that he was satisfied with the performance accuracy of the AINS.

(ii) He did not regard 6,000 feet as a Company limitation and that he did not believe that descent below 16,000 feet was limited specifically to the TACAN arc if approved by MAC Centre.

Evidence to date has clearly shown that the area of low cloud was confined largely to the McMurdo area on the day and that the weather over Victoria Land was excellent for sightseeing. To my knowledge there has been no evidence presented as to the weather conditions on the alternate route. Therefore the decision to proceed south of Cape Hallett I believe was reasonable.

At 0018:05 Captain Collins received the weather at McMurdo as being low overcast with a ceiling of 2,000 feet and visibility below cloud of 40 miles. At this time, permission to descend to 18,000 feet was given. The weather had been given to all aircraft at 0010:54 in similar terms with the addition of the information "surface fair, horizon fair". It is doubtful whether the significance of this information would have been appreciated in relation to the Antarctic environment and VMC flying. From a distance of some 200 miles out it must have been apparent to Captain Collins that a VMC descent to 6,000 feet over McMurdo was not possible but that sightseeing could be accomplished below cloud at 2,000 feet with conditions of excellent visibility and in accordance with what he believed he was entitled to do. Therefore I believe the decision to descend to flight level 180 was reasonable.

At 0018:52 the forecaster further advised that the clear areas around McMurdo were to the north west. This would have included the McMurdo Sound. Less than two minutes later Captain Collins was advised that within 40 miles radar could let him down to 1,500 feet. This would tend to confirm to him that a flight at 2,000 feet would be in clear conditions. Questioning of Captain Collin's decision to accept the radar let-down depends largely on the interpretation of the Memorandum dated 8th November 1979. If the understanding was that the only let-down permitted under any circumstances was the one contained in the Memorandum then clearly the decision to accept a radar let-down was an error. However, if the understanding was that the Memorandum let-down was in fact not the only let-down available because an alternative radar let-down had been offered and provided that the crews had not been specifically directed at their briefing not to accept radar guidance, then I believe that the initial decision to accept the radar let-down was reasonable. It should also be noted that there was no hesitation in accepting this let-down offer by Captain Collins and at no time was this acceptance questioned by any other crew member present.

Flight level 180 was maintained until 0032:07 when further descent VMC was requested and approved and descent from flight level 180 was commenced at 0032:10. This was at a reported forty-three miles out, that is still outside the maximum radar range advised. The decision to descend VMC was clearly influenced, I believe, by the difficulty in establishing proper VHF communication. It was probable that by 0031 Captain Collins had concluded that relying on a radar let-down with unsatisfactory VHF communications was unacceptable. Dr Thompson stated in his evidence that failure to establish VHF communications should have started the alarm bells ringing. I agree. We know subsequently that the crew was puzzled by this failure to communicate but there is no indication as to their reasoning for it. Four possibilities are —

(i) They believed there may have been a blind spot due to terrain even on the Cape Hallett to 77.53S 164.48E track.

(ii) That as the primary communications at McMurdo were UHF, the VHF communications equipment was maintained to an inferior standard.

(iii) An erroneous assumption that Antarctic conditions could create large scale interference with VHF frequencies.

(iv) The low power output of the ground transmitters.

Be that as it may, provided that VMC conditions could easily be maintained to position the aircraft for its sightseeing run down McMurdo Sound and evidence would indicate that this was so, I believe that the decision to descend VMC was reasonable.

The level of 10,000 feet was maintained until 0042:01 when a decision to let-down on a grid of 180 degrees and proceed visually to McMurdo was made. This was approved and instructions given to maintain VMC and advise altitude. The aircraft was below MSA but both the crew and MAC Centre appeared to believe that conditions were suitable for visual flying. At approximately the same time the aircraft was asked to report 10 DME from McMurdo. I believe that the decision therefore to continue descent VMC was reasonable.

The final decision to descend to 1,500 feet has already been commented on and I believe that the evidence shows that when it became apparent that the conditions encountered were deteriorating substantially below those forecast or advised, the decision to climb out was made.

Conclusion

I believe that provided the assumptions as stated are accepted, the progressive decisions made by Captain Collins based on the information and knowledge available to him at the time, were reasonable and consistent with the conscientious and professional approach that was the hallmark of Captain Collin's reputation.

Appendix 2

Purpose:

To show that the F/E's were not displaying any undue concern with the flight progress by taking an arbitrary point at 0046:14 and following through with the comments and/or conversation directly attributed to the pilots.

0046:14	CAM 1	Two nine three oh . . . right
	CAM 2	Yes
0046:21	CAM 2	A thousand to go.
	CAM 1	O.K.
0046:24	CAM 2	Alt, Nav track, vert speed.
	CAM 1	Speed.
0046:28	CAM 2	Speed I see.
0046:43	CAM 2	Yep Yep.
0047:06	CAM ?	Down to two thousand feet **
	CAM 1	Yes
	CAM 2	Yes
0047:20	CAM 1	You've got speed set up there anyway haven't you
0047:23	CAM 2	Yes alt cap (nav) track
0047:28	CAM 1	Speed nav track alt . . .
0047:43	CAM 1	We might have to pop down to fifteen hundred here I think.
	CAM 2	Yes O.K.
0047:47	CAM 2	Probably see further anyway.
0047:49	CAM 2	Its not too bad
0047:55	CAM 2	I see vert speed for fifteen hundred feet
0048:10	CAM 1	Right *
0048:12	CAM 2	(Terrain) fifteen hundred
0048:22	CAM 2	Alt hold
0048:30	CAM 1	We didn't get that TACAN frequency did we?
	CAM 2	No
0048:38	CAM ?	What's the frequency one oh nine two? **
0048:40	CAM 2	Well we think thats what it is but its channel twenty nine
0048:46	CAM 1	Actually those conditions don't look very good at all do they
	CAM 5	No they don't **
0048:50	CAM 5	You're down at one one four now are you? **
0048:51	CAM 1	Fifteen hundred
0048:55	CAM 1	Have we got them on the tower? ***
0048:59	CAM 2	No . . . I'll try again ***
0049:04	CAM 1	Try them again
	CAM 2	O.K.
0049:24	CAM 3	I don't like this ***

* Indicates comment not recorded in American transcript
** Indicates comment from another source to which pilot(s) reply to
*** Indicates where first indication of concern was noted by CAMS 1, 2 and 3.

Appendix 3

To establish the occasions where the operating crew converse with non-operating persons.

1) The occasions where Captain Collins addresses other than the operating F/E or the F/O
   To person/persons unknown
   0037:06 "Well no, the cloud base is low and ah . . .
   0037:54 "Well we're having trouble with communications right now so if you'll just wait there and see what we come up with.

   To P. Mulgrew
   0042:59 Sorry haven't got time to talk but . . .
   0043:02 . . . both the VHF channels that they have here, we're not picking them up at fifty miles
   0045:26 We had a message from the Wright Valley and they are clear over there . . .
   0045:31 . . . So if you can get us over that way . . .
   0045:34 . . . Right

2) The occasions where F/O Cassin addresses other than the operating F/E or the Captain. None apart from a brief conversation with F/O Brooks relating to depressurised operation.

3) The occasions where F/O Molony whilst occupying the F/E operating seat addresses other than the F/O or Captain.
   To P. Mulgrew
   0042:05 (In American transcript) Yeah: you can get a mike there

4) The occasions where F/O Brooks whilst occupying the F/E operating seat addresses other than the F/O or Captain.
   To person/s unknown
   0044:06 About up to here now
   0045:18 Yes we'll be going down to five thousand feet and then to McMurdo . . . We're going down to two . . . we're going down to two

Appendix C

Evidence presented to the Royal Commission by Professor Ross Henry Day.

I am the Foundation Professor of Psychology, Monash University, Australia (since 1965). I hold a Bachelor of Science Degree with First Class Honours in Psychology, University of Western Australia, (1949) and Doctor of Philosophy, University of Bristol, (1954). I am a Fellow of the Australian Psychological Society and a Fellow of the Academy of the Social Science (Australia).

I am the author of *Perception* (1967), *Human Perception* (1969) and the editor of *Studies in Perception* (1978). I am the author of about 125 papers in technical and scholarly journals. I am a member of the Board of Editors of *Perception*, the leading British journal in perception, and of *Perception and Motor Skills*, an American journal. I have been the editor of the *Australian Journal of Psychology* since 1977.

My special qualifications with regard to human factors in aviation are:

(i) About five years as a member of the Panel on the Presentation of Information of the Flying Personnel Research Committee, Air Ministry, U.K.;

(ii) About five years as Consultant to the Human Factors Group, Aeronautical Research Laboratories, Melbourne;

(iii) About four years as a member of the Australian Aeronautical Research Committee;

(iv) About ten years as a member of the Human Factors Committee of the Australian Road Research Board, four years as Chairman.

While Consultant to the Human Factors Group, Aeronautical Research Laboratories, I was engaged in the research projects that led to the development of the T-VIS visual glide path system originally termed the TVG system.

I was approached earlier this year by Captain Vette and subsequently by counsel for the Collins' estate with a request that I consider and advise on Captain Vette's views on the visual perception and the efficiency of visual functioning of the crew of TE901. I have studied the Accident Report and Captain Vette's evidence and have discussed the more general aspects of the accident with counsel. I consider I have sufficient information to form an opinion and reach conclusions. In studying this accident I have discussed the matter with Doctor J. C. Lane, Director of Aviation Medicine, Department of Transport, Commonwealth of Australia. Doctor Lane is regarded as one of the leading world authorities on human factors and air accident causation. Doctor Lane generally concurs with my views on this accident having also studied the Accident Report.

Whiteout

The conditions for the occurrence of whiteout, are described in detail in the Accident Report. It will be noted that loss of perception is among the main effects of whiteout conditions. However, it should also be noted that the cause of this loss is the absence of contrast occasioned by the high, diffuse illumination and by high reflection from white surfaces. I have acquainted myself with those portions of Mr Sinclair's evidence concerning the meteorological conditions that he said were existing in the Lewis Bay area of Ross Island at the time of the accident, and I have studied the passenger photographs that are numbered one to eleven in Mr Sinclair's evidence.

It is apparent from this information that the necessary conditions for the occurrence of degraded visual functioning (i.e. loss of depth perception, and lowered threshold for contrast) existed during the final period of the flight. This loss of efficiency would mean that the crew would be impaired in their capacity to accurately judge distances ahead, and any features contrasting with the white background could have been misperceived.

I have made a careful study of the literature on whiteout and it is my view that the loss of visual efficiency can range from a mild impairment to one which is totally disabling. Whether or not the impairment has serious consequences is bound to depend on the degree of contrast in the visual environment and the amount of information for distances. If in viewing outside the cockpit contrast was poor and distance cues minimal, then the effect of whiteout on visual efficiency would be very considerable. On the other hand, because of the high contrast in the cockpit itself, and the familiarity of the crew with the instruments and controls, it is likely that visual efficiency for the 'near space' would be relatively unaffected. Since the visual functioning in the near space would have probably been unaffected, *there would have been no appreciation of the extent to which vision was impaired in looking outside the cockpit.*

It cannot be emphasised too strongly that the effects of whiteout are insidious in the extreme. Even on the ground, the effects are not recognised by the affected individual until a gross error has been made, such as walking into a snow bank, or falling into a hole. The effect occurs quite rapidly under the conditions of intense light stimulation and white surfaces above and below. There is no way of knowing the visual system is grossly affected until an untoward event occurs.

I should add that I have not observed whiteout nor experienced it in Antarctic conditions. However the conditions for its occurrence are now well established. They can be reproduced in the laboratory, although there have been no systematic laboratory studies of it. It seems to me that the conditions which existed during the final stages of the flight were sufficient to produce a significant degree of visual impairment when looking ahead from the cockpit.

Mental Set

It is generally accepted that human perception is determined in large part by expectancies on the part of the observer, set up by prior experience including experience gained from instructions. Such a 'mental set' determines in large part what aspects of the total scene are selected for more detailed attention, and how what is selected is interpreted by the observer.

There is a continual interplay in perception between the stimulus conditions themselves, that is the pattern of physical energy impinging on the sense organs, and the higher-order expectations that exist. Thus, if the visual scene is of high contrast, clearly patterned, and in relatively high illumination, it is likely to be seen accurately even though expectations to the contrary exist. Alternatively, if the external scene is poorly defined due to low contrast, poor illumination, or visual impairment then the mental set of the observer is likely to play a much stronger role in determining what is perceived, and how it is interpreted by the individual.

There is now a considerable body of convincing evidence to show that the mental set of the individual can almost completely determine what a person sees when the stimulus array is ambiguous, poorly defined, or when the visual efficiency of the individual is impaired. In point of fact, this is one of the major conclusions from research studies of the last twenty years or so.

In studying the Accident Report and giving careful consideration to the briefing procedures described in the Report it seems reasonable to conclude that the crew were 'mentally set' for an approach west of Ross Island down McMurdo Sound. This conclusion is based upon the following:

(i) The audio-visual presentation —

   a. Whilst I have not seen the audio-visual presentation, it seems clear from the Accident Report and from information shown to me by counsel, that the slide taken from the rear of a passenger brochure shows a track to the west of Ross Island down McMurdo Sound.

   b. A slide which gave an indication of a track passing to the 'west of Mt Erebus'.

(ii) The passenger information map which was given to the crew and clearly showed vividly marked in blue the route to the west of Ross Island.

(iii) The strip map showing a route from Christchurch to McMurdo with two clearly marked tracks of a passage to the west of Ross Island.

(iv) The RNC4 chart which also indicated a track to the west of Ross Island.

(v) Counsel has informed me that the evidence of pilots who

attended the briefing on 9 November 1979 is that their attention was not drawn to the fact that the route would pass over Mt Erebus and that they left the briefing having concluded the RCU briefing and the simulator exercise with the clear impression that their route was down McMurdo Sound to the 'McMurdo position'.

These aspects of the briefing procedure lead me strongly to the conviction that a mental set was induced in the crew by both audio-visual and cartographical information that the route was to the west of Ross Island and Mt Erebus. I am also convinced that the failure to stress that the route was directly over Mt Erebus would have contributed to the state of expectancy set up in the minds of the crew that the route was westward of Erebus. Given this strong expectation, it seems to me most unlikely that the crew would have been searching for any features of a route east of that which they believed themselves to be following. Their states of mind would have been such to search out route features associated with the track west of Ross Island. My conviction about the inductions of this mental set during the briefing process gains considerable support from the flight crew of an earlier flight who were present at the same briefing as Captain Collins and First Officer Cassin and who have indicated that it was their belief that the route was to the west of Ross Island.

It is characteristic of human perception that expectations that have been formed from past experience are continually confirmed by available stimulus information. Thus, in following a route, a pilot who was 'set' for a particular direction would be searching out geographical features to confirm the validity of his expectations. It is to be emphasised that the business of human perception involves not only states of expectation but frequent and repeated checking that those expectations are valid or invalid.

After reading Captain Vette's evidence and having studied the passenger photographs, it seems to me reasonable to suppose that certain features on the route being followed could be taken as confirmation by the crew that the route being flown was to the west of Mt Erebus. For example, the cliffs of Cape Bird to the right of the aircraft in Lewis Bay could easily have been taken for those of Cape Bernacchi, and the cliffs ahead and to the left of Cape Tennyson mistaken for those of Cape Royds. In making this point, I wish to refer back again to the strong possibility that visual efficiency when looking ahead of the aircraft was impaired even though the visual angles subtended by the cliffs of Cape Bird and Cape Tennyson may have been different to those of Bernacchi and Royds respectively, and even though the distances might have been different, the impaired perception of distance would have made these differences of little or no consequence. In short, prominent features of one route could easily have been mistaken for those of the other, given that discrimination of contrast and of distance would have been rendered inefficient in consequence of the impaired visual functioning.

Beaufort Island

My attention has been drawn to the fact that on the approach to Ross Island (Lewis Bay) and during the let-down procedures, Beaufort Island would have been clearly in view and therefore would have served as a clear indicator that the route was towards Mt Erebus rather than to the west of it.

I have had the opportunity of inspecting photographs of Beaufort Island showing it as a darkish mass in a white surround. It is to be noted that in the same photographs there are a number of other areas formed by discontinuities in the ice showing the dark water. Given the low expectation of Beaufort Island appearing on the route that the crew believed they were following, and given that the mass of the island bore a close resemblance to those dark areas formed by breaks in the ice, it seems to me not at all unlikely that the island would have been identified or perceived as another break in the ice revealing the dark mass of the sea. Seen through cloud, this likelihood in perceptual terms would have been enhanced.

(See note 9 on page 345 headed *Significance of Beaufort Island.*)

Illusory horizon on approach to Ross Island

In reading Captain Vette's submission I have paid close attention to his suggestion that because of the likely existence of sea fog obscuring the Lewis Bay ice cliffs and the existence of regular cloud base at a height of 2000 or 2500 feet cutting across the slope of Mt Erebus, an illusion of a distant horizon would have occurred. In addition, Captain Vette has taken the view that the presence of the fog would have blended with the ice slope of the mountain and would have given the perceptual effect of continuous flat terrain stretching ahead and beneath the aircraft. I have noted also that Doctor R. B. Thompson in his evidence has supported Captain Vette's contention that a higher horizon and continuous flat terrain would have resulted from the fog and cloud. Again, given the visual functioning was quite probably impaired when looking ahead at the relatively featureless scene, and given the strong mental set for an obstacle-free route, it is reasonable to suppose that the effect of low cloud cutting across the mountain and the obscuration of the ice cliffs by sea fog would have contributed to the recognition of a false horizon. Again I would wish to stress that poor visual functioning and strong expectation that the route to the west of Ross Island was being followed would be strong contributing factors in the misrecognition of the horizon.

Summary

In summarising my comments, I wish to state again that the total act of visual perception not only consists of stimulation at the eyes themselves and the efficiency of the visual system, but also the confirmation or otherwise of certain mental expectations.

In the circumstances under review, the visual stimulus from outside the cockpit was adequate but of relatively low contrast. The level of efficiency of the visual system was probably markedly degraded through the phenomenon generally described as 'whiteout' — the high intensity stimulation of the eye by rebounding light between the white land surfaces and cloud. Among the main consequences of this lowered visual efficiency would have been the loss of sensitivity to contrast and greatly impaired distance perception.

The mental set of the flight crew is almost certainly to have been for the route west of Ross Island, a set induced by the verbal and audio-visual briefing prior to the flight. This expectation would have been confirmed by visual cues occurring in the expected positions on the right and left of the flight path. With poor vision and inappropriate mental set, and confirmation of the latter, the perceptual experiences of the individuals were quite erroneous.

Conclusion

I would like to emphasise that the visual perceptual system can for present purposes be regarded as a piece of biological machinery. If it is forced to function under conditions of stimulation which render it highly inefficient as in whiteout conditions, or if it is wrongly programmed by prior information, and if this programming is supported by incoming stimuli, then the machine simply fails to operate in a fashion appropriate to the circumstances. This is not 'human error' in my view, but a system failure. It is analogous to placing a metal structure under stresses which it was never intended it should withstand. In short, it seems to me, that the visual system of the flight crew was rendered biologically and cognitively inefficient through the stimulus conditions and the prior briefing. This failure of the visual system could not be said to be culpable or blameworthy. Nor is it a result of a failure to be vigilant. It is simply a failure of the system itself.

The strongest evidence in support of this view is that it was not the visual perceptual system of one member of the flight crew that failed, but that of five persons including an experienced Antarctic observer and commentator. This seems to me to indicate beyond doubt that the conditions were such as to lead to a complete breakdown of efficient visual perceptual functioning.

Appendix D: Evidence of Chief Pilot Wilson

I am Chief Pilot with Helicopters New Zealand Limited, which is a company based in Nelson operating helicopters. I have been with the company for six years.

In mid 1979, the company was approached by the West German Government to see if contractural arrangements could be made for the company to assist a German expedition to Antarctica in late 1979. Arrangements were concluded, the plan being, that the operation would take place from a ship, Schepelsturm which was anchored generally in the Yule Bay area in North Victoria Land. Three helicopters were used in the operation of which two were provided by Helicopters New Zealand Limited, and were Hughes 500D's.

In preparing for the Antarctic expedition, I took a helicopter to Lake Tekapo and took part in the survival course run by the Antarctic division of the D.S.I.R. Mr Peter Tait, the general manager of the company and I went to McMurdo on a U.S.A.F. Starlifter for familiarisation purposes. We wanted to know what information could be obtained from the Operation Deep Freeze personnel. Both Peter Tait and I had had extensive helicopter flying experience in Artic regions, operating with oil rigs, but we found that conditions in Antarctica were quite different from what we had experienced in the North. One of the most significant differences as far as flying is concerned is that the Antarctic is mountainous, whereas Arctic areas are generally flat. During our time at McMurdo, we received a briefing from Commander A. Coltrin. One of the most important matters to come from that briefing as far as helicopter operations were concerned was Commander Coltrin's advice that we monitor H.F. Radio Frequency 8997 which was the frequency used by civil aircraft so that we would be aware of their presence in Antarctica. Our concern was not of colliding with civil aircraft but rather avoiding the severe wake turbulence created by jet aircraft.

Mr Tait and I went to Antarctica in early November 1979. Before going there, but after the contract with the West German Government had been concluded, I prepared an operations manual for the use of helicopter pilots involved in the Antarctic operation. I annex photocopies of Section 2 of the manual dealing with operating procedures to be followed in the peculiar weather conditions experienced in Antarctica. This manual was submitted to the Civil Aviation Division for approval which with minor amendments was granted in mid November 1979. The C.A.D. Officer I dealt with was Mr Evans, Controller of General Aviation.

The expedition sailed for Antarctica on 6 December 1979. I was not involved in the recovery operation after the crash of the DC10. The expedition carried out its operations throughout December. At the end of December during a violent storm, one of the expedition members was badly injured and I flew him to McMurdo for treatment. Some of the helicopters were also damaged and the ship sailed into McMurdo Sound, and anchored off Cape Royds, so that repairs could be affected to the helicopters. The ship was anchored from 23 December 1979 to 3 January 1980. During that time I was able to observe the pattern of the weather as it affected Ross Island. Every morning the weather was completely clear of cloud. About 11.00 am fog would form over the lower slopes of Ross Island. During the course of the morning and early afternoon, the fog would rise higher over Ross Island and up the slopes of Mount Erebus. At about the same time cloud would begin forming a layer across the middle section of Mount Erebus. As the morning and early afternoon progressed the cloud would extend to cover the summit of Mount Erebus, and the fog and cloud base would merge at a point 1,000 feet to 1,500 feet above sea level. At times the fog would extend out over the ice shelf and would obliterate the rock cliffs at the head of Lewis Bay. On some days this fog and cloud formation did not take place but when it did the general pattern is as I have described it.

In preparing for the expedition, I obtained without any difficulties all the maps I required from the Antarctic division of the D.S.I.R. I derived great assistance in setting up the expedition from the D.S.I.R., Operation Deep Freeze at Christchurch and in Antarctic and from my visit to McMurdo.

From my experience of flying in polar regions, I know that it is impossible at all times to fly visually and because of difficulties of perception that can be encountered by aviators in these regions. I specified in my operations manual that all flights would be V.F.R. By this I mean that pilots fly at all times by reference to the ground with continual monitoring of instruments in particular the radar altimeter. I have experienced whiteout conditions many times and have experienced similarly difficulties with horizon and surface definition. These conditions occur most frequently in cloudy conditions.

Definition problems are at their worst with a cloud cover of medium density. Thin or whispy clouds can affect surface and horizon definition but there is some definition obvious to an experienced pilot. Very heavy or thick cloud can often be distinguished from terrain and although definition problems occur when thick cloud is present, it is obvious to an experienced pilot that he is flying towards cloud and a false horizon or other illusion is not created to the same extent as occurs with cloud of medium density which is lighter in colour and indistinguishable from ice. Surface and definition problems can occur when the cloud is at varying heights and can occur even in the presence of cirrus cloud at between 20,000 feet — 30,000 feet. I have been shown prints of photographs I am informed were produced by Mr Sinclair of the

Meteorological Office being prints from film recovered from passenger's cameras. In my opinion the cloud depicted in those photographs is of medium density and is the type of cloud I have previously described as being in that category. That is to say the cloud appearing in these photographs would be of a type most likely to cause surface and horizon definition problems, particularly to pilots inexperienced in flying in those conditions.

One further factor which can confuse all those who live and work in Antarctica is the position of the sun. It maintains the same general position in the northern aspect of the sky, and is difficult to use by pilots as a rough fix of their position at any given time of day. One is unable to compare time with the position of the sun, readily as a pilot flying in New Zealand would be able to do. Pilots unconsciously use the sun to maintain basic orientation and this can be lost in Antarctic latitudes.

Even on bright sunny days, ground definition can be a problem particularly in the absence of shadows, and if a helicopter pilot does not pay constant attention to his radar altimeter it is virtually impossible for him to operate because he is unable to obtain any sure sighting of the ground.

Appendix E: Evidence of First Officer Rhodes

I am employed by Air New Zealand Limited as a First Officer and for the last five years have flown DC-10 Series 30 aircraft.

My aviation career commenced in the RNZAF in 1956. I have flown fighter and bomber aircraft and carried out flying instructor duties in the RNZAF, the Royal Malaysian Air Force and the Royal Australia Air Force.

In 1968 I was trained by the RAAF in the flight safety and accident investigation duties. I left the RAAF shortly after this and joined the Air Safety Branch of the Australian Department of Transport where I received further theoretical and practical accident investigation training at Melbourne University, and the Aeronautical Research Laboratories.

I was an accidents investigator in Papua New Guinea until the time of joining Air New Zealand in March 1973 as a First Officer on DC-8 aircraft.

During my flying duties in Air New Zealand I have received further accident investigative training, including a short course at the University of Southern California and attendance at several flight safety conferences.

During my flying and accident investigation duties in Australia and New Zealand I have received technical training on Boeing 707, 727 and DC-8, DC-9 and DC-10 aircraft types.

I am a member of the International Society of Air Safety Investigators, an Accidents Investigator for the New Zealand Airline Pilots Association and accredited to the International Federation of Airline Pilots to act in that capacity for any member country.

During my flying duties with Air New Zealand I have kept contact with the Australian and United States investigative organisations where libraries and technical facilities are available to a degree not practicable in New Zealand. In particular I have studied the use of recorded data (both CVR and DFDR) as a preventative function in aviation safety.

Accident Investigation of ZK-NZP

On the 28th of November 1979, I was advised by Air New Zealand of the overdue nature of flight 901 from Antarctica.

I discussed the problem with Mr George Oldfield (Air New Zealand Safety Manager) and, with Mr Wylie (Accidents Investigator, Ministry of Transport) we met in the Auckland Search and Rescue Centre. I then assisted Mr Wylie in obtaining technical data from Air New Zealand, weather and communications reports from the Civil Aviation Division and fuel samples from the bulk depot, all at Mangere.

The I.C.A.O. Annex 13 (reflected in part by Civil Aviation Accident Investigation Regulations 1978) gives guidance to the Investigator in Charge as to the conduct and composition of his team. I believe in this case the Chief Inspector of Air Accidents accepted the N.Z.A.L.P.A. request for my presence in the team because of my accident investigation background and working knowledge of the DC-10 aircraft.

Like other members of the investigation team (from Air New Zealand, F.A.A., National Transportation and Safety Board, Douglas and General Electric) I was responsible directly to the Chief Inspector (Mr Chippendale).

Mr Chippendale directed me to examine air traffic and meteorological aspects until the accident site work could commence when the weather improved. This involved taking statements from those A.T.C. and Meteorological personnel available. I transcribed the tapes of the radio conversations betwen McMurdo Centre, Ice Tower and the aircraft in the area. This data was then used by Mr Chippendale in the production of his report.

Together with U.S. Navy Flight Surgeon Lt Cmr's Deegan and Goodrin I maintained a daily photographic reconnaissance of the crash site to attempt a wreckage identification plot and check the position of significant parts of the wreckage should the vagaries of the weather cause any change.

Accident site investigation was at the direction of the Chief Inspector.

On my return to New Zealand I was assigned to assist Mr Wylie with the analysis of the CVR and DFDR data and the plotting of this information on a chart of common time base and scale.

In February 1980 I went to Longbeach, California with the accident team to discuss the conclusions reached and to examine the various experts available.

Functions of McMurdo and Scott Base

An outline of the McMurdo and Scott Base functions is necessary to explain certain equipment and procedures. The two bases are one mile apart on the south western tip of Ross Island.

McMurdo and Scott Base are scientific in purpose, the latter being run by the New Zealand Department of Scientific and Industrial Research, the former being run by the United States National Science Foundation.

Doctor Todd of the National Science Foundation runs McMurdo Base and has in support on a contract basis the United States Navy, which provide catering, accommodation and air transport support for the various scientific projects. The National Science Foundation actually own all the C130 (ski equipped Hercules) and helicopters, pay the salaries, fuel and running costs of all the U.S. Navy involvement.

This contract arrangement provides for the maximum availability of aircraft hours for the support of the scientific tasks with the minimum of aircraft, crews, fuel and support equipment as a matter of budgeting. Although the A.T.C. procedures and facilities used by the Navy personnel at McMurdo are in accordance with the TERPS manual common to all U.S. Forces this manual is acknowledged in its preface as being of equivalent standard to F.A.A. procedures.

Aeronautical facilities at McMurdo Search and Rescue

The U.S.N. staff at McMurdo consider ditching between New Zealand and McMurdo as non-surviveable due to the water temperature and therefore have no over water rescue capability. Land rescue is limited and based on Antarctic personnel always carrying survival kits as a time buffer for a suitable rescue attempt, rather than a permanent search and rescue organisation.

Doctor Todd and Captain Westbrook (Command U.S.N. McMurdo) were concerned when they became aware of the missing Air New Zealand DC-10, that their field parties may be in jeopardy, if the search for survivors and subsequent rescue was going to place a large demand on their limited airlift resources. The field parties rely on air transport for food and fuel. In the event there were no survivors. The flying tasks were switched to bringing in the various field parties to base. This released flying hours in support of the accident investigation and body recovery.

Communications Commander Coltrin and Chief Petty Officer Priest (U.S.N.) both expressed concern over the unreliable nature of communications with the New Zealand Civil Aviation links. There has been correspondence on this subject over the last two years (see Exhibit No 101). There was doubt as to when the flight plan of TE901 was received at McMurdo on the 28th of November but the controllers at MAC Centre said that they were in radio contact with the aircraft before they received the flight plan.

Non-directional beacon — At the time of the accident the NDB identification M.C. on a frequency of 518 KHZ was transmitting but had been withdrawn from publication as an aid. It was found that the TACAN is a more precise aid for radar pick-up than the NDB and is a better aid for IMC approaches in the event that the radar is not available. This is particularly true of higher speed aircraft such as the C141 feeding into a radar pick-up. The TACAN bearing data is not usable in DC-10 aircraft but the distance information (D.M.E.) is.

Ground Radar — The radar installed at McMurdo is adequate for the needs of the U.S.N., R.N.Z.A.F., and U.S.A.F. but is primitive by modern ATC standards. The equipment is called a type AN/FPN/36 Gilfillan QUAD radar. The aerial rotates through 360 degrees in azimuth and is capable of being elevated by the operator from minus one degree to plus thirty degrees. A pencil beam of

1½ degrees diameter is transmitted from the aerial to an effective range of 40 nautical miles. The narrow beam requires the operator to know fairly accurately where an aircraft is, to enable him to elevate the aerial in order to get a return on his screen. This problem was accentuated with higher speed aircraft and resulted in a retrofit of transponder facilities to extend the detection range to a nominal 200 nautical miles. In practice the transponder range of detection is usually about 150 nautical miles. The transponder is not usable within the 40 nautical mile range of the primary radar aerial. The operator has a single screen. When he switches to the precision functions (Ground Controlled Approach) the surveillance function is lost. In the precision function both the azimuth and vertical paths to landing are displayed on the one screen. This technical detail is available from the C.A.D.

Use of radar for high level approaches — In discussions with the U.S.N. controllers it was apparent that they could and were willing to elevate the aerial to 30 degrees above the horizon to scan for aircraft approaching from overhead Mt Erebus but would prefer an alternative approach because: —

a. Low level surveillance of local traffic would be lost and elevations above 15 degrees are not normal.
b. Ground clutter in vicinity of Mt Erebus obscures aircraft targets on the screen.
c. Monitoring of safe terrain clearance is not possible if precision approaches are in progress.
d. Calibration of radar returns above 20 degrees has not been carried out.
e. A knowledge of the route to the McMurdo airfield is essential, for radar pick-up, and in the case of TE901 on the 28th November 1979 the previous flights had flown down the middle of the McMurdo Sound to a latitude and longitude which resembled the final path of their aircraft. Further a path over the top of Mt Erebus was not considered a likely direction of approach. TE901 was not expected from over Mt Erebus.

Monitoring of descent in TACAN Arc — A descent within 20 nm of the TACAN is possible but would be a precision exercise requiring close pilot and controller monitoring from 16,000 feet. As the aircraft descended the problem would become easier but was not within the knowledge of the controllers as a procedure for Air New Zealand aircraft. The blind overhead cone of the radar is displaced some 7 nautical miles from the TACAN and coupled with the turning circle of a DC-10 at 260 knots (340 TAS) at 16,000 feet there is little room for manoeuvre to stay within the Air New Zealand TACAN descent area. (See Annex 'A'). The situation improves at lower altitudes but the controllers emphasised that they would have to continually change aerial elevation to try and keep contact with the aircraft. It was further pointed out that at the time of arrival of TE901 a C130 from the southeast and a C141 from the north were arriving, each required a precision approach to landing during which surveillance was not possible.

It was the factors outlined in paragraphs 22 to 29 that I understood was the reason for the limited support available signalled from the U.S.N. to C.A.D. on 13th October 1977 (Ex. 1 Doc: 21).

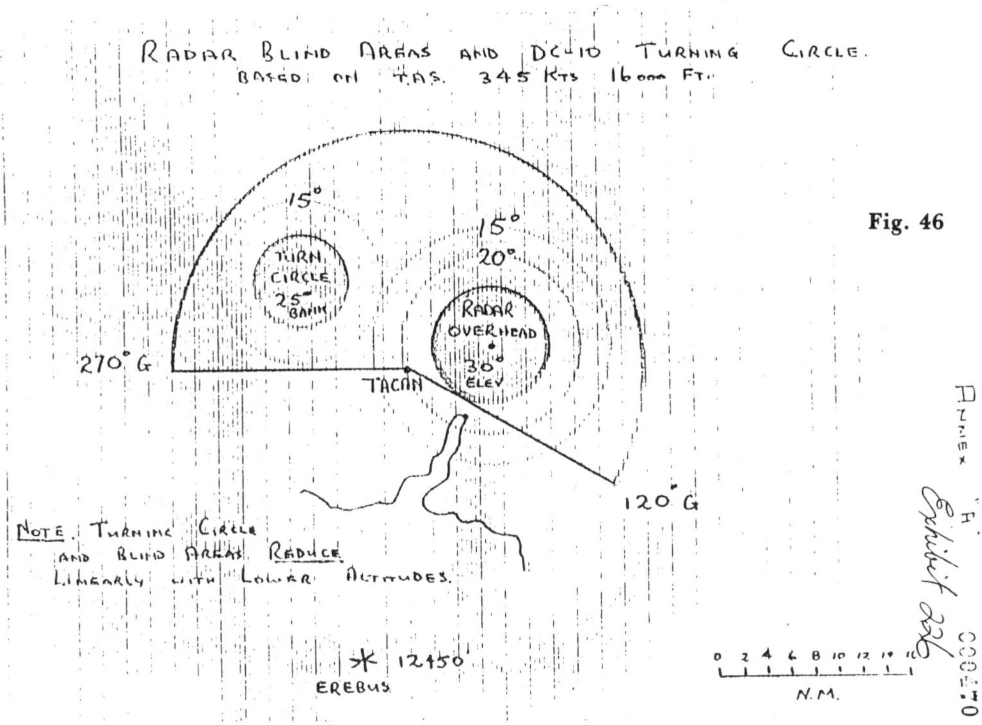

Fig. 46

Visual phenomenon in the Antarctic

My experience of flying in snow and ice conditions prior to this accident has been limited to search and rescue flights in the New Zealand Southern Alps as a member of the RNZAF.

The flights in helicopters in Antarctica during the course of his investigation exposed me to new flying conditions. I believe the term whiteout is inappropriate to describe the various visual illusions experienced in the Antarctic.

It was my experience that the 'flying in a milk bottle' effect was worse in conditions of an overcast sky but the insidious nature of not seeing a snow bank in front of one's path on a clear day, no cloud, bright sunlight, was disconcerting. The matt surface of the snow gives no depth perception even in conditions of 50 miles visibility and causes the wall of snow ahead to appear as a flat plateau with a distant horizon.

In discussing the problem of visibility in the Antarctic with the two U.S.N. Flight surgeons and pilots of the U.S.N. and RNZAF at McMurdo I obtained the following: —

a. Flight by visual reference is permitted only when both the surface and horizon definition are good.
b. The technique for visual flight appears to be a three point visual reference used by both pilots to maintain depth perception and in the event that this reference cannot be kept instrument procedures are followed.
c. The visual illusions are experienced at any height when flight is by visual reference but the degree of illusion is greater at lower levels.
d. Specific training is required.

On the accident site I carried out location and recovery of certain equipment and I took photographs of a specialist nature but the appearance of a surface fog and visual changes each day warranted additional attention. It appeared that at around midday and for the following two hours the accident site was obscured by a shallow fog that moved rapidly up the slope making helicopter landing impossible even though the aircraft could be seen clearly overhead at about twenty feet. The following slides have already been viewed and have been made available as prints (Exhibit No 123). It should be remembered however that the view of a picture at leisure may be different to that seen by aircrew at a single glance in a moving aircraft. The fog rolling up the slope would cause a diffusion of the surface in the direction of travel even though visibility to the sides was in excess of 20 nautical miles.

I concluded that a radar monitored descent in the TACAN arc was technically possible but very difficult to achieve without special briefing and close pilot and controller understanding of the procedure. In my opinion the limitation of 6,000 feet would not have prevented ground collision if the aircraft strayed outside the arc with the knowledge of the visual phenomena given in training and briefing at the time. It need not, however, be unsafe given the knowledge we now have.

Use of weather radar for navigation

The use of weather radar was discussed with experts from Collins and Bendix in the United States during February 1980. These experts could not be sure of what one could be expected to see over the Antarctic sea and ice. They had experience of testing their equipment in the Arctic and for iceberg detection in the North Atlantic. The conclusions so far are that a trained radar operator would be able to have a good idea of what he was getting returns from once he had gained experience on the patterns of returns from wet ice, snow covered ice, ice covered rock, but it was a specialist task. The view on the weather radar approaching Ross

Island would be inconclusive as the outline of the sea ice may mask the coastline. A shadow effect from the mountain ahead would probably be seen but as the correlation of this pattern would not be a usual view for pilots coupled with the possibility of it being confused with a break in the sea ice makes useful conclusions from this aid doubtful. We have already heard from the Director of Civil Aviation that its use for navigation is prohibited. Even as a guide, it would be difficult to relate the ice coast, land coast and clutter at low altitude when the visual activity should be concentrated outside the flight deck.

General Items

Crew Training — As a First Officer I am subject to periodic training and testing by Air New Zealand in accordance with C.A.D. directions. In the course of this training the emphasis is on crew activity and cross monitoring. We practise the incapacitation of the crew members on the basis that there may be no warning of such incapacitation. I cannot therefore accept the comments made during this Commission that co-pilots in general and First Officer Cassin in particular would be reluctant to comment on any practice they felt uncomfortable with before it got to the unsafe stage. The Engineer is an essential part of this communication loop and even though he has a different background to that of the pilots his aims are the same, the safe operation of the aircraft.

AINS — This area navigation system is a navigation aid that is different to earlier navigation systems. It is not a single aid but is made up of three inertial platforms each feeding their accelerations into two independent navigation computers. The navigation computers assess the accelerations and are capable of picking one platform out as being faulty. The reception of ground navigation aids further refines the accuracy of the data. It is important to note that since the advent of AINS equipped aircraft there have been numerous occasions when ground navigation aids have been found to be faulty or in the wrong geographical position as a result of the AINS rejecting or questioning the validity of the ground data. Crews using this equipment are aware of the limitations as laid down in the operations specifications but the equipment is approved as a sole means of fixing the aircraft position above the safety height and I believe Captain Collins and his crew would have known that the equipment was accurate from visual checks over Cape Hallett and perhaps earlier. The argument that the latitude and longitude readout could be used is one for extensive debate but it is not a commonly used method of defining one's position, rather the distance to run and distance off track.

Appendix F: Memorandum of the Royal Commissioner (Mr Justice Mahon) as to Evidence and Information Obtained Overseas

Evidence from United States Navy

One of the questions which I am required to answer under the Terms and Reference is whether there was any act or omission on the part of Air Traffic Control at McMurdo Sound which was a contributing factor to the disaster, such act or omission being one arising out of a duty owed by Air Traffic Control.

In order to answer this question, it was necessary for evidence to be taken from the Base Commander and the Air Traffic Controller and his staff who were deployed at McMurdo Sound on 28 November 1979. The persons who were engaged in Air Traffic Control at that time are all now in the United States. I originally asked whether these witnesses could come to New Zealand in order to testify before the Commission, but following representations by the New Zealand Ministry of Foreign Affairs and discussions between that Ministry and the American State Department, it transpired that the State Department will not permit American military personnel to testify outside the United States' jurisdiction in relation to litigation or inquiries which do not directly involve American military personnel as parties. But the American State Department agreed that there was no objection to depositions being taken from United States Navy witnesses within the United States' jurisdiction, and it was finally agreed that Mr Baragwanath and I could obtain these depositions in California and in Washington.

Appointments for the taking of the evidence were duly made and the evidence taken was reduced to the form of depositions. The depositions could not be completed during the limited time available to us in the United States as they needed to be engrossed and then sent away to the different locations of the witnesses within the United States for signature and attestation. We expect, however, that these depositions will arrive in New Zealand at a fairly early date.

The general tenor of the evidence thus taken will be, firstly, to describe the exact nature and location of the radio transmitters and their respective frequencies at McMurdo Sound, and secondly, to describe the nature of the radio contact and radar contact between McMurdo Air Traffic Control and Flight TE 901.

The gist of the radar contact evidence will be that the DC-10 was at no time observed on the radar screen. When a transmission was made from the ground to the aircraft requesting the crew to activate the aircraft's transponder, this had reference not to identification of the aircraft on the radar screen at the ice tower but to the radio link between the aircraft and the TACAN. The United States Navy evidence therefore is that there was at no time any radar contact with TE 901.

It will also be said that Air Traffic Control did not anticipate the arrival of TE 901 over the top of Mt Erebus, nor that its track would lead it towards Ross Island, Air Traffic Control expected that TE 901 would arrive down McMurdo Sound, in conformity with previous 1979 flights and the flights of 1978.

When the aircraft was asked to report its position when 10 miles out, this was intended to give the observers at the ice tower the opportunity to make visual contact with TE 901, bearing in mind that the paramount duty of Air Traffic Control was to separate air traffic in that area. As well as the C-141 arriving from the true north, there was also the C-130 arriving from the true south, and in addition there were helicopters operating in the near vicinity. In addition, it will be said that although the ordinary approach of TE 901 was expected down McMurdo Sound, the exact time of its proposed arrival could not be predicted because the aircraft, which was announced by the Captain to be flying VMC, might have deviated towards the Admiralty Range for sightseeing purposes. It was not until the aircraft had failed to report by radio at the due time that Air Traffic Control began to inquire as to where the aircraft was, and for some time after the occurrence of the crash a number of VHF transmissions were initiated in unsuccessful attempts to make contact with TE 901.

The United States Navy Controllers had no authority over civil aircraft. They did not believe themselves to have any function except on a purely advisory basis insofar as the movements of civil aircraft were concerned. However, McMurdo Air Traffic Control, acting in the interests of its own aircraft, did have the paramount duty of separation of traffic and this would include separation of civil aircraft from military aircraft. The effect is that leaving the separation function on one side, the Base Commander at McMurdo acted on the principle (which was one of long standing) that they could not direct the flight path of any civil aircraft.

It was said by United States Navy witnesses that it was the regular practice of Air New Zealand pilots to divert from their nav. track for sightseeing purposes as the weather might be good in one place and poor elsewhere. There were occasions, so it was said by one witness, when Air New Zealand aircraft were observed to be flying in and out of cloud although on the basis of their radio transmissions they were flying VMC. However, they said that the normal pattern of cloud over the Ross Sea is thin strato-cumulus, and it is unusual to encounter any thick layers of cloud.

The flight plan for the fatal flight arrived at 6.40 a.m. (McMurdo time) on 28 November 1979. The information conveyed in the flight plan reached the Communication Centre at the above-stated time, and was then passed across to Mac Centre, Operations Department, which was commanded by Commander André Coltrin.

Commander Coltrin told us that on 8 October 1979 Mr Jones of

Qantas got in touch with him concerning an up-to-date briefing on Antarctic flying. There was a protracted discussion regarding radio facilities, search and rescue capabilities, clothing, altitudes and weather phenomena, including a discussion on horizon and surface definition problems. The Commander produced to Mr Jones the United States Navy Survival Manual, and made it available to Mr Jones. Commander Coltrin was surprised that Air New Zealand had not joined in with Qantas on this up-to-date briefing procedure.

The United States Navy view is that it is of prime importance for a Pilot-in-Command in these latitudes to have had previous experience of Polar flying and this, in turn, is based upon the unnatural weather hazards and strange definition phenomena which occur in those regions. In the United States Navy it is necessary, before obtaining a Polar aircraft command designation, to have spent 20 hours of co-pilot time on the ice. The United States Navy witnesses all emphasized the dangers of visual deception in Antarctic flying and the conditions which will lead almost inevitably to a dangerous whiteout situation with a white ground surface under unbroken overcast and the sun coming down through the overcast from behind the observer. The multiple reflection of the sun's rays, coupled with the disappearance of distinguishing points on the white surface, leads to a condition which may result in an illusion of flat snow running for miles ahead even though the ground in front may be rising sharply or dropping steeply. All the witnesses said that this condition must be experienced to be fully understood. We were given sundry examples of Polar aircraft accidents where a pilot had been deceived as to approaching terrain even though flying in clear air.

On the question whether the United States authorities would have agreed with the proposal to fly aircraft over the peak of Mt Erebus, Commander Coltrin would have declined to express any view, because he had no jurisdiction over civil aircraft. But the Chief Warrant Officer in charge of Air Traffic Control said that he would have objected to the flight path and to the 6,000 feet descent sector.

The Aircraft Radar in the DC-10

Evidence was given by the Director of Civil Aviation that the pilot of TE 901 ought to have been able to observe the outlines of Ross Island on his aircraft radar. In addition, the Chief Inspector reached the same conclusion in his Report, although he said in evidence that he did not recall the name of the person who told him this at McDonnell Douglas. On the other hand, there was at least one pilot from the executive branch of the airline who gave evidence to the effect that he was not so sure whether a clear terrain picture would have been seen. It appeared that the only people who could answer this question in a definitive way were the Bendix Corporation wbo manufacture the type of weather radar used in the DC-10 and who are in fact the leading manufacturers in the world of this equipment.

Accordingly, on 31 October 1980, I paid a visit to the Avionics Division of Bendix situated at Fort Lauderdale, Florida. I there saw the Director of Engineering and the Manager of the Design Systems. I raised with them the theory advanced by the Director of Civil Aviation, the information apparently given to the Chief Inspector, and the current doubts expressed by one or two of the operational pilots. I also referred to the fact that Major Gumble, who was the pilot of the C-141 which was following TE 901, had said that he got a good picture of the terrain from his weather radar when set in the mapping mode although it seemed that the radar return on his aircraft had been interpreted by two experienced navigators who were on the flight deck. It seemed also that these navigators were familiar with the region.

The answer I got from Bendix seemed to me to clarify the situation. First of all, with regard to the C-141, I was told that the weather radar on this aircraft was not as sophisticated as the radar installed in the DC-10, but gives a better mapping return. In other words, although the C-141 radar was not as efficient at detecting clouds containing rain precipitation, it gave a better terrain picture than the DC-10 radar therefore the C-141 would get a better terrain return than the DC-10.

I was then given an explanation as to the function of weather radar in general. The primary purpose of this radar is to detect the level of rain precipitation in cloud because it is the water content in the cloud which warns a pilot of impending turbulence. The radio waves emanating from the aircraft's radar system are directed towards ascertaining the presence of moisture. When they strike raindrops in the cloud the radar set on the aircraft receives a clear echo. The radar beam will give a good return from rock or earth but the return which it gives from the sea will depend upon whether the sea is calm or disturbed. If there are waves then the return from the sea is quite good because the beam strikes the angled surface of a wave and a reasonably good echo is received. On the other hand, if the water is calm then the radar beam tends to slide off the calm water and travel onwards, and the return received on the aircraft radar is correspondingly blurred and uncertain. When the radar beam comes upon a conjunction of land and sea, it readily distinguishes between the water and the ground and a good terrain outline is obtained, and also if there is a hill behind the shoreline the beam will produce a shadow effect on the screen which will indicate the presence of that hill. However, in the present set of circumstances, the conjunction of land and ice, or shoreline and ice, raises a special problem and evidently causes a drastic reduction in the quality of the return from the radar in its mapping mode.

When the radar beam strikes ice the quality of the return will depend upon whether there is a water film on top of the ice caused by some degree of melting. If the beam strikes ice with water on its surface then a reasonably good return will be received. If, on the other hand, the beam strikes dry ice then the beam, or rather the radio waves which comprise the beam, will be absorbed by the ice surface and will penetrate the dry ice. The more they penetrate the dry ice the more power they lose. If the radio waves strike a damp layer somewhere in the ice, then they will impart an immediate return to the aircraft's radar, but it will be a fairly weak return. If, however, there is no damp ice layer beneath the surface then the radio waves will continue on onto the ice and be absorbed by it, and the ultimate return will be blurred and attenuated.

The reason for the difference between a return from rock and a return from dry ice is that the radio waves act rather like light waves. A light wave will not penetrate rock, but it will penetrate ice. So with a radio wave. Since there is no humidity in Antarctica, it follows that both the ice and the snow will normally be quite dry.

If, therefore, one recalls the type of Antarctic terrain over which the radar beam in this case was travelling, then the radar beam would penetrate pack ice and would slide over any intervening flat water which it then encountered. Then, as the aircraft got closer to Ross Island and a solid ice shelf was encountered, the radar beam would penetrate the solid ice just as it had penetrated the pack ice. Then, when the radar beam struck the ice-covered slopes of the mountain, it would again be absorbed by the dry ice and in the result the pilot of the DC-10 would get approximately the same return from the mountainside as he had been getting from the pack ice and from the ice shelf itself. In other words, the return would be substantially the same as he had been receiving from the time when the pack ice first came within range of the radar beam. Therefore, the pilot would not detect from his radar that he was approaching solid terrain.

The explanation above given is the reason why radio altimeters are unreliable in Antarctic and Arctic regions. The radio waves descending vertically will be absorbed by snow and ice, and in an area where there is very thick snow the radar beam will penetrate the snow and will give a false reading on the radio altimeter. This is also the reason why there is a special warning to pilots contained in the Bendix Handbook, which deals with the operation of the DC-10 weather radar. The warning relates to the possible presence of ice crystals in the air.

The pilot may see on his weather radar a clear picture of clouds ahead and he will estimate that he can climb over the clouds. But there is a danger that the area above the clouds may be filled with ice crystals formed by the freezing of raindrops as they are propelled upwards by the wind inside the cloud. Ice crystals in the air are productive of substantial turbulence, but the radio waves from the radar will travel through the ice crystals and not produce any return on the radar set. The radar beam will therefore travel on, disregarding the ice crystals, until it reaches some cloud far ahead which is within its range. So, unless the pilot is alert to the ice crystals danger, to which I have referred, he can fly into apparent clear air above clouds and encounter severe turbulence.

The result of all this is that in the opinion of the Bendix experts, relating to the case of TE 901, the pilot would have received some kind of return on his radar (if set in the mapping mode) but the return would be so blurred and so attenuated as to give no reliable indication of terrain. If it were not for the preceding pack ice and ice shelf, then the pilot might see that there were some solid structures far below him and in his path. But, as stated previously, the prior returns off pack ice, calm water, and ice shelf, would mask the return received from the mountain because the latter would look like the previous returns from the pack ice.

It would be possible, so the experts said, for a pilot to note a slight change in the return received from the ice-covered slopes of the mountain and the return received from adjacent shelf and pack ice by reason of the shadow effect, but the latter would be distorted. If the pilot had been in the area before, he might be able to discern that there was either some type of terrain or at least suspected terrain ahead. But he would only deduce this by reason of the fact that he had flown over the area before. That is, although his eyes would see the same type of blurred return which he had been obtaining from pack ice, his pre-existing knowledge of the terrain would cause him mentally to reject those parts of the picture which did not resemble the known terrain, and his identification of terrain would therefore depend not upon his view on the screen but upon his prior knowledge of the area which he was approaching.

The same principle, so it was said, would apply to the terrain mapping described by Major Gumble. His navigators had flown towards Ross Island before. Their particular set would give a rather better terrain return than the DC-10's set, but nevertheless it would not be very satisfactory. However the navigators, being aware of what they were approaching, would again be able to interpret what they were seeing as solid terrain, providing they disregarded those aspects of the map which did not coincide with what they knew was there.

So in the result, the effect of the Bendix evidence was that not only would the DC-10 weather radar (set in the mapping mode) give a return hard to distinguish from pack ice, but that type of return would tend to confirm in the Captain's mind that he was in fact flying over pack ice in the centre of McMurdo Sound, if indeed that is where he believed he was.

The Bendix people also made this point. They said that in all probability the radar set on the aircraft was either in the weather mode or was on stand-by at a time when the aircraft was still a long way out from Ross Island. Then, when the Captain saw the gap in the clouds and the sea below and began his orbiting procedure to fly down to the height recommended by Air Traffic Control, there would be no point in switching the radar over to the mapping mode. He would not be interested in the mapping mode if he could actually see the area of pack ice and water toward which he was descending. But suppose that he switched the radar on to the mapping mode once he had levelled out at about 3,000 feet or thereabouts. Then he would be flying too low for the mapping mode to be of any assistance because all he would get would be an insignificant return at the very bottom of his radar screen. So in the end, even if one presumed that the radar was set in the mapping mode, as from a long way back in the approach towards Ross Island, a Captain who had not been in the area before would not receive any radar echo clear enough to warn him that there was any terrain in his path.

The Whiteout Phenomenon

As counsel will recall, we have had considerable evidence on this point in Auckland, and in particular we have had evidence illustrating the way in which the apparent validity of a visual reference will be reinforced to the point of being considered conclusive having regard to the mental set or mental condition of the air crew. I have already referred to opinions of United States Navy witnesses who have referred to the whiteout phenomenon, and one can say that they are in unison as to the type of visual phenomena which is experienced in Polar regions.

The point was made by these witnesses that there is, of course, a difference between a whiteout caused by the air being full of fine powdered snow driven before a wind, and perfectly calm clear conditions with a light overcast being penetrated by the sun. In the first case the effect is no different from flying through cloud, but in the second case it appears to coincide with the weather conditions prevailing at about 12 noon near McMurdo on 28 November 1979. The pilot is looking ahead through clear air towards a scene which may be what he sees but which may be totally different from what he sees. We were advised in the United States to make a point of seeing four designated experts in this field. They were Commander Goodson and Captain Briska at the big Naval Base at Pensacola, Florida, Captain Arthur P. Ginsburg of the United States Air Force stationed at the Wright Patterson air-field near Dayton, Ohio, and Mr Shannon, a Canadian pilot with extensive Polar experience.

Mr Baragwanath's interview with Commander (Doctor) J. E. Goodson, M.S.C., N.A.R., at Pensacola, was as follows:

Dr Goodson has had over 20 years experience in the study of vision as a Psychiatric Physiologist. He has supplied substantial literature which he has researched for our purposes and as one would expect there is a continual of changing visual preception as one moves from normal conditions into a condition of true whiteout. As to this last, he accepts the description contained in Exhibit 228 at paragraph 5.9. In a true whiteout (by which he means the "milk bottle" situation) there are numerous particles in the air which cause the light to diffuse in all directions, as the result of which visual stimuli were completely absent. The experiments establish that one simply fails to see rather than sees a positive problem — in this regard he qualifies what is said at page 46 of Exhibit 228. Eyes require good stimulus to operate normally. As these are lost, the eye becomes much less efficient. At an extreme there are no shadows and no visual cues.

Long distance judgment is heavily dependent upon these, as is well-known from experiments. However, an observer tends to judge size in accordance with his past experience. Whether the case be one of whiteout or not, the visibility of an object depends upon its contrast. If one has scattered light between the observer and the object, contrast is reduced. As you reduce the visual cues you reduce the ability to perceive something and its size and distance. It is a well-known visual phenomenon.

If one has no density gradients, one has no perception. The eye is automatically part of the brain and plays a part in the processing of data received. Changes in stimulation are required to give perception: if the field is largely empty the eye is less responsive to such stimuli as are there.

Experience and expectation are important to perception; as in the case of a man who for a sustained time looked through inverting prisms: for the most part (with an unexplained occasional exception) his brain sees the inverted object as the right way up. A famous experiment is the use of the ambiguous figure comprising two curved lines joined at the end. After a period of fasting, some saw the object as a hot-dog; others after being outdoors saw it as a canoe. The passage at page 46 of Exhibit 228 should be read as referring to the reduction of forward visibility to the observer. If one is not looking for obstruction to one's view, the only clue is a gradual reduction in visibility. Whiteout and its variants are very different from a well formed and well recognized cloud. (This observation is made not from experience but as a matter of logical opinion).

Perception depends upon the gradient of the onset of a stimulus. If gradual, one would not perceive the onset until a later stage than if there were a steeper gradient.

Of possible relevance is the phenomenon known as "empty field myopia". One needs a good contrast of contours (distant stimuli) to focus at a distance. Without them, recent evidence leads to the conclusion that the eye focuses very closely — at some 6–14 feet. In the absence of distance stimulus an observer tends to become myopic: what was at a distance is not in clear focus, and there is therefore a blurred ridge on the retina and limited perception of it. The viewer is slow to realise that he is in fact myopic.

It should be noted, however, that a professional should be quicker to react to any change in conditions: it is inappropriate to conclude too rapidly that he has been deceived. He should preserve a sceptical mind and seek to confirm his perceptions.

Nevertheless, there can be difficulty even with experience in appropriate conditions. For example, as documented by Conrad Kraft of Boeing — there can occur, with regard to certain landing fields, a compelling visual phenomenon which can give rise to bias which leads to landing short. This relates to a visual perception of distance and approach angle, and an array of stimuli (including known airfield lights) when the background lights are consistent with an earlier landing.

It is unwise to be too dogmatic that an attempted reconstruction is correct. Take, for example, the case of the helicopter simulator operation which was perfectly satisfactory for inexperienced crews and of value to them; but experienced men possessed susceptability to stimuli which were at odds with those constructed by the simulator. As a result, they became sick. It is very hard to determine, on reconstruction, what are clues that really matter in the real world.

Conclusion

For any estimate of distance one needs to have cues. It is certainly true that one looking at rising snow slopes with the sun behind and cloud above could think that he perceived a 40 mile expanse of level ice. Without textures or contours he could see it as far, instead as close. He could also think that a snow slope was ground rising in perspective and therefore level to a horizon if he saw it rising. This would in part be due to his expectations if he did not see any rising which would purely be a matter of expectation. With cloud cover it could give the appearance of a horizon (and thus a rising). Of course, to the extent that cues actually existed the foregoing premises would progressively disappear.

Captain Philip T. Briska was interviewed on 31 October. He is a Flight Surgeon and Chief Opthalmologist at the Institute and is also a naval aviator with over 2,000 hours mostly on fighters, and over 150 carrier landings. He also has expertise as an accident investigator. A clear view was formed that he is a perceptive and practical man and his opinion is worthy of weight. (Doctor Goodson's view is that of a pure scientist and is limited to what *can* happen).

Captain Briska has expressed the view that the flight crews must have seen Beaufort Island (unless it was concealed from them by cloud, of which there is no evidence). This would have occurred at the stage of their left hand turn and at the end of the northern leg. Such a turn would have been relatively slow and decorous for the comfort of the passengers, unlike the type of rapid turn which can be accomplished in a fighter, and he commented that the flight crew must have had their heads in the cockpit or asleep not to have seen it. He observed that they would be looking for land after coming down from altitude. He expressed the thought that what they saw was Dunlop Island, adjacent to the mainland, which would, of course, account for the observation "land ahead" and also reinforce the fallacious impression of their being further to the west. He appears to be of the school of thought that thinks the "bit thick here" is referring to somewhere other than the aircraft's proposed track. He emphasized that the essence of VMC flying is to have adequate visual cues. If the Captain saw the land displayed in photographs 10 and 11, that would perhaps be consistent with his mental picture of his track as falling between Cape Bernacchi and Cape Bird. Seeing the cloud above the land to the left and right — the land being in appropriate directions on his assumed track — this would confirm his sense of being on the appropriate track. If the cloud base ahead was consistent with the 500 feet to left and the 1500–2000 feet to right described by the Meteorologists, and the cliffs were obscured by a fog ramp, thee may well have been ahead of him similar cloud layers. If the top cloud came down progressively in step with that to right and left (here it is essential to have reference to the local patterns in this regard) and if there was a fog ramp obscuring the cliffs, there would have been a cloud-free area ahead; though this would be difficult to distinguish from actual cloud. On the basis of a belief to be overflying sea ice and that there was a 40 mile visibility below the overcast from McMurdo in his direction, it is possible that he would believe himself to be seeing a sea ice ahead with a gap above the cloud; this being, in fact, the slope of Mt Erebus.

The 'light bounce' or (less probably) empty space myopia, may have aggravated such misconceptions; to which the absence of adequate visual cues would also contribute.

Because of the factors inducing belief in a westerly path his brain could well be affirmed in its belief, even if there were some non-confirmity in what he saw.

There is, however, a great degree of uncertainty when one approaches a cloud overcast as to its altitude and distance. This observation is not confined to Antarctic aviation but is quite general. Consequently it would be difficult to know — even in temperate conditions — whether VMC was being maintained or not.

This difficulty was no doubt aggravated by the Antarctic factors just referred to — that his absence of visual cues and (at present) the 'light Bounce' phenomenon or (if present) empty space myopia.

Our interview with Captain Ginsburg was highly instructive. He is a world expert on illusory visual phenomena and was awarded a Ph.D. by the University of Cambridge in the United Kingdom for his work in this field. His work for the United States Air Force is, not unnaturally, directed towards, the special visual problems which may occur in the handling of fighter aircraft travelling at supersonic speeds.

We demonstrated to Captain Ginsburg a summary of the total information which we had obtained regarding the weather conditions at McMurdo at the relevant time, and he made himself thoroughly familiar with all this information. We also asked him to take into account, but only as an hypothesis, that Captain Collins had flown on nav. track southwards from Cape Hallett until orbiting through a cloud break and had armed the nav. track again as the last orbit concluded because of his belief that the nav. track would lead the aircraft down the centre of McMurdo Sound with many miles of flat sea or ice on either side.

First of all, Captain Ginsburg deferred his consideration of this latter evidence suggesting Captain Collins believed that by maintaining the nav. track he would be keeping the aircraft many miles away from any high ground. Captain Ginsburg concentrated his attention upon what the pilot and co-pilot were likely to have seen at the conclusion of the second orbit when the plane was locked back on to nav. track as it approached Ross Island. In Captain Ginsburg's opinion, having regard to the known height of the overcast — which, judging by the passengers' photographs was still well above the aircraft at the moment when it struck the mountainside — and having regard to the position of the sun and its 34° inclination, then the pilot would have seen a white expanse of flat terrain extending forwards for an unlimited distance. His point of visual reference would only have been the shallow strips of black rock some miles to the left and some miles to the right of the aircraft, representing Cape Tennyson and Cape Bird.

Looking forward, there would be no points of reference over the ice and snow. Not only would there be no points of contrast but there would be no perception of depth. The fact that the flat white carpet in front was in fact rising upwards at an inclination of 13° and then 19° before meeting the overcast would not be perceived. There would be no shadows and no points of reference to terrain in a forward direction, and Captain Ginsburg expected that a pilot not familiar with this type of visual illusion would merely fly straight on.

We referred to the undisputed evidence that no one on the flight deck ever saw the snow-covered slopes into which the aircraft flew. Captain Ginsburg said that this was not a surprising feature at all and indeed he would expect, in the conditions prevailing, that no one on the flight deck, even flying in clear air, would detect that the aircraft was about to strike a rising expanse of white terrain.

He said that the only pilot or pilots who would detect that ocular phenomenon would be people who had flown in Polar regions before. He said that pilots with Arctic or Antarctic experience would also not see the mountainside but, having noted the overcast, would be aware that there might be something in front of them which they could not see. The two or three pieces of rock face in the ice cliff directly ahead of the aircraft would not be observed as anything but the black strips of sea of the type previously encountered.

We then asked Captain Ginsburg to consider the factor previously mentioned — namely, that the pilot may have believed himself to be flying over a very wide expanse of flat ice in the approximate centre of McMurdo Sound. Having studied the maps, Captain Ginsburg expressed the opinion that the two thin strips of dark rock to left and right of the approach to Lewis Bay would coincide, in the pilot's opinion, with the entrance to McMurdo Sound, and if the Captain's nav. track confirmed the pilot's belief that he was in the centre of McMurdo Sound then the totality of the illusion would be complete. He said that the pilot, upon levelling out after the second orbit, and upon looking far ahead along the flat white surface, would be expecting to see the high terrain 20–30 miles away which lies approximately to the true south of the head of the Sound, and when he could not see it he no doubt decided that it was safer to climb away. Overall, Captain Ginsburg was of the opinion, having studied all the evidence which we were able to give him, that the absence of depth and contrast definitions would have produced what he described as a characteristic example of visual deception.

Captain Ginsburg placed very considerable emphasis on the same point as had been made by Professor Day, namely that everything turns on the mental pre-condition of an observer. He stressed that the eye is not a camera. He said that the observation of a particular object necessarily requires a combination of the function of sight with the function of mental activity associated with the process of observation. Discrepancies between what appears to be seen and what is known to be visible are automatically cancelled out by the mind in favour of a picture of what is known to be there.

If Captain Collins believed, on various grounds, that he was flying down the approximate centre of McMurdo Sound then he would, as a necessary function of his intellect, relate whatever he saw to what he expected to see, and would co-ordinate objective and subjective perception. But this would only occur if he was certain of his position. If he were in any way uncertain of his position then his subjective perception would be disengaged, so to speak, and he would be guided by visual perception alone. If certain of his position, and his course, he would automatically discount minor variations in the visual perception as opposed to what he expected to see, but only up to a certain limit of tolerance, that is to say, if visual perception suddenly appeared to present a picture which was markedly different in some respect from his expected observation, then that factor would intrude upon the pre-condition of certainty of position and, for the first time, a state of mental uncertainty would arise as to whether he was in fact upon the course or in the position previously assumed.

In this respect, according to Captain Ginsburg, the similarity in the approach to Lewis Bay and the approach to the head of McMurdo Sound had constantly to be borne in mind because, judging from the passengers' photographs, it was in all probability a factor confirming the mental set of Captain Collins that he was certainly in the centre of McMurdo Sound. It seemed clear from the passengers' photographs that the tip of Cape Tennyson, as seen from an approach to Lewis Bay, and the tip of Cape Bird, as seen from the same position, each revealed a very shallow line of black rock surmounted by snow. If the appearance of Cape Bird from the centre of the Sound also presented a narrow strip of black rock at sea level, and if Cape Bernacchi presented a similar picture, then the inequalities of distance would not matter. There would not be any sufficiently obtruding difference from the expected vision sufficient to cause any doubt to arise.

Although out of sequence in the narrative, I should here refer to another expert opinion on this topic. On 10 November 1980 we visited Farnborough and, having there listened to a reproduction of extracts from the cockpit voice recorder tape (as hereinafter mentioned), we were asked to see Mr Roger Green who is a Psychologist employed in a civilian capacity with the Royal Air Force as a specialist in flight skills, including visual illusion. His attendance is required at about one-third of the Boards of Inquiry held by the Royal Air Force involving incidents in which the presence of human factors appears to have been an operative cause.

Mr Green laid stress upon the guides provided by visual cues, and emphasized the point that without visual cues the factors of depth and contrast substantially disappear. He gave by way of example a sudden snowfall over one of the military air-fields in England which resulted in three fighter pilots landing well short of a runway of which the location was very familiar to them. What had happened was that the snow had obliterated the visual cues by which they had been guided in previous approaches, and Mr Green stressed the point that in many cases people, and especially pilots, are not necessarily aware of the fact that they are using visual cues so that the disappearance of the latter passes unnoticed.

In snow-covered terrain, a pilot is deprived of texture information which will alone acquaint him with slope and distance. In bright sunshine he is only deprived of that information to a partial extent. But even so, his normal appreciation of variations in terrain is adversely affected. Mr Green also stressed the importance of the mental set of a pilot, and believed that Captain Vette's comparison between Lewis Bay and the approach to McMurdo Sound was a good example.

Mr Green said that stereopsis and binocular cues are only of help up to relatively short distances, perhaps not exceeding 100 feet or so, and that thereafter perception depends very largely upon experience. It is for this reason that Mr Green placed the greatest stress upon the principle that there can be no substitute, in the aviation field, for past experience of terrain. An R.C.U. briefing should be confined, in Mr Green's opinion, to purely conventional briefings. It cannot implant knowledge in the memory in the same manner as actual experience. He pointed out that a civil pilot does not often fly purely visually, and Mr Green was of the opinion that when this occurs and a pilot is unfamiliar with the terrain, a number of dangers will instantly arise.

The third of the experts whom we were advised to see was Mr G. W. Shannon, Vice-President of Operations for Bradley Air Services Limited of Ontario. Mr Shannon's company flies both passenger and freight schedules up to North Canada and the sub-Arctic. He was also retained some years ago to carry out a commercial contract in Antarctica. He flew from the southernmost point of South America across to Shackleton Base in Antarctica, and then across the Polar Continent to McMurdo. This flight, and other operations in the Antarctic, was carried out in a de Havilland twin Otter, Mr Shannon's work being connected with the operations of a United States drilling site. He is reasonably familiar with the McMurdo region by reason of that particular contract which he carried out in Antarctica. Mr Shannon was recommended to us as being an expert whose knowledge and experience of flying in snow conditions is exceptional.

We saw Mr Shannon at his company's location at Nepean, some miles out of Ottawa. Mr Shannon had the advantage of having no prior knowledge of the DC-10 disaster, except that he naturally knew of the occurrence. He had not read the Chief Inspector's Report, and had no detailed knowledge of the circumstances. Over a period of between two and three hours we displayed to Mr Shannon all the relevant maps and diagrams, weather information, cloud locations, passengers' photographs and so forth. We also showed him the Chief Inspector's transcription of the cockpit voice recorder. Mr Shannon noted all this material and paid close attention to the cockpit voice recorder transcript which he read and re-read on a number of occasions, particularly the closing stages.

In Mr Shannon's opinion, the prevailing weather and the location of the sun and the other factors previously mentioned would present to the pilot and co-pilot of the DC-10 a forward vista of flat snow and ice extending away to the far distance, and he had no doubt that a pilot unfamiliar with Polar conditions would believe that he was flying forward with clear visibility over flat terrain for many miles. Mr Shannon believes that the pilot and co-pilot would have therefore an apparent flat and clearly visible terrain definition, whereas in fact there would be no terrain definition at all.

We asked Mr Shannon whether the overcast extending forward would form an illusory horizon in the distance at a point where it met the snow-covered rising ground. Mr Shannon said he thought not. He said that in such conditions the almost invariable effect is that the under-side of the overcast turns white so that there would be no horizon at all. He said that there was a possibility of a false horizon, but he regarded that possibility as remote. His own years of experience of flying in such conditions led him to the conclusion stated above, namely, that the overcast in front of the pilot would seem to disappear by reason of the fact that its grey under-surface would become white in colour through the multiple light reflection provided by overhead sun behind the aircraft.

Mr Shannon gave as an example an occurrence which often takes place at their own airport when the ground is covered with snow. He said that if there is a light overcast overhead then in daylight the underside of the overcast turns white and it is not possible from the ground to discern the height of the overcast. He said you know the overcast is there because you cannot see the sun, but it is not possible to say whether the overcast is 1500 feet high or 5000 feet high when looking upwards from the ground. He said that they are dependent upon reports from pilots as to the height of the overcast.

Mr Shannon said that having regard to the known weather conditions which we had exhibited to him, he would expect that as Captain Collins levelled out following the second orbit, and having dropped height to 1500 feet to try and see something in the distance but without success, that Captain Collins would then have elected to climb away because he could not see any landmarks in the distance. Mr Shannon said he noted, from his study of the cockpit voice recorder transcript, that Captain Collins decided, very soon after having levelled out, that he should fly away and he attributed that decision to the fact that although the aircraft was flying under the overcast, and although the ground seemed to stretch away for miles, nevertheless there was no terrain anywhere to be seen.

As with other witnesses whom we saw on our travels, Mr Shannon placed primary significance upon the adherence by Captain Collins to the nav. track. He said that if Captain Collins had plotted the nav. track on a map, then he would obviously have believed that there was no danger of any kind ahead, and that he was many miles away from any high ground. Mr Shannon said that one of the reasons why he had studied and re-studied the closing stages of the transcript was to try and see whether there was any expressed concern or doubt on the part of the pilot or co-pilot in relation to the course or position of the aircraft. Mr Shannon was not very interested in the cross-talk which was taking place behind the pilots. He said that he drew the conclusion that neither the pilot nor the co-pilot entertained the slightest apprehension at any stage, and he drew the further conclusion that each of them was perfectly satisfied as to the course and position of the aircraft.

We raised with Mr Shannon the theory that a pale fog may have drifted off the ice and covered the ice cliff. We told him that a helicopter which had landed on the ice shelf below the cliff about an hour later saw no sign of the ice fog although it was observed on subsequent days to be coming off the ice. Mr Shannon said that the winds in that area are very fickle, and even a temporary breeze from the true north would instantly form ice fog the moment it reached the ice shelf below the cliff. The fog might persist with a steady breeze, climbing upwards, but if the breeze died away then the fog would disperse. Mr Shannon believed that if there were patches of black rock visible forward of the aircraft on that part of the ice cliff not covered by glacial ice, then this would have no significance to Captain Collins or the co-pilot as, from a distance, any shallow patches of black rock would merely resemble the patches of black water which they had previously observed.

Mr Shannon said that the situation confronting Captain Collins as he levelled out after the last orbit would have signalled a red light to the experienced Antarctic flier who would immediately have flown away. Photograph no. 10 illustrates that the weather ahead was getting more solid, and in Mr Shannon's view any experienced pilot would have realised that conditions were no longer appropriate for VMC flying. But Mr Shannon again pointed out that it had taken Captain Collins only a brief interval to reach that conclusion, and he had then attempted to fly away once he could not discern the clear visibility which Mac Centre had told him would be apparent once he had descended to 2,000 feet.

Transcription of Cockpit Voice Recorder Tapes

It will be recalled that a question arose during the hearing as to the accuracy of two particular extracts from the transcription of the cockpit voice recorder tapes, as incorporated in the printed report of the Chief Inspector. These two extracts were:

"Bit thick here eh Bert"

and

"You're really a long while on . . . instruments at this time are you "

It was contended that each of these extracts had been regarded at Washington as being unintelligible, or alternatively, not sufficiently intelligible to be of any assistance, and that the Chief Inspector had thereafter gone over to Farnborough in the United Kingdom and that the two extracts had been considered intelligible at Farnborough and had therefore been printed as part of the transcript.

It was decided by Mr Baragwanath and myself that we would take the tapes to Washington, and also to Farnborough, and listen for ourselves to the rendition through the special filters which are used in the equipment at each of these locations. We had heard the tape in Auckland without these electronic aids, and although the comments of the pilot and co-pilot were clear enough to be unmistakeable, the conversation involving the flight engineers and Mr Mulgrew and other persons had been so garbled as to be, for the main part, unintelligible.

We were particularly interested in what was said on the flight-deck behind the two pilots because the pilots themselves had not seemed to express any doubt as to their position and it was only the cross-talk behind the pilots which seemed to exhibit doubt on the part of various speakers as to exactly where the aircraft was. The tape was first of all played at Washington by Mr Turner of the National Transportation Safety Board. He had some new equipment there, and the relevant sections of the tape were played through on his original equipment and then again on his new equipment. The extracts above-quoted were played over and over again, and it was evident that there was room for considerable difference of opinion as to what exactly was being said.

I deal first of all with the reference to it being "a bit thick here, eh, Bert?" In the first place, Mr Baragwanath and I were quite unable to pick up the word "here". The sequence of words seemed to be "thick eh Bert". There was no hesitation or pause or auditory gap between the words "thick" and "eh" into which any other word seemed capable of being interposed. One had to bear in mind that the word "here" contains no hard consonant and represents merely an aspirate mono-syllable. Just the same, we never heard the word "here". There was also doubt, shared by Mr Turner, as to whether the word "thick" had in fact been used. It might have been another word. Then there was the fact that there was apparently no one on the flight-deck named "Bert". If that is so, then the mono-syllable "eh" appeared suspect, as it could not have been an interrogative word addressed to a person. Mr Turner believed the entire sentence or phrase to be quite unintelligible, and Mr Baragwanath and I agreed.

As to the second disputed sentence referring to "instruments", it seemed clear enough that that word in fact was used. But whereas there was the expression "this time", I kept hearing it as "that time". It also seemed to me that the words after the word "instrument" might have been from a different speaker and dealing with a different subject. Mr Turner said that in his opinion the word "instruments" marked the end of a sentence and that the following words, whether spoken by the original speaker or not, appeared to relate to a different topic. In the result therefore, although the sentence as appearing in the Chief Inspector's report may possibly have been correctly transcribed, it was impossible to be sure. In view of the doubts as to whether one was hearing a single sentence or two parts of different sentences, possibly uttered by different voices, it was Mr Turner's opinion that the sentence should be classified as either not intelligible or not sufficiently intelligible to be given any reliable translation.

At Farnborough the same extracts were played over and over again through a different variety of filters by Mr Davis, and we listened to them in the ordinary way and also through ear-phones. Once again, we could not discern the word "here" as following the word "thick", and it appeared that Mr Davis had been unaware that there was no person called "Bert" on the flight-deck.

As to the second sentence involving the word "instruments", I came to the same conclusion as at Washington.

Those listening to the tape at Farnborough were Mr Davis, Mr Tench (Chief Inspector of Air Accidents for the United Kingdom) and Mr Shaddick, together with Mr Baragwanath and myself. Mr Davis did not venture any opinion as to the intelligibility of the two extracts. He merely played the part of expert technician (which he clearly is) in reproducing the two extracts in a variety of forms from his variety of filters. The consensus of opinion among the four of us (I am excluding Mr Davis) was that the extracts were either unintelligible or not sufficiently intelligible to be given any reliable meaning.

My own impression was that every attempt had been made by Mr Chippindale and Mr Davis to attribute some bona fide meaning to these two passages, but the result was in the end only the best possible construction which, in their opinion, the garbled speech appeared to present. I was also left with the impression that a very considerable proportion of what was said on the flight-deck (excluding the remarks of the Captain and the co-pilot) could not be reliably transcribed. There were words and phrases which were clear, especially those of a technical kind, but the transcript appearing in the Chief Inspector's Report of what was said on the flight-deck behind the two pilots seems to me to be overshadowed by the extreme difficulty in picking up what really was being said, and in attempting to separate different voices and different subjects of conversation. Mr Chippindale had warned me about this when the tapes were played at Auckland.

It may be worth observing that this difficulty of getting an accurate impression of what was said on the flight-deck had been foreseen by the experts at the Bendix Corporation when I saw then on 31 October. Although, as previously stated, I was only enquiring into the radar question, it appeared that the Director of Engineering and the Manager of the Design Systems had both studied the Chippindale Report, and had read the transcript. They warned me to expect to hear only a garbled version of what was said on the flight-deck behind the pilots because, from their understanding of the position, there were on the flight-deck, in addition to the pilots, two flight engineers and an official commentator and apparently, on occasions, one or perhaps two passengers.

They said that, being familiar with the cockpit voice recorder system installed in this aircraft, they were well aware of the difficulties of transcription. The microphone is fixed in the roof between the two pilots and just behind them, so that when the pilots turn and speak to each other there is a reasonably clear recording of what they are saying. If there were a single engineer who sat just behind the pilots then it is quite possible that a fairly satisfactory recording might be obtained of what was said by that flight engineer. But if there are three or four people on the flight-deck behind the pilots then the voice recording system becomes very difficult to interpret. Conversations tended to cross with each other and, in addition, the system picks up voices from the galley.

I said that Mr Chippindale had pointed out these difficulties before we heard the tape-recording in Auckland. I said that we had heard the entirety of the recording twice, and apart from what was said by the pilot and co-pilot — which was uniformly intelligble — the balance of the various comments and discussions were most difficult to interpet. I said, however, that I expected that with the use of the sophisticated filtering apparatus at Washington and at Farnborough, Mr Baragwanath and I would get a much clearer version of what was being said. The Bendix experts warned me not to be too confident about that. They said that the filtering, especially if aided by computer programming, would tend to clarify certain words and could be used to give a louder rendition of words or phrases previously indistinct but that, generally speaking, the filtering systems would not help us a great deal owing to the very poor quality of this recording system. I said I understood that at Washington some 5 days had been spent in endeavouring to transcribe 30 minutes of recording, and that even then there were a number of passages thought to be unintelligible. They said they were not surprised.

Both at Washington and at Farnborough the problem of understanding what was being said in the area behind the pilots seemed to be primarily created (as Mr Chippindale had warned us) by the extent of cross-talk. But there was a further difficulty. One of us would believe that he could identify, say, six words out of eight, but someone else would identify only three words out of those eight and would also be inclined to think that the other words (out of the six) were different from what the first person thought they were. Then the same type of differing view would be expressed by the third person, and perhaps by the fourth person. So, the ultimate result was that, despite a prolonged effort to produce a joint decision to try and give effect to all opinions, a transcription could not be decided upon with which everyone agreed.

In other words, although it might be popularly thought that the official published transcript represents what actually was said, the transcript in fact does not produce that result. What it produces, to a considerable extent, is merely a consensus of opinion as to what may have been said at any given moment. And where the words or phrases are reasonably clear, one must beware of the absence of proof as to any accompanying directional pointing of hands towards the landscape, and so on.

Visit to Antarctica

We left for Antarctica on 26 November 1980 at 6 a.m. We flew down in an RNZAF C-130 Hercules aircraft. The pilot was Flight Lieutenant Russell and the Commander of the flight was Wing Commander Gayfer. The aircraft was loaded with 14 tons of cargo for Scott Base.

With about 400 miles to run the Wing Commander suggested that I go on the flight-deck for the remainder of the journey. At that point, the aircraft was flying at 29,000 feet and had crossed the Admiralty Mountains and was flying over the Victory Mountains. There was some light cloud ahead which disappeared as it was overtaken by the aircraft. From that time onwards the range of vision over Victoria Land to the right and due south was about 300 miles. There was no cloud and the view of the Continent was composed entirely of snow-covered mountains as far as the eye could see.

I looked at the aircraft radar from time to time and some distance further on, the outline of Ross Island could be picked up with the radar set in the mapping mode. We eventually came out over the sea about 40 miles to the true west of Coulman Island. At about 150 miles out the Wing Commander took over the co-pilot seat and said that it was proposed, with my approval, to now bring the aircraft over to the track followed by the DC-10 and to execute the orbit to the right and then the orbit to the left which the DC-10 had followed. Thereafter the Wing Commander said he intended to fly directly at the mountain-side along the exact track taken by the DC-10 and he would pull away at a fairly late stage. I said I agreed with all this.

First of all, the aircraft flew to the Byrd Reporting Point to establish its position with ground control, and then we flew over the crash site, where parts of the wreckage are still visible. The aircraft was then flown away to the true north, reaching the same altitude as the DC-10 before it had commenced its first orbit.

The Wing Commander had a map upon which had been plotted the exact location of the figure-8 manoeuvre adopted by Captain Collins as he accomplished his descent to 2,000 and then 1,500 feet.

Thereafter he directed the course, speed, altitude and rate of descent so as to follow exactly the flight of the DC-10. At the appropriate moment he ordered the commencement of the right hand orbit and he controlled the descending altitude and angle of turn at each stage, with the Flight Lieutenant carrying out these instructions. After completion of the first orbit the aircraft commenced the second orbit to the left and the Wing Commander kept the aircraft on its descent at the same descending altitudes as had been followed by the DC-10. From time to time, if there appeared to be slight deviations from the actual track of the DC-10, the appropriate modifications were made to the heading.

Eventually, the second orbit was completed and the aircraft was straightened out for the run towards the mountain. The Wing Commander suggested to me that he flew in at 2,000 feet at 250 knots with the result that we would be flying at the same approximate altitude and speed of the DC-10 and along its exact course. I agreed to this and Flight Lieutenant Russell was directed to drop to a flight level of 2,000 feet and hold the speed at 250 knots. Wing Commander Gayfer then instructed the pilot that at the appropriate moment he would direct a left-hand rate one turn which meant a banking angle of 25 degrees.

As we approached the mountain there was bright sunshine and the ice cliff could be seen in the distance without any difficulty. There appeared to be only two shallow patches of black rock visible at about the level of the ice shelf at the bottom of the cliff. These were in line with the crash site. These were not very long and on my estimation about 20 or 30 feet high. Apart from these two dark areas, and possibly one or two other smaller but similar areas well to the true east, the ice shelf in front of us was uniformly white. We could see the crash site without difficulty in the clear air. The Wing Commander warned the Flight Lieutenant that he was about to issue the order to bank left and then at a point two miles from the crash site, he gave the necessary order. The aircraft banked sharply left and held a 180 degree turn so as to adopt a heading of 180 degrees grid. Since the approach speed was approximately 300 miles per hour, we had turned away when about 30 seconds from the crash site, and consequently the reconstruction of the final approach of the DC-10 was quite realistic. The aircraft was then flown on a heading of true north and then turned left around Cape Bird on to the military track.

At this point the transmissions from Air Traffic Control became very faint, and it was not possible to hear what the ground operator was saying. The Wing Commander therefore instructed the Flight Lieutenant to dispense with a ground control approach and said that the run into the ice runway would be made visually. The aircraft eventually touched down on the ice runway after a flight time of 8 hours 18 minutes. The landing at McMurdo took place at 2.15 New

Zealand time.

On 27 November we were taken on an extensive tour, in a tracked vehicle, of the McMurdo area. The weather was quite fine in the morning with bright sunshine. In the afternoon over by the Scott Hutt, a northerly front was seen to be approaching. Standing on a height near Scott Hutt we could look out across the ice at the Ross Sea and at the mountains along Scott Coast, and the visibility range was more than 100 miles to the most northerly aspect of the mountain ranges along the Scott Coast. On the right there was seen the black outline of Tent Island and behind Tent Island, but obscured by it was Cape Royds. When the United States Air Traffic Control say they can see 40 miles, they mean that they can see the mountains. Visibility is then usually 100 miles or more.

In the afternoon, the northerly front enveloped Ross Island, in accordance with an extremely accurate forecast by Sir Rochford Hughes, but still left a clear area looking down the Sound a few points to the west of true north, and Scott Coast was entirely clear. Mt Erebus then totally disappeared in pale cloud and as one looked at it the cloud pattern which obscured the mountain kept shifting and changing. The plume of steam at the top of the crater was no longer visible. Sometimes a long streamer of cloud near the top altered from a horizontal angle to a downwards angle pointing to the true east. The light cloud patterns around the mountain kept changing both in contour and in colour throughout the afternoon. The conditions then began to resemble what they had been like on 28 November 1979, that is, a low overcast had come in with the northerly front, but as stated previously, the vision out towards McMurdo Sound into the north-west was clear and bright. To the true east, that is to say looking to the right of Ross Island, the snow now blended with the horizon so as to make the horizon invisible, and it was impossible to say where the snow ended and the cloud began.

The plan was to fly by helicopter on to the mountainside the next day, but this was thought to be a doubtful procedure if the cloud covering Ross Island continued to persist as it would be impossible to fly in conditions where both ground and horizon definition had entirely disappeared. In such circumstances, the helicopter pilot would not be able to tell whether he was flying into or over the snow-covered ridge. In the meantime, however, a programme was scheduled whereby the helicopter would leave at 11 a.m. next day, and there was to be a Memorial Service at Scott Base at 3 p.m.

28 November 1980

The overcast which had been produced by the northerly front now spread over Scott Base and well to the true south. But out over

McMurdo Sound a fraction left of true north, visibility was as clear as it had been on the previous day and particularly over towards the true easterly border of Victoria Land. The helicopter flight to Mt Erebus had been scheduled for 11 a.m. but had to be cancelled because the overcast and the cloud concealing Mt Erebus made it impossible to fly. Radio communication with a party of people at Cape Bird indicated that the overcast was lifting over Lewis Bay. The flight was put off to 12 noon, but by then the weather was still impossible for helicopter flying.

The Memorial Service to be held at Scott Base was therefore brought forward to 2.30 p.m. and the next flight time for the helicopter was fixed at 3.30 p.m. At this time the helicopter arrived on the pad at Scott Base, and in view of the latest information received it was decided to proceed with the flight, but whether we managed to fly into Lewis Bay would depend upon the degree of cloud. We took off at about 4 p.m. and flew towards the saddle which runs between Mt Bird and Mt Erebus. Heavy cloud concealed Mt Bird and the cloud extended across the saddle. There were, however, certain thin breaks in the cloud which could be discerned by strips of sunlight on the snow.

The pilot first attempted to fly through the narrow cloud breaks as revealed by the sun, but as this seemed slightly hazardous he elected to turn right and fly over the saddle but under the cloud base. The gap available for this purpose was minimal, but there was enough clearance between the cloud base at the top of the saddle for the helicopter to fly through.

It had been decided that if Lewis Bay was either in cloud or covered by low overcast, that we would have to fly away to the true north and return to the Base. However, the weather over Lewis Bay was free from cloud and there was bright sunlight, so we were able to carry on. The helicopter then flew towards the true north and turned and came back on a heading of 357 degrees grid which put it upon the track of the DC-10.

As we approached the ice shelf at about 75 knots, the latter could clearly be seen as on 26 November, and the rising ground which commenced at the ice shelf was also clearly apparent in the sunlight, but the mountain itself was totally invisible being enveloped in pale cloud. Even though the mountain slope began only some 2 or 3 miles ahead of the crash site, no part of the mountain could be seen.

Over to the true east we could see a narrow strip of black rock at about the level of the sea ice which marked the western border of Cape Tennyson. However, in towards the Bay from Cape Tennyson there could be seen an ice fog lifting off the solid ice which was drifting over the ice shelf and which was entirely concealing 3 or 4 miles of the ice shelf. Over to the right we could see the narrow strip of black rock which marked the tip of Cape Byrd. The approach towards the ice shelf was being made over solid ice, covered with snow.

We flew over the crash site and made several passes back and forth at a low altitude and the pieces of wreckage previously mentioned could now be seen more exactly. We then flew upwards to a point about 3,000 feet up the mountain and the helicopter was then landed on a rock outcrop after some delicate manoeuvring of the landing gear so as to avoid boulders. About 200 yards down the slope from where we landed was the small cross which had been installed there in the previous year. Air New Zealand's representative, Mr Davies, had with him a wreath and also four containers of ashes of victims which the relatives desired to be scattered on the mountainside. These victims were one American, one Australian, and two New Zealanders. The wreath was duly placed, and Mr Davies scattered the ashes. We went back through the volcanic rock and snow to where the helicopter was waiting, and took our departure by flying away to the true west around the steep slope of Mt Erebus and down towards the flat land to the true south, and after a flight of about 20 miles landed at Scott Base.

It was then decided that Air Commodore Crooks, Dr Thomson, Mr Davies and myself would leave that night for New Zealand by a Royal Australian Air Force Hercules which was flying out at about 6.30 p.m. local time. The Australian flight crew asked me to sit on the flight-deck during the take-off as there was something which they said they wanted me to see. The overcast was still low over the whole area and they confirmed that the conditions were virtually identical to those prevailing on the same day one year previously.

The pilot told me that he would fly out to the true east and attain a height of 1,000 feet, and then he would turn back and fly to the true west and pass Scott Base at 500 feet, before commencing the climb away to New Zealand. He asked me to look out for a snow ridge which we would encounter as we approached Scott Base. Near the left-hand end of the ridge was a black outcrop or rock, and the ridge then ran off to the right from that point. The pilot said that it would be difficult to discern in the diffused light under the overcast where the slope began and where the top of the ridge was located. But he said that I would be able to ascertain its approximate position by reference to the black outcrop which formed the edge of the ridge.

The pilot suggested to me that as we approached at 500 feet I should lift up one hand so as to obliterate the black outcrop to my left, and I would then notice that the snow-covered ridge would entirely disappear. That is to say, without any geographical reference point it would be impossible to observe the rising ground.

The pilot radioed his intended departure course to Mac Centre and obtained their clearance. He then flew off in the indicated direction for a few miles, then reversed his course and dropped to 500 feet. I could see the snow ridge lying ahead in that I could make out the top of the ridge, though not its base, and when I blacked out the view of the black outcrop of rock, the ridge immediately disappeared. All that could be seen was a flat expanse of snow-covered ground running on for some miles ahead. The flight captain said that in his opinion this would be an exact replica of the visual deception to which Captain Collins would have been exposed as he flew under the same level of overcast approaching Mt Erebus one year ago.

After passing Scott Base the aircraft flew out into the Sound, turned right and began its climb to its cruising altitude. I kept an eye on the altimeter and noticed that we entered the bottom of the cloud base at 3,000 feet and that we emerged from the cloud, which was light coloured and not very solid, at 5,000 feet. Looking over to the right, the top 7,000 feet of Mt Erebus was clearly visible. Having regard to the complete disappearance of any ground and horizon definition in the overcast as observed by us during the climb out from the air field, the flight captain and the other members of his crew were satisfied that the visual conditions were identical with those experienced by Captain Collins when he levelled out at 2,000 feet on his approach to Ross Island.

It was noticeable that radio transmissions which we were receiving from Mac Centre were fluctuating and uncertain as opposed to the very clear transmissions being received from the R.A.A.F. Operations Centre at Sydney, even though we were only 100 miles from Mac Centre. The aircraft was flown upwards to its cruising height of 31,100 feet, and then held that altitude all the way back to Christchurch. We landed at Christchurch International Airport at 4 a.m. New Zealand time, after a flight of 7 hours.

Notes on Text

by Stuart Macfarlane, Senior Lecturer in Law,
University of Auckland

These notes commence with an analysis of the Chippindale Report, comparing it with *Impact Erebus* and the Mahon Report. The following nine numbered notes expand certain issues mentioned in the text of *Impact Erebus*. The notes end with a conclusion which sets out the views of the note writer.

Abbreviations

B	Brief of evidence before Royal Commission. The page numbers relating to proceedings before the Commission refer to those in the Case on Appeal before the Court of Appeal C.A.95/81.
C	The Chippindale Report. Aircraft Accident Report No. 79-139 of the Office of Air Accidents Investigation.
The Commissioner	Mr Justice Mahon, the sole commissioner of the Royal Commission.
The Inspector	Mr R. Chippindale, Chief Inspector of Air Accidents.
M	The Mahon Report. Page numbers from the Report of the Royal Commission into Mt Erebus Disaster.
PR 3 Feb.	Press release of 3 February 1982 being comments by Mr R. Chippindale of 6 May 1981 on the Mahon Report.
PR 18 Jan.	Press release of 18 January 1982 being comments by Mr R. Chippindale on the publicity following Court of Appeal decision.
S	Final submissions of counsel to Royal Commission.
T	Transcript of evidence before Royal Commission. Page numbers from C.A. 95/81.

Notes

The Chippindale Report

'It has been my experience that we have as many interpretations of the Report as you have parties.'

R. Chippindale (T. 169)

The Chippindale Report is that of Mr R. Chippindale, Chief Inspector of Air Accidents ('the Inspector') into the crash of the Air New Zealand DC10 on 28 November 1979 on the slopes of Ross Island approaching Mt Erebus proper in Antarctica.

Because he holds the position of chief inspector of air accidents, Mr Chippindale's views have been given weight in the community. For instance *The New Zealand Herald* on 29 January 1982 wrote:

"The Government can scarcely ignore some of the criticisms directed by the chief inspector of air accidents, Mr Chippindale, at the Royal Commission's findings on the Mt Erebus crash. The chief inspector may be supposed to possess knowledge that gives weight to his statements on technical matters."

Another newspaper, the *Auckland Star* put the issue in a nutshell:

"In essence, Mr Chippindale said the DC10 crew must be responsible for the decision to descend the aircraft in an area on the opposite side of Mt Erebus to that approved for any descents.

Mr Justice Mahon, in his report, said the dominant cause of the crash was the act of the airline in changing the computer navigational track of the aircraft without telling the aircrew.

"I believe, if informed people read both reports [Mr Chippindale's and Mr Justice Mahon's] in their entirety, there is sufficient information for them to come to their own conclusion," Mr Chippindale said. . . .

. . .now that they [Mr Justice Mahon's findings on the crash] have been challenged by an accident investigator of the standing of Mr Chippindale, an attempt must be made to settle the matter.

The public and the claimants, frustrated and unhappy in the present situation, will expect an exhaustive effort to decide between the Mahon and the Chippindale views, or at least to reconcile them.

Until that effort is made, the Erebus affair will not go away."

Impact Erebus in essence sets out the same analysis as that given by Mr Justice Mahon in the Mahon Report. Opposed to these views in almost every respect is the analysis of Mr Chippindale in the Chippindale Report.

Between them *Impact Erebus* and this note on the Chippindale Report represent "an exhaustive effort to decide between the Mahon and the Chippindale views."

Though the primary aim of *Impact Erebus* is the advancement of aviation safety, an incidental consequence is to explain why the Mahon Report is right. This note explains why the Chippindale Report is wrong.

How sound are the conclusions of the Chippindale Report may be judged firstly by considering the evidential basis upon which it is founded and secondly by examining to what extent evidence supports or opposes the five major propositions which constitute the elements of the inspector's analysis of the accident.

Evidential Basis of Chippindale Report

In his press release being his memorandum dated 18 January 1982 Mr Chippindale says:

"The fact of the matter is that [Mr Justice Mahon] has only expressed an opinion and it is the opinion of one man who does not appear to have accepted sound advice on aviation lore..."

"The official aircraft accident report published by my Office [the Chippindale Report] was the result of the detailed investigations and deliberations of some 18 highly qualified international experts and aircraft accident investigators. It had also been considered before publication by the legal representatives of the deceased pilots, Air New Zealand and the Civil Aviation Division whose representations had been carefully considered prior to the publication of my report."

"The sadness is that [the Mahon Report] abounds in errors which indicates that it was not reviewed by any uninvolved technical assessor or the arguments tested by anyone familiar with this particular accident or the investigation of large aircraft accidents. Certainly it was not checked for accuracy by any of my investigators."

The Inspector made similar comments in his press release of 3 February, where he says:

". . .I consider it manifestly misleading for [Mr Justice Mahon] to state as fact so many assumptions on the conduct of this flight which were deduced from the statements, not subject to the rules of evidence, made by so many interested parties, and information gathered in private but not tested by experiment or cross-examination.

"In contrast [The Chippindale] Report was the considered assessment of a group of extremely well qualified DC10 experts, pilots, and accident investigators who decided this accident could have been avoided if the crew had conducted the flight in accordace with the briefing in relation to minimum safe altitudes."

"Justice Mahon has described his version of the events clearly but has not refrained from expressing as facts his hypothesis of a sequence of events which should have happened but arguing strongly against material which did not conform to this pattern and on occasion omitting evidence which supported a contrary view."

Through these press releases the Inspector has made the evidential basis of his own report a public issue. Can his claims be substantiated?

The fundamental strength of the Royal Commission, upon which the Mahon Report is based, derives from the facts that it was conducted in public, the evidence was subject to cross-examination, was taken down in writing and independent researchers are able to subject to lengthy and detailed examination the 284 exhibits which were produced, the prepared briefs of evidence and the oral examination, comprising 3,083 pages and the 368 pages of analyses of it made by counsel.

The corresponding weakness of the Chippindale Report is the lack of all those features. That investigation was conducted in private, the evidence was not subject to cross-examination, with few exceptions (e.g. three statements from U.S. navy personnel returning to the States) it was not taken down in writing, and it is not available to independent observers. In short it is not possible to make any internal scientific assessment of it, since the data on which it is based are not available to public scrutiny. The only means of checking is externally by utilising the data made public by the Royal Commission enquiry. Not even the names are revealed by the Inspector of those who are alleged to have provided stated information.

When considering the Inspector's investigation of the crash one should bear in mind Captain Vette's reference in his author's note (page 13) to the Inspector's lack of experience with large civil airliners.

Who are these "some 18 highly qualified international experts and aircraft accident investigators" who support the Chippindale Report? The Inspector says (PR 3 Feb) that his report was reviewed in detail and agreed by members of CAD,

Air New Zealand, New Zealand ALPA, the United States Transportation Safety Board, the U.S. Federal Aviation Administration, McDonnell Douglas and the General Electric Company. For this assertion to represent more than mere unverifiable assertion, of no evidential value, would require the open production of statements setting out in precise detail why those parties consider the Chippindale analysis is correct and why the analysis contained in both the Mahon Report and *Impact Erebus* is incorrect.

The first named by the Inspector is the Civil Aviation Division. They are an "interested party", since their actions were involved in the crash and were a subject of the Royal Commission inquiry.

Air New Zealand, also an interested party, naturally would support a report which in substance accepted their case, and virtually exonerated the airline of all responsibility, and held the pilots alone responsible for the crash. The Inspector had as a technical advisor Captain Gemmell, the chief pilot of Air New Zealand (B.592), who was one of the principal witnesses for the airline, and who suggested throughout that the accident was caused by pilot error. The captain was one of the few permitted to listen to the CVR tape (T.1777). He did this at the Inspector's residence in Wellington (T.1797).

Next named by the Inspector as agreeing with his report in detail is the New Zealand Airline Pilots Association. Their support will doubtless come as a revelation to that Association whose members clearly put a great amount of time and money into opposing, at the hearing before the Royal Commission, the Inspector's propositions.

Unless the Inspector produces statements from the U.S. Transportation Safety Board and the U.S. Federal Aviation Administration his assertion of their support must be treated as unsubstantiated.

Appendix 4 to Annex D of the Chippindale Report purports to be upon an NTSB form. However that appendix is a graph showing grid headings, indicated air speed, and mean height above sea level plotted on a time scale. If the support of the NTSB referred to is merely the provision of such technical data, then it is irrelevant to the disputed conclusions of the report.

McDonnell-Douglas would undoubtedly have agreed that the crash was not caused by failure of the aircraft body. Likewise the General Electric Company would have concurred that engine failure was not to blame. Since those who oppose the Inspector's conclusions have never suggested mechanical failure, the views of such bodies are not to the point. I cannot trace the anonymous McDonnell-Douglas radar expert whose views conflict both with the Bendix view and with pilots' evidence and so must treat his opinion as unsubstantiated. See note 4.

In his brief of evidence to the Royal Commission the Inspector names a number of persons who assisted with his enquiry. Most of these belong to the categories of support which the Inspector mentioned. Are the some 18 international experts whom the Inspector says support his report among these? They are:
Crawford, Grossi (NTSB), Chesney (USFAA), Moehring (Gen. El.), Hammett, Lund, Wright (Douglas), Wylie, Graham (inspectors), Gemmell, Wood, Wyatt, Olliff (Air N.Z.), A. Cooper, Foley, Rhodes (ALPA), R. Cooper, Spence (C.A.D.), Davies (U.K. CVR lab).
Others named, but outside nominated categories, include: Fink (L&S), Thomson (D.S.I.R.)

The Inspector elsewhere (T.123) explained that he had thoroughly discussed the accident with the "leaders of the various fields of expertise" whom he names as: A. Crawford, P. Chesney, Wright and Lund, J. Moehring, I. Gemmell, P. Rhodes, M. Baker (CAD), and a representative of the meteorological service. He advised them of the general tenor of the report and the facts he intended to include in it. Their consensus of opinion was, the Inspector said, that with few exceptions there were no other areas to be explored. He then gave notice to the parties he considered could be held blameworthy.

The Inspector's statement that the Chippindale Report had been considered by legal representatives of deceased pilots whose representations he had carefully considered, implies in its context that those representatives supported his report. Otherwise the Inspector should have said his report represented his own opinions which were contrary to the views of those representatives.

In actual fact, did those persons support the Inspector's report? Junior counsel for the pilots association acting also for Captain Collins' family said (Submissions p.24):

"...the investigation of Mr Chippindale previously in the drawing of his conclusions lacked objectivity and impartiality."

Counsel for First Officer Cassin (Submissions p.3) strongly contested that the aircraft was flown or manoeuvred in a manner that was unsafe.

The following two extracts from the transcript of evidence show that the Inspector's policy is not to take statements.

Relating to aircraft radar showing Erebus ahead the evidence reads:

Q. Now, can you tell us who the pilots were who reported this correlation of contours on the screen with what they'd seen visually?

Chippindale: No. The discussions that we held with most of the crews who had participated in previous Air New Zealand Antarctic flights were specifically informal and designed to achieve a complete frankness between the investigators and the pilots concerned. Therefore no specific records were kept of much of the discussion which took place.

Q. Do I take it from that, that you do not have a record or recollection of who gave you this information?

Chippindale: I have no recollection at this time whether or not, in the brief note I took, there is a record. (T.76)

Comment: It later transpired the Inspector had taken a record, but even that record did not substantiate his report. See note 5.

Q. You said that several members of earlier crews were of the opinion that the inbound track to McMurdo was intended to be on an alignment which was over the sea level ice to a point adjacent to McMurdo but to the west of that base. Are you in a position to tell us who these members were?

Chippindale: No, the same considerations apply to that statement as the one which I mentioned earlier, in that we spoke informally and off the record to these members, in an endeavour to obtain the maximum amount of information in order to establish the cause of this accident. (T.88)

The Inspector's evidence quoted above appears to imply that he believes that if proper evidence were taken, then pilots would not reveal the truth. What of those pilots and others who gave evidence on oath before the Royal Commission?

The ordinary private individual, whether a reporter or researcher, holds no power to require any person to appear before him and to give evidence on oath; and he is obliged to rely upon the goodwill of those of whom he makes inquiry to provide any information at all in whatever form his informant chooses.

Those appointed to make official inquiries possess whatever authority the law provides in their particular case to call witnesses and to examine them on oath or affirmation in order to carry out the purpose of their inquiry. The Civil Aviation (Accident Investigation) Regulations 1978 gives an Inspector of air accidents the same powers and authority to summon witnesses and receive evidence as are conferred on Commissions of Inquiry. Thus Mr Chippindale possessed the same statutory powers, to all intents and purposes, as did Mr Justice Mahon to summon witnesses and receive evidence.

The inquiry into the Erebus crash demonstrates the unsatisfactory results which follow from the failure of the Inspectors of air accidents to take evidence in accordance with the powers they have been given for that purpose.

The results are unsatisfactory to those whose conduct is being examined, to those who rely upon the conclusions of the inquiry, to the witnesses who provide the information, and to the Inspectors who carry out the inquiry.

Speaking informally off the record and with no specific records of much of the discussion run contrary to the accepted principles of law applicable to gathering facts for a decision which is to be made public. It is my submission that if such procedures are to be continued at all, then the results of such inquiries should neither be published officially nor otherwise made available to any other than the parties involved.

The confusion and misunderstanding which may arise from failing to take statements are demonstrated by the Inspector's handling of the incident which led to the airline altering the Antarctic computer flight path by 27 miles in the early hours of the morning of the fatal flight.

The flight manager, line operations DC10/DC8 wrote on 14 December that the pilot of the 14 November Antarctic flight rang him "and said that the McMurdo position was in error and should be 166°58' ". The Inspector received and assimilated that letter and discussed the matter with the pilot. The Inspector said that the pilot told him that as a result of the error which the pilot reported, "he believed himself to be well to the west of what he had anticipated as the waypoint for McMurdo" (T.199).

When the pilot gave evidence, he contradicted what was said by the flight manager and what the Inspector said the pilot had told him. (note 3 mistake 45). The pilot in question, Captain Simpson, was questioned before the Royal Commission concerning what he told the Inspector (T.1414):

> *Q.* When and how did you first learn that you were supposed to have reported an inaccurate position?
>
> *Simpson:* I first learnt of this reading a newspaper report about the time that Mr Chippindale's report was made public and it surprised me to learn that I supposedly had reported an error when at my interview with Mr Chippindale on the 14th December 1979 and again on the 19th March 1980 I had told him information similar to that that I had described to this Court.

Perhaps the pilot altered his story, but he had no motive for doing so. It is more likely the Inspector either misunderstood or forgot what the pilot said. The moral is that the failure by the Inspector to take statements of evidence leads to unsatisfactory and unreliable conclusions.

The widow and daughter of the pilot said they told the Inspector that Captain Collins spent the evening before the flight working on his atlas and charts in preparation for the flight. The Inspector publicly denied this and by implication accused them of inventing their evidence to the commission. Had the Inspector taken statements such a situation would never have arisen.

Had the Inspector taken a proper statement from the anonymous McDonnell-Douglas radar expert, then the controversy over radar showing the mountain should not have occurred. The former Royal Commissioner told me that it was because of the unsatisfactory nature of the evidence presented that he and counsel assisting the Commission felt it advisable to inquire direct from the radar manufacturers in the U.S. Of course, as noted previously and in note 5, the Inspector's report did not correctly represent the note he took from the pilot on the subject. I cannot reconcile the Inspector's assertion that the expert's name was known to airline counsel with counsel's denial. Nor can I reconcile the Inspector's denial that he knew DC10 radar was said not to give a dry ice return with the Commissioner's statement that he received a memorandum from the Inspector which mentioned the point.

Another instance of unsatisfactory evidence underlying the Inspector's report was his treatment of Annex J (printed on page 44). Of this the Inspector said in paragraph 2.4:

> "One track and distance diagram issued at the route qualification briefing showed that the track from Cape Hallett was direct to the McMurdo TACAN But this did not show the location of any topographical feature."

There can be no doubt that Annex J is referred to, since paragraph 2.4 is almost the same as paragraph 1.17.5, which identifies the diagram.

The implication is that had Captain Collins plotted Annex J onto a topographical map he would have learnt that the route ran across Mt Erebus. What does the evidence reveal? As noted elsewhere the Inspector was mistaken in that Annex J was not issued at the RCU briefing. The only issue remaining was whether it was instead distributed by Mr I.A. Johnson in the Antarctic envelope handed out at pre-despatch briefing. The Inspector in his press release (3 Feb. page 15) said:

> "Annex J was listed in the contents of the Antarctic envelope and Mr I.A. Johnson stated on oath that he checked it was there before handing the envelope to the crew."

The Inspector's statement is mistaken. Mr I.A. Johnson stated (T.1138) that both Annex J and exhibit 164 (printed above on page 45) came within the same description of a track chart, and that he could not say whether either of them was in the envelope. He did not even know whether the item represented three different charts or three copies of one chart.

The Inspector as one of his criticisms of the Mahon Report said (PR 3 Feb.), "The Judge does not appear to believe that the co-ordinates shown on the mini-plan recovered from the accident site and dated 1977 were genuine for they showed a longitude of 166°E."

The Inspector on page 1 of his press release refers to "the errors in the prologue catalogued below" and on page 19 he wrote: "There are other points of inaccuracy in the Report of the Royal Commission which I have not enumerated but I feel I have detailed a sufficient catalogue of errors to make my point that the Judge should have had the services of a suitable assessor to assist him."

Owing to the obscurity of the Inspector's mode of expression it is unclear what error he is here asserting has been made by the Commissioner. The supposed error is meant to be made in paragraph 229 of page 82 of the Mahon Report, but reading that paragraph in conjunction with the Inspector's criticism does not clarify what the Inspector means.

Does he mean that the 166°E co-ordinate was genuine, but that the Commissioner was in error in believing it was not genuine? The mini-plan was exhibit 3A produced by the Inspector who said (T.39): "We also have a document which was recovered from the aircraft and which is on the obverse of [the Radio Navigation Chart being exhibit 3], which I believe is known colloquially as a mini-plan."

It was marked 10.10.77 presumably meaning 10 October 1977 and is not in computer format. The longitude of the waypoint referred to as McMurdo is 166 degrees 41 minutes E. The Mahon Report in paragraph 227 says the longitude of the waypoint was 166 degrees 48 minutes east for the first two flights of 1977. These were the flights of Captain Gemmell on 15 February 1977 and Captain Grundy on

22 February 1977. The Mahon Report says that for the flight of 18 October 1977 the McMurdo waypoint was altered to 166 degrees 41 minutes east. The Commissioner's statement co-incides with the longitude of the waypoint on exhibit 3A. If the Inspector means what he says viz. that the waypoint on exhibit 3A was 166° E then he is mistaken; and the Commissioner is correct in that the waypoint was 166 degrees 41 minutes east.

Does the Inspector mean that exhibit 3A was recovered from the ice, as he said in his evidence, but that the Commissioner was mistaken in that he did not believe it was so recovered? The only evidence that it was found on the ice came from the Inspector himself. What is the source of the Inspector's statement to that effect? If the Inspector is correct, then it is puzzling how the crew came to have such a plan. Though Brooks had been to Antarctica previously, his flight was on 22 February 1977 when the longitude was 166 degrees 48 minutes east.

The Inspector's claims that Mr Johnson stated on oath that he handed Annex J to the crew and that a mini-plan showing 166 degrees E was recovered from the accident site form two of the many criticisms he made of the Mahon Report.

Elements of Chippindale Analysis

The Chippindale Report is founded on five major propositions:

- The change in the computer flight path from McMurdo Sound to Lewis Bay did not mislead the crew
- The crash was caused by the pilots descending beneath 16,000 feet contrary to to the airline's instructions
- The crew was not certain as to their position
- The aircraft's radar would have depicted the mountainous terrain ahead
- The captain headed the aircraft toward cloud-covered high ground appearing to the pilot as an area of limited visibility or whiteout.

The Inspector said that his report:

"did not 'whitewash' the airline nor blame the pilots alone..." (PR 18 Jan).

He also said (PR 3 Feb):

> "I was not subject to pressure by the Minister of Transport, the Director of Civil Aviation, NZALPA, Air New Zealand or any other person or organisation."

The *Auckland Star* of 29 January 1982 reported that:

> "Mr Chippindale today said he would only take up the point that Mr Justice Mahon reportedly said he (Mr Chippindale) accepted the Air New Zealand view on the cause of the crash.
>
> 'I believe determination of the cause of the crash was based on the assessment made by all the investigators involved in preparing our report and was in no way affected by Air New Zealand's view on the matter', Mr Chippindale said."

The only criticism of consequence made by him against the airline is contained in paragraph 3.7 which says:

> "Some diagrams and maps issued at the route qualification briefing could have been misleading in that they depicted a track which passed to the true west of Ross Island over a sea level ice shelf, whereas the flight planned track passed to the east over high ground reaching to 12450 feet AMSL."

However, the effect of this paragraph is negated by paragraph 3.26 reading:

> "Despite the shortcomings of some aspects of the route qualification briefing, this flight and Antarctic flights in general were not unacceptably hazardous, if they had been conducted strictly in accordance with the route qualification briefing as presented."

The Chippindale Report in the first sentence of its analysis of the accident (paragraph 2.1) states:

"The *initiating factor* in this accident was the captain's decision to make a VMC (visual meteorological conditions, meaning 20 km visibility or more) descent below the specified minimum safety height while north of McMurdo." (my italics)

Does this not mean that whatever happened before the decision of the captain was irrelevant? The decision was made only minutes before the crash and so, to the general reader at any rate, the Report excludes as a cause the airline changing the flight path 27 miles and not telling the aircrew.

I now examine the Inspector's five propositions by comparing them with the Air New Zealand viewpoints propounded during the course of the hearing, adding to each proposition my own comment.

The Inspector's first proposition is that the change in the computer flight path from McMurdo Sound to Lewis Bay did not mislead the crew.

Air N.Z.

". . .notwithstanding any possible misconception which may have arisen from knowledge of the incorrect McMurdo longitude co-ordinate, the crew would have checked the McMurdo position on their actual flight plan (either pre-departure or at some stage during the flight as I would expect them to) and noted that the aircraft was planned to fly to the McMurdo TACAN." (Gemmell B.601).

Chippindale

After a formal introduction the Chippindale Report commences in paragraph 1.1 by relating the history of the flight. It reads, inter alia:

"1.1.1. In preparation for Flight TE 901 two of the pilots attended a route qualification briefing. . . .

. . .The briefing gave details of the instrument flight rules (IFR) route to McMurdo which passed almost directly over Mt Erebus, a 12,450 ft high active volcano, some 20 nm prior to the most southerly turning point, Williams Field."

Comment

The instruments which must be referred to are the computer navigation system and the NDB. The paragraph is mistaken, and mistaken in the vital matter which led to the crash. The briefing did not give details of an IFR route almost directly over Mt Erebus. Quite the contrary. The briefing officer handed out a computer flight plan which showed the route passed down the centre of McMurdo Sound to the Dailey Islands, some 27 miles away from a Mt Erebus route. Apart from the crew who were killed, the three surviving pilots had all been given the impression by the briefing that the route lay down McMurdo Sound.

After an earlier reference in paragraph 1.17.7, the Inspector in paragraph 2.5 wrote:

"2.5 The flight plan was printed for each flight from a computer stored record which, until the night before the flight, had the longitude for the McMurdo destination point incorrectly entered as 164°48'E. The error in longitude had persisted for 14 months and was not corrected on the sample flight plan shown to this crew at their preliminary briefing. The error had been discovered 2 flights earlier but neither the crew of the previous flight or that of the accident flight were advised of the error by the flight dispatcher prior to their departure. The error had placed the destination close to the longitude of the Byrd Reporting Point (165°E) had this aligned the track close to that displayed on the RNC chart and to that used by military aircraft. The fact that the computer error of over 2° of longitude to the west could exist for 14 months indicates there was no regular valid comparison between the topography and the geographical

> co-ordinates by briefing officers or flight deck crews. As all previous flights to McMurdo had approached the area in VMC earlier crews had not adhered to the flight plan track and hence had not detected the error. In the case of this crew no evidence was found to suggest that they had been mislead [sic] by this error in the flight plan shown to them at the briefing."

Comment

This paragraph obscures the simple point which was that at the briefing a computer flight plan was distributed which showed the programmed nav track ran down the centre of McMurdo Sound. Before they boarded the aircraft the aircrew were handed a flight plan which lay across Erebus. The flash ops line on the plan, the purpose of which was to draw the crew's attention to anything of note, was blank. The crew were not told in any other way that the plan they had been briefed on (with the waypoint underlined or highlighted) had been altered in one vital leg.

The Inspector's conclusion that no evidence was found to suggest that the crew had been misled was itself misleading. The Inspector said in the course of his cross-examination (T.242) that by "no evidence" he meant only documentary evidence; his office had not come across any documents held by the crew of the fatal flight which showed a route to the Dailey Islands.

The Inspector went on to add that "we had found no one who attended the briefing who was misled by those co-ordinates." He interviewed First Officer Gabriel. He believed he asked the first officer about this point (T.242). First Officer (by now Captain) Gabriel said (B.530) that during the briefing he compared the TACAN co-ordinates with the co-ordinates of the "McMurdo" waypoint on the flight plan which he had in front of him. He realised the waypoint was to the west of the TACAN. He therefore believed the navigation track ran to the west of Ross Island though he did not know its precise location (B.1713). Either Captain Gabriel must have altered his evidence between what he told the Inspector and what he told the Royal Commission or the Inspector must have misunderstood or forgotten what he was told.

Another pilot at the briefing was Captain Simpson. He said (B.422) that he looked at a flight plan produced by Captain Wilson at the briefing and noticed that the "McMurdo" position on it was further south-west than the Byrd reporting point, being at the head of McMurdo Sound. Thus by inspecting the co-ordinates Captain Simpson was clear the computer flight path ran down McMurdo Sound and not across Mt Erebus. The Inspector's conflict with Captain Simpson's evidence is referred to elsewhere. In the present context the Inspector's statement that he found no one misled by the co-ordinates into believing the track lay down McMurdo Sound is mistaken (or the pilots are mistaken) in two out of the three surviving pilots who attended the briefing.

Only one other surviving pilot attended the briefing, First Officer Irvine, who said (B.553):

> "When I left the briefing I had a clear understanding that we were flying into the McMurdo area up the McMurdo Sound with Ross Island and Mt Erebus well out to our left."

He derived this understanding (T.1733 and T.1734) from having seen a number of visual displays of tracks into the area, all of which indicated a track up the McMurdo Sound itself and none over Ross Island. Thus although the briefing led him to believe the track lay down McMurdo Sound, the belief did not derive from the computer flight plan showing the Dailey Islands waypoint. The Inspector is correct that it was not the co-ordinates which "misled" this pilot: one correct out of three.

The Inspector's use of the word misled is curious. The computer flight plan shown at briefing to Simpson, Gabriel and Irvine ran down McMurdo Sound to the Dailey Islands. They were given at despatch a plan with the same track with the result that their actual track did follow that same path down McMurdo Sound. The Inspector said that showing Simpson, Gabriel and Irvine the McMurdo Sound flight plan did not mislead them into believing that the computer flight plan ran down McMurdo Sound; which of course was the actual track, and had been for 14 months. This terminology was devised by the airline shortly after the crash and it enabled them to maintain in public that no error had been made in the flight plan given to the crew of the fatal flight.

Despite the importance of navigation procedures in general, and the change in navigation co-ordinates (which altered the flight path) in particular, the Inspector never interviewed the navigation staff of the airline (T.201). The Commissioner believed this to be one of the most surprising facts to emerge from the hearings, as when the Inspector had returned from the ice he had said in a press statement that the key question was the navigation system.

The Antarctic envelope handed to the crew at pre-despatch briefing had a check list of contents (produced by the Inspector as exhibit 38. T.133) which listed, inter alia: "spare flight plans — NZAA — Cape Hallett — McMurdo Sound 2; spare computer plans — NZAA — Cape Hallett — McMurdo Sound 2." Such notation would confirm the belief of the crew that the flight plan ran to McMurdo Sound.

Descents beneath 16,000 feet

Air N.Z.

Pilots were forbidden to descend beneath 16,000 feet until they reached McMurdo Base. If Collins had obeyed his instructions he would not have crashed on Erebus at 1500 feet, and he would have discovered the change in co-ordinates, since the pupose of flying to McMurdo Station was to check the AINS (Gemmell, T.1788).

Chippindale

The Inspector says that the pilots were forbidden to descend beneath 16,000 feet until they reached McMurdo Base (paras. 2.1, 2.11, 2.20).

> "2.1 The initiating factor in this accident was the captain's decision to make a VMC descent below the specified minimum safety height while north of McMurdo.
>
> "2.11 ...This descent was made despite the safety requirements to maintain a minimum sector altitude of 16,000 feet until overhead McMurdo TACAN and to descend below that height only in a specified sector and in weather conditions of 20 km visibility and no snow showers, and after contacting the radar controller.
>
> "2.20 The pilot-in-command did not comply with the company's requirement to limit the descent to 16,000 feet until overhead 'McMurdo Field'.
>
> "3.20 The descent was intentionally continued below the VMC limit specified by CAD and Air New Zealand Limited, of 6000 feet to an indicated 1500 feet."

Comment

Firstly, Captain Vette, in chapter 9, his epilogue dealing with low level Antarctic flights, has disproved the quoted propositions stated by the Inspector in all of those paragraphs 2.1, 2.11, 2.20, and 3.20.

Secondly the proposition in paragraph 2.11 conflicts with the affidavit subsequently given to the Royal Commission by the Chief Traffic Controller and MAC Centre Supervisor at McMurdo Sound. The Inspector ought to have checked with the controller and discovered and resolved this conflict. In fact the Inspector did contact the controller in question (T.46) but evidently failed to note the inconsistency between what the airline said were its let down instructions in the so-called safety arc and the controller's view that such a procedure would be absurd.

The controller said:

> "We were not aware at the time of the sector within which the Air New Zealand DC10 flights were said to be required to fly. This required sector as now identified to me by representatives of the Royal Commission would in my opinion have been absurd. Any radar controls contemplated by those planning such flights for purposes of assisting in descent manoeuvres would have been virtually impossible since Ice Tower radar does not function above 30 degrees and since without vertical aiming radar or an Azimuth Beam on the TACAN, there could be no adequate ground based control should it have been required. I re-emphasize that to the best of my knowledge, Operation DEEP FREEZE was unaware of any officially approved Civil Aviation Division (CAD) or Air New Zealand flight plan or descent approach. However, had such a flight plan or descent approach been provided for DEEP FREEZE consideration and had DEEP FREEZE agreed to provide radar let-down assistance, I would have regarded such a plan as extremely ill-advised for the reasons stated."

The Inspector lays great stress, both in his report and in his subsequent press releases, upon the pilots breaching their briefing instructions in regard to height. Hence the opinion which other pilots derived from their briefing as to permitted heights is important and is the third issue for consideration. If a majority or even an appreciable number of pilots concluded that the passage in their briefing referring to 6000 feet was only a termination of a letdown from 16,000 feet and that further descent was permitted, then Collins and Cassin could hardly be blamed for taking the same view. Unfortunately for them, it appears from the Inspector's evidence which follows that only two pilots (or possibly one pilot) from the previous three years' flights held that view. The questions to and answers by the the Inspector follow (T.176):

Q. Now, you have said in evidence that some captains saw 6000 feet as the termination of the letdown from 16,000 feet and interpreted the briefing notes to allow a further descent at the captain's discretion. How many of the captains that you spoke to held that view?

Chippindale: I believe two captains expressed that opinion.

Q. Are you prepared to name them?

Chippindale: I would prefer not to under the provision I made this morning.

Judge: You had better tell us the range of captains among whom you enquired.

Chippindale: The ones on the previous flights, do you mean? At the stage I enquired into the minimum height to which the aircraft had descended we were talking to pilots from the three years' flights.

Comment:

In the context of radar the Inspector referred to (T.76): "The discussions that we had with most of the crews who had participated in previous Air New Zealand flights. . ."

Thus his evidence is detrimental to Collins and Cassin and correspondingly supports the paragraphs quoted above from his report. But to what extent is his evidence supported by the evidence which the pilots themselves subsequently gave? The short answer is that the pilots' evidence contradicts the impression given by the Inspector. Their evidence is summarised, in sequence, crew by crew, commencing with the flight before the crash.

White (B.546-7) believed he was not restricted from descending below 6000 feet. Irvine (B.556) did not regard 6000 feet as a minimum. Simpson (B.427-8) regarded the reference to 6000 feet as ambiguous, but descent was to be co-ordinated with ATC, and said it was normal for a pilot to negotiate his own terms and clearance with ATC. (On his own flight Simpson descended to about 2500 feet — Gabriel B.536).

Gabriel (B.536) considered 6000 feet to be for cloud break and further descent in VMC permitted if approved by ATC. Dalziell (B.574) said the briefing officer R.T. Johnson led a discussion on Antarctic flights below 6000 feet and the briefing officers made no criticism of this practice. Dalziell knew from other personnel that previous flights had flown round Erebus below 16,000 feet.

Ruffell (B.504) was under the impression 6000 feet was the minimum descent permitted. He got that impression partly from a letter he received relating to his particular flight. Calder (B.510) did not consider 6000 was anything other than cloud break procedure and understood descent below 6000 feet was at pilot's discretion.

McWilliams (B.517) understood 6000 feet was a minimum for cloud penetration only, and understood he was not prohibited from descending below 6000 feet in VMC.

Vette (*Impact Erebus,* passim) believed 6000 feet represented cloud break procedure, and lower descent permitted subject to ATC approval.

R.T. Johnson (B.389) was one of the briefing officers. He said he descended to 3000 feet while aware his briefing stated 6000 feet was the minimum. He himself led a discussion at briefing on flights below 6000 feet and offered no criticism of that practice (B.574).

Hawkins (B.639) was part of the airline's executive, being flight manager training. He said he declined to descend below 6000 feet on 18 October 1977 because the minimum briefed altitude was 6000 feet. Captain Vette attended the same briefing and Vette's view of the briefing conflicts with that of Hawkins. Hawkins' co-pilot was A.A.E. Lawson. On Hawkins' flight was Graeme Kennedy, a reporter who knew both pilots. Kennedy wrote an article in the *8 O'Clock* (exhibit 83) that the flight descended to 200 metres, a misprint for 400 metres or about 1300 feet (T.816-817).

The instructions given to counsel for the passengers' consortium were that either Hawkins or Lawson told Kennedy that when over McMurdo the aircraft was between 1200 feet and 1300 feet.

Lawson denied he told Kennedy the flight descended to between 1200 and 1300 feet and said that to the best of his recollection, Hawkins' flight did not descend beneath 16,000 [sic] feet. (T.817-818). Hawkins said at (B.638): "I continued my descent visually to 6000 feet", at (T.1882): "I remained at 6000 feet. . . 6000 feet was my minimum altitude and that's where I stayed." Hawkins said (T.1878) he knew Kennedy but he did not give Kennedy the information. Counsel said (T.1879) a passenger stated he understood the flight descended to approximately 1000 feet. Hawkins answered that he was aware of the passenger who was the source of that information and it was totally incorrect, and it would be very difficult for a passenger to judge whether it was 6000 feet or 16,000 feet.

Grundy, flight operations manager, said the early flights had no discretion to descend beneath 16,000 feet. (T.432 D41) and that his flight on 22 February 1977 did not descend beneath that height. (T.432 D30)

Gemmell, chief pilot, who flew the first Antarctic flight on 15 February 1977, said that on that flight he did not descend beneath flight level 160 (16,000 feet) at any time. (T.462, T.463, T.490, T.505). His co-pilot, A.A.E. Lawson, is recorded as saying that to the best of his recollection the flight descended below 16,000 feet (T.817). Later he said: "If I could remember descending below 16,000 I am sure I would indicate it to this commission.'' (T.819). Bob Thomson of the DSIR Antarctic Division believed the flight descended to 9000 feet.

Evidence was not given by the pilots of 28 November 1978 and 1 November 1977.

We have it on oath from the Inspector that his office had discussions with most of the previous crews of the three years' Antarctic flights but that only two ''captains'' expressed the view they were entitled to descend below 6000 feet. The fact is there were 13 Antarctic flights, three pilots on each flight i.e. 38 pilots (Lawson went twice). I believe that counsel when using the word ''captain'' meant ''pilot.''

The Inspector may instead have given a literal answer meaning that of the 38 pilots only two of those who held the rank of captain told his office they believed they were permitted to descend below 6000 feet while omitting to mention the views of the 20 first officers. Eighteen of the pilots were captains.

Confirmation that the Inspector means pilots, not captains, is given in the following extract from his evidence. But he now says that he knew of only one pilot (not two captains) who did not share his interpretation of minimum flight altitudes.

> *Q.* But we know, don't we, that there are, and have been, numerous or a significant number of pilots who did not share your interpretation of that minimum flight altitude, don't we?
>
> *Chippindale:* I know specifically of one pilot and a sequence of reports of lower flights. (T.215)

Eight line pilots (i.e. not members of the administration) said in evidence they believed descent was permitted, one said he did not believe he was so permitted. The other five pilots who said descent was not permitted were all executive pilots or part of the airline administration. To recapitulate, of these five Lawson, formerly a briefing supervisor, knew of low level flying but did nothing about it, Johnson, a briefing officer, led briefing discussions on flights beneath 6000 feet but without criticising it, and descended below that height himself. Johnson's co-pilots were Captain M. Buchanan and First Officer O. Faulls. Either here are two pilots who believed descent below 6000 was permitted, or they concurred in a breach, by their pilot in command, of briefing instructions. Likewise Captain A. Potts and First Officer Clapshaw either believed Captain Vette's descent was authorised or in breach of briefing. McWilliams (B.518) descended to 2000 feet, so First Officers G. Roud and W. Woodhams believed descent was permitted or breached briefing instructions. The flight of First Officers M. White and P. Wilkins descended to 2500 feet. (B.510). First Officers J. Eaton and N. Pullin descended to about 4000 feet (B.576). First Officer Woollaston descended to 2500 feet (B.536) and First Officer T. Shepherd to approximately 2000 feet (B.550). The Commissioner, Mr Justice Mahon (M.75) concluded that all Antarctic flights from 18 October 1977 descended considerably beneath 6000 feet, generally to between 1500 feet and 3000 feet.

The end result is that most Antarctic pilots believed descent beneath 6000 feet was permitted, or descended contrary to briefing. How is it that out of most of the Antarctic crews the Inspector's office discovered only two captains (or one pilot) who believed such descent was permitted? He spoke to both Eaton (T.189) who gave no evidence and Simpson (T.189 and T.199); were those the two he refers to?

But Captain Vette tells me he was interviewed by the Inspector and that he made it clear that he did not believe there had ever been any prohibition on descent beneath 6000 feet, so evidently Captain Vette is the one pilot who told the Inspector there was no such prohibition.

Captain Vette adds that a number of other pilots from previous Antarctic flights were interviewed by the Inspector. Prior to the interview the subject of a supposed prohibition on descent beneath 6000 feet was raised. None of the pilots had heard of such a prohibition and did not know what it referred to.

Apart, then, from the five executive pilots, and Ruffell, Vette and possibly Eaton, what of the other 30-31 pilots, of whom seven testified they believed descent was permitted, and most of the remainder either shared that belief or breached their briefing instructions?

As noted previously, the Inspector wrote (PR 3 Feb): "(the Chippindale) report was the considered assessment of a group of extremely well qualified DC10 experts, pilots and accident investigators who decided this accident could have been avoided if the crew had conducted the flight in accordance with the briefing in relation to minimum safe altitudes."

Who are the group of pilots to whom he refers? By a process of elimination are they reduced to the executive pilots of the airline's administration: Gemmell, Lawson, Grundy, Hawkins and Johnson together with the sole non-executive line pilot Ruffell? If such be the case, it is perhaps not surprising that the Inspector's report reflected the airline views.

Finally, when the Inspector referred to "two captains", what are the implications if the phrase means "two pilots with the rank of captain" so excluding pilots who were first officers, or if it means "two pilots"?

From the quotations it will be noted that in answer to the judge's question concerning captains the Inspector replied using the term "pilots" and at T.76 he referred to "crews".

By the hearing at least two first officers had become captains: Eaton and Gabriel. The Royal Commission procedings were not intended to be adversarial, but investigative; the Inspector's function was to assist the Commission to ascertain the truth. If his answer of "two captains" excluded the opinions of 18 or 20 first officers on the topic it was not only misleading and obscured the facts but it also conflicted with the evidence of seven captains that they believed descent was permitted. Is the Inspector implying they changed their evidence?

Consideration of his press release as a whole suggests the Inspector may be implying they did. He said his report was the considered assessment of, inter alia, pilots "who decided this accident could have been avoided if the crew had conducted the flight in accordance with the briefing in relation to minimum safe altitudes." (The plural must refer to 16,000 feet and 6000 feet.) Surely this means that, among others, pilots told him the briefing prohibited descent.

In the same press release (see note 1) he said he felt he "had a right to legal representation during the inquiry. . . The lack of representation placed me at a distinct disadvantage in not having any avenue for questioning witnesses who had led me to believe that a different set of circumstances existed after the accident to that which they described in evidence."

Is the Inspector implying the pilots told his office they believed the briefing prohibited descent beneath 6000 feet, and then gave evidence to the contrary before the Commission?

If by "two captains" the Inspector means two pilots, then is he implying that over 20 pilots either changed their evidence after their interviews with his office or that they knowingly breached a briefing prohibition on descent?

Another hypothesis is that the Inspector's officers misunderstood or forgot what pilots told them in more than 20 instances, reporting only one or two correctly. If this is so, then my thesis that written signed statements should be employed in order to avoid misunderstandings finds further support.

The Inspector was asked (T.138) whether any of the three earlier flights had descended in the McMurdo area beneath what was printed in the transcript evidently incorrectly as 1600 feet. From two references on T.139 it appears that the question actually asked concerned descents beneath 16,000 feet.

In his answer the Inspector said that he was given the opportunity to speak to all the previous crews relating to this point. The question materially affected their conduct as captains of aircraft with Air New Zealand.

They were unable to speak for themselves, he gave them no specific caution, and they were not on oath. He had indicated to the crews that his discussion with them was informal in order to get to the bottom of the cause of the accident rather than establish liability or blameworthiness.

The Inspector asked the Commissioner whether, in view of these facts, it was necessary for him to answer the question. The Commissioner gave him permission not to answer, so that there is no record what the pilots told the Inspector about actual descents beneath 16,000 feet.

The Commissioner went on to add that Mr Chippindale could hear what the pilots had to say and that if his impression of their evidence was not quite the same as what he understood that they told him, he could testify about that (T.139).

If the pilots told the Inspector they did not descend beneath 16,000 feet, then it is unfortunate the Inspector did not take up the offer to testify as to what they told him. On the other hand if all the crews told him they descended beneath 16,000 feet (with the exception of some executive pilots) why did the Inspector not answer the question? If virtually all pilots descended beneath 16,000 feet in the so-called McMurdo area, which the airline later said included McMurdo Sound and the Ross Sea, how could any one of them be prejudiced by the Inspector revealing the fact? The answer to the question is crucial since upon it depends whether Collins and Cassin were blameworthy in almost alone of all Antarctic flights descending beneath 16,000 feet outside the arc around McMurdo Station, or whether it was the universal practice among crews.

The Inspector said (T.144 and T.145) that he requested and was given a repeat of the RCU briefing which had been given to Captain Collins and First Officer Cassin. The repeat was organised within a day or so of the Inspector's return from the Antarctic. In connection with that briefing the Inspector was asked (T.154):

> "Did you hear any statement to the effect that the captain had a discretion to go lower than 6000 feet?"
>
> *Chippindale:* "No."
>
> *Q.* "Anything to that effect?"
>
> *Chippindale:* "Nothing to that effect."

Captain Gemmell said (T.1835) that Captain Wilson, the briefing supervisor, gave a sample briefing to himself and the Inspector which indicated that Captain Wilson at RCU briefings mentioned, without any criticism, the fact that flights had descended below 6000 feet with the approval of American ATC.

If one puts together the evidence of the Inspector and the airline's chief pilot, the conclusion is that two or three weeks after the crash the briefing officer repeated for the Inspector the briefing which he gave Collins and Cassin during which he mentioned without criticism that previous flights descended beneath 6000 feet. First Officer Irvine said (B.553 and B554) such descents were mentioned at the briefing of Collins and Cassin.

I cannot reconcile the Inspector's denial with Captain Gemmell's evidence. It is also puzzling that the Captain, as technical advisor to the Inspector, did not remind the Inspector that what was said at the briefing as to heights was contrary to what the Inspector was inserting in his report to the effect that descents beneath 6000 feet contravened briefing instructions. The Inspector comments upon this issue (PR 3 Feb):

> ". . .the briefing officer apparently advised some captains that they could

decend to any level approved by the USN Air Traffic Control but the airline's policy as stated in the taped RCU briefing and correspondence with the Civil Aviation Division (CAD) did not disregard the [altitude] limits mutually agreed."

The briefing officer, Captain Wilson, said he did not tell the Inspector he had discussed without criticism flights below 6000 feet. He did not believe he told the Inspector that the aircrew could negotiate descents below 6000 feet with the McMurdo ATC. The Inspector did not ask him his view on whether descents below 6000 feet were authorised. (T.1236)

The Inspector agreed (T.215) that a briefing officer would have sufficient standing in the airline to indicate to pilots that they were permitted to descend beneath 6000 feet.

The evidence given on behalf of the airline appears to conflict. How can Captain Gemmell's evidence be reconciled with Captain Wilson's? Perhaps Captain Gemmell is correct and Captain Wilson did discuss without criticism flights beneath 6000 feet during the briefing but that Captain Wilson meant only that he had never personally explained this to the Inspector, as opposed to merely revealing his discussions during the course of his sample briefing.

Whatever one makes of the evidence, it is clear that the Inspector's investigation of what descent restrictions, if any, were imposed on the pilots by their briefing instructions was inadequate to support his unequivocal conclusion that their breach of instructions caused the crash.

Fourthly, for the Inspector to have stated his conclusion as to height in the way he did, requires that like CAD and the airline he too must have been ignorant of low-level Antarctic flying. See note 4 on page 335.

The fifth reason why the Inspector's conclusions cannot be supported is that the briefing officer himself believed that flight beneath 6000 feet was permitted by both the airline and CAD, provided the McMurdo authorities concurred. He discussed such flights at a number of briefings, probably including the briefing of the crew which crashed.

Sixthly, there was brought to the attention of the Inspector a report of an aircraft having been observed 1000 feet above glacier level. In view of the importance of low level flights in relation to the crash, this incident should have induced the Inspector to investigate thoroughly the incident and its possible broader implications. Instead, he merely exchanged pleasantries (T.175).

Seventhly, the McMurdo authorities told the Inspector that previous flights flew at low levels. The McMurdo ATC supervisor Charles Priest in his statement to the Inspector (exhibit 55) said:

> "On numerous previous Air New Zealand flights, I have observed VMC altitudes as 500 feet (estimation) in the McMurdo Sound."

The statement to the Inspector of Air Controller Kenneth Kondewich (produced by the Inspector T.344 as exhibit 58) was to like effect. He said that previous flights descended to about 1500 feet. The hills at the back of Scott Base are approximately 1000 feet above the sea/ice and a flight above Scott Base camp would clear the nearest hills by about 500 feet.

Knowledge by the Inspector of these statements seems hardly consistent with the criticisms and comments which he made in his report.

My conclusion is that the views on descents beneath 16,000 feet being permitted which are expressed in the Mahon Report and in *Impact Erebus* are correct and that the contrary views in the Chippindale Report are mistaken.

Crew uncertain of position

Air N.Z.

The crew did not identify or mistakenly identify their position (Gemmell T. 1775). The crew did not give their actual precise geographical position much thought as they descended in orbit toward Erebus (Gemmell T. 1779).

One possibility, which has received little attention at this Commission, is that the crew knew that they were flying on the direct track between Cape Hallet and McMurdo Station which was over Ross Island (Gemmell B.601).

Any misconception the crew may have had as to where the navigation track from Cape Hallett lay in relation to the high terrain based on knowledge of the incorrect McMurdo waypoint would not have been the deciding factor in the decision to descend below MSA [16,000 feet] (Gemmell, B.610.)

Chippindale

The Inspector when stating the probable cause of the accident in paragraph 3.37 wrote: ". . .the crew was not certain of their position. . . ."

Comment

For the Inspector's conclusion to be correct requires two conditions to be satisfied. Firstly, what was said by the two pilots on the flight deck on the approach to impact must indicate uncertainty. Whether the other crew member, the flight engineer, was uncertain is not relevant. In Chapter 4 *Flight to the Ice* and in the explanation of the azimuth map from page 49 to 60 Captain Vette explains the meaning of the CVR transcript and shows that the crew were certain in their own minds of their position, but the un-notified change in co-ordinates and the effect of sector whiteout resulting in Lewis Bay taking on the appearance of McMurdo Sound, caused their belief to be mistaken. The Mahon Report comes to the same conclusion. Both believe that the Chippindale Report misinterprets the CVR as to certainty of position.

Cockpit Voice Recorder

In the Chippindale Report the Inspector makes four statements which he appears to infer from his understanding of what was said on the flight deck before the crash.

> "There were discussions on the flight deck indicating that some of the speakers believed they were to the west of Mt Erebus, but the two Flight Engineers on the flight deck had voiced frequent queries about the procedure and expressed their mounting alarm as the approach continued on at low level toward the area of low cloud." (paragraph 2.20)
>
> "The apprehension expressed by the flight engineers indicated that these members of the crew were endeavouring to monitor the flight responsibly but their suggestions of caution, as with the captain's decision to climb out of the area, were overtaken by the speed of the sequence of events." (paragraph 2.25)
>
> "The flight engineers endeavoured to monitor the progress of the flight and expressed their dissatisfaction with the descent toward a cloud covered area." (paragraph 3.24)
>
> ". . .the crew was not certain of their position. . ." (paragraph 3.37)

Of these views, the Commissioner states in paragraph 116 of the Mahon Report:

> "In my opinion none of these views expressed by the Chief Inspector in his Report is substantiated, either by the transcript of the CVR (cockpit voice recorder), or by the process of listening to the playing of the tapes."

To arrive at his conclusions, the Inspector relies upon correctly interpreting a tape-recording from the CVR, the quality of which was so bad that it took the National Transportation Safety Board five days to prepare a 30 minutes' transcript.

There was only one microphone, and this picked up simultaneously what was said by the two pilots, the two flight engineers, the flight commentator, cabin crew and sight-viewing passengers at the rear of the flight deck and, so long as the flight deck door was open, persons in the galley. It also recorded the plane's public address announcements and air to ground, and ground to air communications. (M.36).

The Commissioner found that (M36-37):

> "the volume of conversations and cross-talk on the flight deck behind the two pilots made it difficult in the extreme to ascertain what exactly was being said. Conversations between different people tended to run together. A sentence uttered by someone would be interrupted mid-way through by a sentence spoken by someone else who was evidently closer to the microphone. Someone would give an answer to an indecipherable question."

It should be made clear that no techniques of voice-prints were employed to identify speakers. The Commissioner says (M.43):

> "It requires no expert skill to listen to a tape-recording."

The only expertise available was filters, and the Commissioner says of this device (M.43):

> "There were very few cases indeed in which an indecipherable comment was made decipherable by use of the filters."

The Bendix experts (M.44):

> "pointed out that filtered [sic] devices were only useful in eliminating to some extent background noise. Such devices in general either confirm, by the medium of increased clarity, what the listener thought had been said, or confirm the inherent unreliability of a transcript sought to be produced from listening to particular recorded comments and remarks."

The Commissioner's summary was:

> "Five days of debate and discussion, followed by transcript partly based on guesswork."

In essence, the Commissioner's opinion was that in key passages of the transcript, which the Inspector relied on for his opinions, the Inspector had held that words had been said which the Commissioner believed were unintelligible, the Inspector had taken chance snatches of conversation from passengers and cabin crew, and wrongly attributed them to the flight engineers, and pieced together words coming from different persons to form some purported connected commentary by the flight engineers.

The weakness underlying the transcription was not revealed by the Chippindale Report, so that the resulting transcript has been taken by the community at large to be accurate. For instance it was broadcast on television, and Tom Scott, in *The Listener,* wrote:

> "What makes the (Chippindale) report particularly grisly reading is the abridged transcript of the cockpit voice recorder tape. It sounds so nonchalant and unconcerned, except for the flight engineer attesting from time to time that he is unhappy about what's happening. My first reaction was one of horror. . ."

Mr Chippindale took the tapes to Farnborough and from listening himself to the tapes he added certain sections to the transcript which tended to support his theory that the crew were uncertain of their position and the flight engineers were expressing concern at the Captain's conduct. He also omitted certain portions of the transcript from his final published version, one of which was notable and which, contrary to the Inspector's theory, tended to suggest the crew were certain of their position.

The Mahon Report found that some extracts from the CVR included in the Inspector's Report had been regarded at the transcription in Washington as being unintelligible or, alternatively, not sufficiently intelligible to be of any assistance, and accordingly these extracts had not been included in the transcript which had been agreed and settled by all parties in Washington.

When all the differences of the Chippindale transcript from the Washington team transcript are compared, it is clear that they differ greatly and would lead to different conclusions. Some examples quoted in the Mahon Report were the following:

"*Unidentified:* Bit thick here eh Bert?

Flight Engineer Maloney: Yeah my. . . oath

Maloney: You're really a long while on. . . instruments at this time"

The Inspector writes in his press release:

> "At no time did I attribute any comment to any person. I relied totally upon the recognition of the voices made by the team in Washington."

There is a clash between the two Reports, since the Mahon Report said that the extracts were believed to be unintelligible by the Washington team. The Chippindale Report transcribes these extracts to appear as if there is a question asked of Maloney, referred to as Bert, which he answers.

However, Maloney's name was not Bert, nor was anyone of the flight deck crew named Bert.

The Inspector says:

> "We were aware there was no crew member on the flight deck called 'Bert' but that is far from proof that no one called Bert was on the flight deck, neither is it in any way proof that the words were not spoken. I knew they were from the many hours spent listening to the tape."

However, by placing these extracts together, the Inspector implies a question is asked of Maloney, who answers. That is misleading when the Inspector knows his name is not Bert.

That in itself is no proof the words were not spoken, but the fact is the Inspector is relying on his opinion alone in inserting remarks into the transcript when the passages had been agreed to be unintelligible by those who heard the recording in Washington.

The Inspector incorporated into his transcript of the CVR a reference to instruments which was detrimental to the pilots that was not included in the Washington transcript. The Commissioner's comments on the incident are reprinted above in appendix F pages 291 to 293.

Impact Erebus at page 55 explains that the American transcript at 0046.39 records the rustling of paper, which was obviously Collins' chart being refolded as the crew monitored their pre-plotted track, and a subsequent explanation of not wanting to block the window. These references which favoured the crew were omitted by the Inspector from his report.

Further treatment of the CVR transcript by the inspector which he handled in a manner detrimental to the crew is explained in note 6 Could "Wilson" mean Wilson Piedmont Glacier?; note 7 Location of Taylor Valley; and note 8 When did aircrew first express anxiety?

The American transcript of the CVR from 0039.49 GMT to 0039.58 has an equivalent in the interim report prepared by the inspector relating to the squelch:

Brooks: Having a bit of trouble with radio contact at the moment?

Maloney: Have you got the squelch off?

Brooks: Clearance to go down.

Cassin: Say again.

Maloney: Have you got the squelch off on that?

Cassin: Yes, on both.

The Inspector omitted all these references from his final published report. He agreed that the squelch is a filter used on VHF radio. If you take the squelch off, you are likely to increase the force of the pick up range of your signal and also get a noisy and/or staticky pick-up. The exchange between the crew shows that Flight Engineer Maloney was monitoring the actions of First Officer Cassin as part of the crew loop procedure. The crew were doing their job properly, in that they were doing all that was within their knowledge to obtain VHF communication. Their thinking was directed to technical difficulties preventing reception.

The inspector omitted from his published report these references which are favourable to the crew and which provide further proof, pace Captain Kippenberger, that they were not all suffering mental problems. The Inspector, with logic which is not entirely clear, said (T.263) that his one and only reason for editing the transcript in this manner was to maintain the co-operation of the Airline Pilots Association.

The Inspector says of these matters:

> "One thing I will state solemnly is that no attempt was made by me to tailor the CVR readout or insert comments which were not clear to me". (PR 3 Feb.)

Plotting Co-ordinates

Secondly, for the Inspector's conclusion of uncertainty to be correct requires Captain Collins not to have noted and plotted on a topographical map the co-ordinates given to him at briefing, and not to have used his navigation track as a secondary means of monitoring his position.

The Inspector said (PR 3 Feb):

> "As Captain Collins was a trained navigator and experienced DC10 pilot familiar with the AINS I submit he cannot possibly have been certain of his position. To be certain he would have checked his destination on a map then considered his heading towards it or plotted his position from the latitude and longitude display. Despite the inroads computers have made into navigation procedures and the accuracy of the AINS the crew must know where each waypoint is from an approved checking reference such as an RNC chart or topographical map."

Collins' widow and daughter say they told the Inspector the Captain spent the evening before the flight working on his maps. The Inspector denies this. In effect he says that he told them on or about 14 December that the co-ordinates had been altered without Captain Collins' knowledge and that the Captain's name would be cleared of pilot error if it could be shown the captain had relied on this information in his flight preparation. The widow and daughters deny all this.

The Inspector's version has several implications. It implies that when he revealed the information on 14 December neither the widow nor daughters thought at the time to invent the story, and still did not think to tell him by 31 May when he filed his report. It implies that they only invented their evidence in time to present it to the Commision. But his version also implies that the widow and daughters decided to remain silent about the change in co-ordinates from 14 December when they knew about a major blunder committed by the airline which, regardless of evidence about map plotting, would have altered the whole perspective of the case of pilot error being currently propounded against the Captain.

There are two possibilities why the widow and daughters would remain silent in such circumstances.

The first, and obvious, possibility is that the Inspector made a mistake, and did not tell them what he now says that he told them. Their lack of knowledge would explain their silence.

The captain's widow gave evidence in person at the Commission but no counsel cross-examined her concerning Captain Collins working on his maps. The captain's daughters gave evidence by affidavit, but the Commissioner told me that he asked all counsel specifically including Air New Zealand and CAD whether any counsel wished them to be called for cross-examination, but that no counsel desired to cross-examine them. He also had the Inspector notified and invited to cross-examine, through his counsel Mr Anderson, the Collins family. Why did the Inspector not accept this invitation?

The captain's widow gave me the names of Air New Zealand officers, including members of the airline executive, whom she told shortly after the accident that her husband had spent the evening before the accident working on his Antarctic maps. One may assume this was the reason why the evidence of herself and her children was not challenged in cross-examination.

I contacted only one of those persons, Captain Ross Gordon, and he confirmed everything that Mrs Collins had told me; including her total ignorance about the change in co-ordinates until he explained this to her from Mrs Cassin's copy of Mr Chippindale's interim report, and her mentioning shortly after the accident that her husband had been working on his maps.

The second possibility is that the Inspector is correct and he did explain the significance of the change in co-ordinates to them. However, this possibility entails a number of improbable consequences.

If the Inspector is correct, then on or after 14 December they must have decided to invent their story, but to say nothing of it until after the Inspector had brought out his report, which he had told them on 14 December would be interpreted as pilot error.

Yet if this is so, how does the Inspector explain why the Collins family told no-one that the co-ordinates had been altered without the Captain being told?

The Inspector's version next requires that the widow and daughters should decide to commit perjury before the Royal Commission, while believing that their evidence would be subject to evidence in rebuttal by the Inspector and to cross examination by counsel for the Inspector, counsel for the airline, and counsel assisting the Commission.

The Inspector's case also requires that the Inspector remained silent regarding this point during the Commission hearings for the reasons that he stated which are set out in note 1 above. However, note 1 explains that the Inspector's reasons are unsupported by the facts. The note is based on a statement supplied to me by Counsel assisting the Commission.

The evidence of the Collins family affects the credence which one can place upon a major proposition of the Chippindale report. The Commissioner had previously invited the Inspector to testify at a later date if his recollection differed from evidence given later by pilots. (T.139). Clearly this would apply in the case of any witness.

I find the implications of the Inspector's statements to be improbable. For these reasons I believe that the Inspector's unsworn statement in his subsequent press release is mistaken. I believe that the sworn evidence given by the widow and daughters and unchallenged until the Inspector's press release, is true.

I believe that the Captain did plot his track using the incorrect information supplied to him at briefing. Captain Vette expresses the opinion that Captain Collins' maintenance of nav track is only explicable if that meticulous captain had pre-plotted his track, monitored his HSI, and kept monitoring his position on the large charts his daughters saw him working on at home, so large that they required spreading on the kitchen floor. The need to keep refolding this chart or charts as the flight progressed is the only credible explanation of the recorded sound of paper being repeatedly rustled. Even his severest critics have not, to my knowledge, suggested Collins was reading a newspaper.

Just before the printing deadline of this book I learnt from the Commissioner of the existence of an extra long chart. This must have been the one which Captain Collins had to spread on the floor. It is the U.S. chart, which is about five feet long, of the military routes from Australia and New Zealand to Antarctica. This would explain why the Captain laid it on the floor at home, why it needed to be repeatedly refolded, why it risked blocking the window, why the Inspector referred to it (see note 2) being available in Los Angeles but not readily available over the counter, and why the Collins daughters could not identify any of the many maps shown to them as being the one which their father was planning his flight with. A copy of this chart had not been produced to them to identify.

I obtained a copy, showed it to the Collins family, and they confirmed it was that chart which Captain Collins was using to plan his Antarctic flight with on the evening before his flight.

The incident involving the Inspector's investigation of Collins working on his charts represents another consequence of the Inspector failing to record his evidence in a proper manner with signed statements.

Radar would have shown Erebus

The fourth proposition in the Chippindale Report is that "The aircraft's radar would have depicted the mountainous terrain ahead". On publication of the report this proposition must have seemed to many, as it certainly did to me, to be convincing proof of pilot error. Note 5 on page 340 shows that opinion to be wrong.

The proposition implies that the pilots either did not watch their radar as they ought to have done, or, less probably, that they were monitoring the radar but that they negligently elected to fly toward the mountainous terrain depicted on it.

During the hearing the Inspector gave evidence that the airline's briefing staff, as a matter of Air New Zealand policy, disapproved of the use of radar as a navigation aid (T.75 and T.186). At that juncture the Inspector's proposition appeared to constitute a criticism of the airline, so counsel for the airline attacked it in cross-examination (T.306-T.309).

Counsel suggested to the Inspector that his proposition contained implicit within it a criticism of the flight crew, since they could have used the radar but obviously did not use it. The Inspector said he did not intend his proposition as a criticism since he had no information that the crew did not use the radar (T.308).

As recorded in note 4 the Inspector subsequently repeated that his expert's view was that the crew would see a shadow on the radar that could be nothing other than high ground ahead.

It seems to follow that the Inspector's proposition in paragraph 3.36 means that the crew saw Erebus on the radar, but elected to fly toward it. The evidence in note 4 is to the contrary and indicates the radar would not have shown Erebus from 1500 feet, or indeed from any other height. Moreover, Captain Vette tells me it would not have been proper airmanship for the crew to have devoted the close and continual attention to the radar to monitor it in mapping mode while flying VMC. If the crew had been proved to have done so, then their conduct would have been subjected to severe criticism for pilot error.

Failure to detect existence of whiteout

Air N.Z.

If Collins had maintained 16,000 feet, whiteout beneath him would not have been a danger. (Gemmell T. 1796). To some extent Captain Collins seems to have been aware of the effects of whiteout (Gemmell, B.616).

Counsel

Page 54 of Mr Chippindale's Report [says that] the company has claimed the whole philosophy behind the Air New Zealand Antarctic flights was for crews to avoid a whiteout situation by remaining strictly VMC throughout the sightseeing part of the flight.

I suggest to you that philosophy is based upon a fallacy. Because, of course, we know a whiteout situation is insidious and the viewer may be in a situation where he believes he has visibility. He may believe that he has VMC conditions when, indeed, that's not the case.

Do you believe that philosophy is ill conceived?

Gemmell: No (T.1791)

Chippindale

"2.16. . . the captain headed the aircraft toward the cloud covered island and no expression of doubt was made by the first officer."

"2.25. . . the first officer. . . should not have overcome his natural caution in relation to cloud covered high ground."

"2.20. . . Observed conditions probably lead [sic] to the particular snow slope and the cloud base appearing to the pilot as an area of limited visibility and this whiteout situation may well have been the deciding factor which made him announce his intention to climb out of the area."

3.37 The probable cause of this accident was the decision of the captain to continue the flight at low level toward an area of poor surface and horizon definition when the crew was not certain of their position. . .

Comment

The impression given by these extracts is that the aircrew knew that high ground covered by cloud lay ahead, yet they flew towards it. They suggest that the snow slope ahead probably appeared to the pilot as a whiteout situation of limited visibility consisting of poor surface and horizon definition, but that he decided to continue the flight towards the area.

The Inspector's explanation of whiteout in paragraphs 1.17.46 - 1.17.58 consists of extracts from a paper by Robert B. Boswell (exhibit 44) but the Inspector did not quote the following passage:

> ". . .the entire visual continuum is erased by whiteout and in such an unfamiliar situation all vestiges of depth perception may be lost without one being even aware of impairment. Naturally, this often leads to confusion and even disorientation."

I have read, as a layman, the evidence on whiteout given to the Royal Commission and this satisfies me that Captain Vette is correct and that TE901 encountered sector whiteout, the existence of which could not be detected. A crew inexperienced in polar flight and misled as to the nature of the phenomenon of whiteout could not even suspect the existence of lowered visibility.

It was said by Air New Zealand that they did not know that form of whiteout which is described in Captain Vette's text and were aware only of the other form of whiteout which is caused by snow showers or blown snow resulting in lowered visibility. This statement appears mistaken. Captain E.T. Kippenberger, Controller of Flight Operations; Captain J. Spence, DC8 Airline Inspector; Captain E. Tredrea, Air New Zealand Fleet Captain DC8 aircraft; Captain P. Grundy, Air N.Z. Flight Instructor DC8 aircraft; and Mr G. Willetts, Air N.Z. Assistant Engineering Manager visited McMurdo from 16 to 23 November 1969 to study the feasibility of the proposed Air New Zealand DC8 operation and to propose operational requirements for it. Captain Kippenberger recorded their visit in minute 98/4/76 dated 25 November 1969 produced as exhibit 97, which clearly distinguished the two forms of whiteout as follows:

"Landing presented two hazards:

(a) White out conditions caused by snow held in the lower levels (10 ft up to 200 ft) by wind turbulence. In such conditions visibility could be reduced to zero.
(b) White out conditions caused by difficulty of depth perception and evident particularly in overcast no shadow conditions with ceiling 3000 ft or below. Markers on runway edges alleviated but did not overcome this problem...

McMurdo terminal weather is generally good between late December and early February with about a 50% possibility of a ceiling 3000 feet or below with nil shadows. White outs due to snow in the lower air layer generally require winds in excess of 25K to lift the snow crystals and 15-20K to sustain the conditions."

Captain Kippenberger agreed (T.536) he had a knowledge of the problem of surface and horizon definition in Antarctica and he believed Captains Spence and Grundy also had that knowledge. He was sure (T.537) Air New Zealand would have given the pilots the training in those problems which would have been necessary if the airline had intended to operate at the low levels at which the aircraft which crashed operated at.

The overall supervisor in CAD of Air New Zealand Antarctic flights was Captain Spence (B.66). His duty was to inspect, evaluate and report on the airline's operational performance, particularly concerning safety and training (B.143). He received (B.146) the airline's written Antarctic briefing, exhibit 134 and he was shown (B.148) the audio-visual briefing on about 12 October 1977, and did not remember having any adverse comments on it (B.149). He said that the briefing which Collins and Cassin received was basically adequate (T.739). He listed the only inadequacies in the briefing as: inaccuracies on spot heights, and there should have been added a warning on radio or altimeter inaccuracies, a reference to visitors on the flight deck, an expansion of survival in the Antarctic, an expansion on the ice runway. The information on whiteout should have been expanded, although white-out was certainly not a factor in the accident (T.740). He did not consider the briefing would have been improved if the McMurdo Sound computer flight plan had not been displayed as that had no bearing on the accident, and there was not a shred of evidence Captain Collins was influenced by it (T.740-741).

Captain Grundy was the airline's Flight Operations Manager (DC10/DC8). He had been Flight Manager Training. He said that the discussions of whiteout in Captain Tredrea's two reports were far more relevant in 1969 than in 1977 because in 1977 landings were no longer proposed (B.24). He and Captain Gemmell discussed their briefing at Deep Freeze and that was the basis of the airline briefing (T.432 D.33). It did not occur to him that knowledge of whiteout and altered perception could be of help to a captain at 6000 feet to decide whether he was in VMC. He was sure the majority of pilots were familiar with whiteout as an aviation hazard in snow conditions (T.433 F).

In his text Captain Vette wrote on page 212 that:

"...had the Air New Zealand operations specification for the Antarctic contained the simple warning that visual separation from white surface should not be attempted, particularly below an overcast sky at any altitude, the accident would not have occurred...

"If [CAD] had adequately researched the route with the ICAO they would quickly have discovered the dangers of attempting any form of visual sightseeing below an overcast sky in Antarctic conditions. They would certainly then not have condoned a descent to 6000 feet to continue visually in an area of the most prolific visual illusory conditions."

Probable Cause

The final conclusion of the Chippindale Report reads:

3.37 Probable cause: The probable cause of this accident was the decision of the captain to continue the flight at low level toward an area of poor surface and horizon definition when the crew was not certain of their position and the subsequent inability to detect the rising terrain which intercepted the aircraft's flight path.

What does the Report mean by probable cause?

The Report does not reveal that it does not really mean what it says. Apart from reading the Royal Commission proceedings, to discover what it means one must read a newspaper:

"The Chief Inspector of air accidents, Mr R. Chippindale, said yesterday *he had not been able to find out definitely what caused the plane to crash* on Mt Erebus last November with the loss of 257 lives."

"It has been hard to establish a definite cause. . ."

"Mr Chippindale said he had had difficulty in finding 'the ultimate cause' for the crash."

" 'Therefore, I have said in the Report what I think is the probable cause — *the last thing that made the accident inevitable'* he said, 'though there were other factors or causes, leading up to the accident' " (emphasis added)

At the end of the Royal Commission proceedings airline counsel said (interpolation E to S.23) that the airline did not espouse a proposition that the accident could be [at]tributed to a sole cause, let alone a sole cause of pilot error; evidently thereby abandoning the Inspector's probable cause theory.

Questions

The Inspector in his press release of 18 January wrote:

"The conclusions in [the Chippindale] report have received favourable comment, since the [Mahon] Report was released, by past and present senior members of the air staff of the Royal New Zealand Air Force, the Civil Aviation Division, senior captains of Air New Zealand and chief inspectors of air accidents overseas."

If the Inspector's statement is not to remain an unsubstantiated assertion, carrying no evidential weight, then the inspector must name these persons and produce statements from them in which they spell out their reasons, field by field, for preferring the Chippindale analysis to that contained jointly in *Impact Erebus* and in the Mahon Report.

I should like to ask these un-named persons to publicly answer the following questions and set out their evidence in support.

Do they assert that the change in the computer flight path from McMurdo Sound to Lewis Bay did not mislead the crew?

Do they deny Collins plotted the co-ordinates which were given to him at his RCU briefing onto a chart or charts which he subsequently used with his HSI to monitor his position, as a backup to his flying VMC?

Do they deny that if the case put forward by the airline is correct, then the airline made 54 mistakes concerning flight path alone? (See note 3).

Do they assert the crash was caused by the pilots descending beneath 16,000 feet contrary to the airline's instructions?

Do they deny that Captain Vette's analysis in his chapter 9 makes nonsense of the airline's claim, which it made after the accident, that the crew's briefing prohibited descent beneath 16,000 feet until over McMurdo Station?

If so, how then for instance do they explain away (see Chapter 9) the airline recommending 1000 terrain clearance over glaciers while never passing over McMurdo Station when on the alternative route to the South Magnetic Pole?

Do they deny that all previous Antarctic flights from 18 October 1977 descended considerably beneath 6000 feet, generally to between 1500 feet and 3000 feet?

Do they deny that if the case propounded jointly by the airline and CAD concerning height is correct, then those two bodies between them made a total of 177 mistakes concerning height? (See note 4.)

Do they assert that the crew was not certain as to their position?

Do they deny Captain Vette's analysis which demonstrates that the crew were certain in their own minds that they were in McMurdo Sound, even though they were mistaken in their belief?

Do they deny, given the unrevealed change in flight path, given the checking procedure then laid down by the airline, given the briefing, given the deceptive similarity of the entrance to Lewis Bay to the entrance to McMurdo Sound, and given the existence of sector whiteout, that the crew had proper grounds for being certain of their position?

Do they assert that the aircraft's radar would have depicted the mountainous terrain ahead?

Do they assert that the captain headed the aircraft toward cloud covered high ground appearing to the pilot as an area of limited visibility or whiteout?

Do they deny that the airline told the crew that a whiteout situation was caused by snow showers or blown snow?

Do they deny that pilots briefed on whiteout as above, with no experience of polar flying, with visibility in excess of 40 miles, would be unable to detect the existence of sector whiteout?

Release of Report

The procedure which the Government laid down for the inquiry into the Erebus disaster differs from that followed in the U.K. The Chief Inspector of accidents at the Department of Trade in the U.K., W.H. Tench, explained that if a public inquiry is to be held, the accidents investigation branch does not prepare a final overall report for public release. Instead the branch merely investigates the facts and prepares reports into the facts for the use of the Attorney General in the public inquiry.

Though the U.K. procedure appears desirable, it was not followed in the Erebus inquiry, in which the Office of Air Accidents Investigation completed a full report in which the Chief Inspector of Air Accidents stated what he believed was the probable cause of the accident. In effect the report blamed the pilots and exonerated the administration of the airline. The shareholder in the airline was reported to be the Minister of Finance, Mr Muldoon. The Chief Inspector's report was dated 31 May 1980, and on 12 June Mr McLachlan, the Minister of Transport, approved it for release as a public document.

Had the Chippindale Report exonerated the pilots, and attributed all blame to the airline's administration, some cynics may wonder whether Mr McLachlan would have decided so quickly to approve the report for public release. More mature consideration may have led to the wisdom being recognised in the U.K. practice of leaving the public inquiry to produce the report for the public.

In commenting on the release of the Chippindale Report as a public document, counsel for the captain's family said:

> "The release of the Accident Report containing such damning comments about Captain Collins in the face of requests that it be withheld until the sitting of the Commission for whatever reason was not justifiable. Its effect was to

promote in the public's mind the cause of the crash as pilot error. The damage that this did is immeasurable. The heart wrenching suffering it caused the Collins family cannot be measured or erased. They were faced with having to weather this ill-founded criticism and massive publicity. I would sincerely hope that those responsible for that adverse publicity are responsible enough to give equal, if not greater publicity to the positive findings of this Commission relative to the conduct of Captain Collins.''

The Inspector's opinions remain unmoved by the facts which I have explained in these notes.

> ''Mr Chippindale said from Wellington today there had been nothing in the subsequent Royal Commission report by Mr Justice Mahon or 18 months of public comment and controversy to change his mind.''
>
> *Auckland Star,* 26 January 1982.

1. Was Inspector represented before Royal Commission?

The statement in Captain Vette's text is denied by the Inspector who wrote in his press release of 3 February 1982 that:

> ''Prior to the commencement of the Royal Commission I advised the counsel assisting that Commission that I considered both my office and myself could well be 'on trial' and although willing to assist them in every way, felt I had a right to legal representation during the Inquiry.''
>
> ''It was arranged that a lawyer (Mr N.C. Anderson) would assist me but he would not be recognised by the Commission or participate in its hearings ... ''
>
> ''The lack of representation placed me at a distinct disadvantage in not having any avenue for questioning witnesses who had led me to believe that a different set of circumstances existed after the accident to that which they described in evidence.''

The witnesses to whom the Inspector refers are evidently the widow and children of Captain Collins. (See following note 2.) Other witnesses, who may be referred to are pilots interviewed by the Inspector's Office. (See preceding note on Chippindale Report.)

I am told by counsel assisting the Commission, Mr Baragwanath, that prior to the commencement of the hearing it was arranged that senior and experienced counsel would be engaged to advise Mr Chippindale and his office. Mr Anderson was accordingly instructed.

At the commencement of the hearing there were no particular grounds to expect an attack upon Mr Chippindale but Mr Anderson was present throughout his examination by counsel assisting and lengthy cross-examination by various parties so as to be able to furnish advice or participate if required. At the conclusion of his eight days of testimony the Royal Commissioner, while reserving his opinion as to the issues, congratulated Mr Chippindale as to the manner in which he had given his evidence.

Cross-examination had not significantly attacked the investigation, although Mr Chippindale's conclusions were resisted by counsel representing the flight crew. Accordingly it appeared that Mr Anderson would be not be required to participate in the public hearing, unless some new issues emerged. Since Mr Chippindale proposed to be present, it appeared appropriate that he should be at liberty to offer to counsel assisting whatever suggestions he might consider helpful, with the right to consult Mr Anderson further, should need arise. A complete copy of the transcript and other material furnished to the parties was supplied to Mr Chippindale and so available to inform him of developments during periods when he was not present.

On two particular occasions written submissions were filed on behalf of Mr Chippindale, one dealing with the term of reference relating specifically to accident investigation, the other with the final submissions of counsel for the parties.

It is thus incorrect to state that Mr Anderson, who was listed in the Mahon Report as counsel for Mr Chippindale, was not or "would not be recognised by the Commission". While one term of reference concerned accident investigation generally, the Inquiry was not into the Chippindale Report but into the accident. There was thus no warrant for Mr Chippindale to be cited as a party.

Mr Chippindale made no request that he be permitted to question witnesses. For him to have done so would, in any event, have been inappropriate, since the subject matter of any question would be within the responsibility either of counsel assisting or, if involving attack on Mr Chippindale, within the responsibility of Mr Anderson.

2. Did Jim Collins plot track before flight?

The Inspector suggests that the incident described in Captain Vette's text did not occur, and that it was a fiction subsequently invented by Mrs Collins and her daughters.

The Inspector (P.R. 3 Feb.) refers to visiting the widow and family of the Captain on or about 14 December 1979 to endeavour to support the plotting of a track by the Captain, and he says:

> "I carefully described that we wanted to locate, if at all possible, the atlas that Captain Collins had used, and asked if she or her family had seen Captain Collins drawing on any maps. I find it remarkable that, knowing the significance of this information, Mrs Collins did not subsequently forward an indication of their subsequent recall of events to me or at least draw this matter to my attention when their solicitor was advised of my preliminary findings ... "
>
> "Again during my investigation I was unable to locate any map supply location with maps readily available which had provided maps for Captain Collins. Although, available in Los Angeles by prior order they were not readily available over the counter and the Air New Zealand briefing officer had to resort to a photostat copy of a map for his RCU (route clearance unit) briefings. Therefore from whence came his large map?"
>
> "Further, although the use of the atlas by Captain Collins was drawn to my attention early in my investigation I think it is sufficiently questionable that he would have drawn a track on a page of a limited edition atlas which was a collectors' item retailing for NZ$250.00 to throw a doubt on the assertion (in the Mahon Report): 'What he did therefore was ... to plot on this map and also upon his atlas the path which the aircraft would take'."

Because of their ambiguity, these passages may be open to different readings, but I think the average reader would take it that the Inspector is saying:

1. He explained to Mrs Collins the flight path had been altered overnight from McMurdo Sound to cross Mt Erebus without her husband being told.

   If her husband had been misled by being shown the McMurdo flight path, this would explain why he crashed on Mt Erebus. Thus she had a motive for inventing a story that her husband worked on maps before the flight. (What else can "knowing the significance of this information" mean?)
2. He asked Mrs Collins if she or her family had seen Captain Collins drawing on any maps, and she and her daughters said they had not.
3. Neither she nor her daughters revealed (either when he visited them or after he informed them of his preliminary findings) that Captain Collins had spent the evening before the flight working on his atlas and maps in preparation for the

flight.

4. He ascertained Captain Collins did not buy any maps which his family subsequently alleged he worked on.
5. Although Captain Collins did possess an atlas it was so expensive that it is unlikely he plotted his flight path on it.

All this is totally at variance with what I am told by Mrs Collins, Kathryn Collins and their counsel. The evidence of Mrs Collins and her two daughters was not challenged in final submissions, nor was it subjected to cross-examination. If the Inspector had indeed believed that their evidence was untrue, then it was open to him to have had the matter raised in public at the enquiry. (See note 1 above.)

I now examine the points numbered above:

1. Mrs Collins says the Inspector did not explain the significance of his inquiries. Nor did she learn about the change of co-ordinates from any other source until much later. She did not even understand at first that the Inspector wanted only her husband's atlas, and while the Inspector was present she sent her daughters to neighbours to find an equivalent atlas.
2. The Collins family cannot now recall whether they were asked by the Inspector if they saw Captain Collins actually drawing on his atlas or maps. But this is a different issue from working on them, or having already plotted a track upon them.
3. Mrs Collins and Kathryn Collins say it was made clear in the Inspector's presence that the last time they saw the atlas was when Captain Collins was working on it the night before the crash.

   Before the Inspector's allegations were known, that is before his press release was published, the counsel for Mrs Collins told me he recalled Kathryn saying the Inspector had been told when he called that her father had spent the evening before the crash working on the atlas and maps.

   The Inspector refers to not having this evidence brought to his attention after he informed them of his preliminary findings.

   It is unclear what he refers to as "preliminary findings". In his letter of notice to the estate of Captain Collins setting out the possible blameworthiness of the pilots there is nothing to indicate that he is going to allege he was never told Captain Collins spent the evening working on his atlas and maps. In his final published Report there is the clear statement that he found no evidence to suggest the crew were misled, so apparently he must refer to this Report. However, this was not seen by the Collins family until it was published. In the earlier versions shown to the family by the Inspector the sentence did not appear, so again they had no cause to suspect in advance that the Inspector was subsequently going to make such an allegation.
4. If this interpretation is the Inspector's assertion, it is contrary to evidence uncontradicted at the Inquiry. It is also unsupported by evidence in its own right. Captain Collins knew, according to Mrs Collins, about the end of September he was scheduled to fly to Antarctica and even if he had not obtained maps before then, he had ample opportunities flying overseas subsequently to obtain them. Mrs Collins said that: "He frequently purchased maps whilst overseas and also had quite a number of old ones from his Air Force days."
5. Mrs Collins did say in her evidence (B.559) the atlas was a copy of the limited edition, but this resulted from confusion between three editions; the paper back, the hard cover, and the hard cover limited edition. It fact, it was the $35 to $36 hard cover edition, and this was confirmed to me by Mrs Collins and by Mrs Collins's mother who was one of the two donors of the atlas, and by the fact that the replacement atlas given to them by Lands and Survey was of that cheaper

edition.

3. The 54 mistakes concerning flight path

Captain Vette's text summarises in a few paragraphs a highly complex chain of events, which are set out in detail as follows.

DC10 aircraft are navigated by the area inertial navigation system, the most advanced system in present use. The system operates by inserting into the computer on the aircraft a series of waypoints located by co-ordinates of latitude and longitude. The system may be locked into the steering controls of the aircraft so that it can be flown automatically from one waypoint to another. The aircraft is then said to be "on nav. track". The course from one waypoint to another constitutes the aircraft's computer flight path. The accuracy of the system is illustrated by the fatal flight when, after a five-hour, three thousand mile flight from Auckland to McMurdo, the aircraft struck Mt Erebus only 1.2 miles east of the nav. track, i.e., there was a "cross-track" error of 1.2 miles.

The Commissioner formed the conclusion that the aircraft was about 27 miles east of the position the crew believed it to be since the destination waypoint before the return to New Zealand had been altered by the airline, without the crew being told, from one location to another. To understand the different destination waypoints one must know something of the local Antarctic geography shown on the map (fig. 47 on page 330). The south west corner of Ross Island comprises a peninsula. On the east side of the tip of the peninsula is Scott Base which is the N.Z. Antarctic base. Two miles from it, on the west side of the tip, is an American scientific base, McMurdo Station (marked "McMurdo STN" on the map). A few miles south east of the two bases is a landing field on the ice, Williams Field. Close to the field were two aircraft navigational aids. One was a non-directional beacon (NDB) (marked on the map "M.F. NDB") and some distance away from the NDB was the McMurdo tactical air navigation system (TACAN). Unknown to the pilots, on the fatal flight the aircraft's computer had been re-programmed to the destination of the TACAN.

The military route used by the U.S. Navy, RAAF, and RNZAF planes lay roughly down the centre of McMurdo Sound to the Byrd reporting point (marked "BYRD R/P"). From that point, military planes normally adopt an instrument landing approach using the ground-based navigational aids, and land on the ice runway of Williams Field. When Air New Zealand commenced its flights to Antarctica in February 1977, its computer flight path did not follow the military route down McMurdo Sound. Instead its destination waypoint was Williams Field. Such a flight path lay almost directly over Mt Erebus, which was an active volcano, 12,450 ft high. From the flight of 18 October 1977, the McMurdo destination waypoint was altered to the NDB. This was about two to three miles from Williams Field. The flight path still ran over Mt Erebus. (B.212)

It was the Air New Zealand case that everyone in the airline continued to believe throughout the whole period of Antarctic flights that the computer flight path ran across Mt Erebus and not down McMurdo Sound. If the witnesses for the airline had admitted they knew that for 14 months prior to the fatal flight the flight path ran down McMurdo Sound, and that a few hours before the flight, they altered this flight path to cross Mt Erebus and they did not tell the aircrew, they may have considered outsiders would believe they were responsible for the accident and not the pilots. To assert they were ignorant of the McMurdo Sound flight path entailed the airline witnesses maintaining they made a series of mistakes.

In 1978 a series of events commenced which the airline subsequently asserted were errors. This note explains the 54 mistakes made concerning course.

Fig. 47 McMurdo Station area.

The figure shows the features of the McMurdo Station area necessary to understand the programming of the waypoint for the computer flight path. The diagram is three dimensional and the descending figure of eight orbits of flight TE901 are seen top right as the aircraft headed toward the high ground of Mt Erebus which lay concealed in sector whiteout. The navigation track of the flight plan programmed into the aircraft lies approximately along the straight line marked A B C on a heading toward the TACAN.

The navigation track of the computer flight plan shown to the crew at briefing which Collins used to plan his flight lay nearly parallel to the track actually programmed into the aircraft but above Beaufort Island and just below the Byrd reporting point, down the centre of McMurdo Sound to a position west of West Dailey Island.

The Air New Zealand Dailey Islands route lay close to the military route, the latter being colloquially called the great circle route or chute.

Source of map is Appendix 3 to Annex D of Chippindale Report.

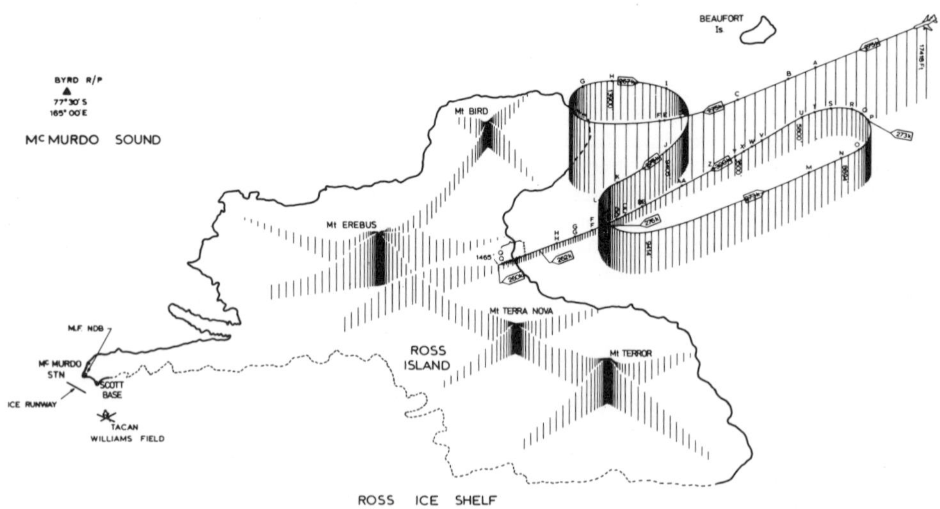

In 1978, when the airline changed its method of producing flight plans, a mistake was made (so the airline told the Commissioner). The chief navigator should have intended to enter the NDB waypoint into the airline's computer, but he made the mistake of believing that the outdated Williams Field waypoint should be inserted. (Mistake 1)

Instead of entering into the computer the Williams Field waypoint (which in itself would have been an error), the chief navigator entered different co-ordinates which gave a destination waypoint some 27 miles to the west. The chief navigator said in evidence he made the mistake in July or August 1978 (B,237) but in cross-examination he said (T.947) he made the mistake in September. (Mistake 2)

The chief navigator failed to detect the error of insertion when he checked the display on the computer screen (VDU) (B.238). By pressing a key the chief navigator could have obtained a printout of what he had entered (T.950), but he did not obtain a printout and merely looked at the screen (T.950,953). This "error" caused the computer flight path to lie roughly down the centre of McMurdo Sound to a destination waypoint close to West Dailey Island (marked on the map on page 134 Fig. 42.) (Mistake 3)

It was this mistake, the shift of the destination waypoint 27 miles west, which is referred to in the Report of the chief-inspector of air accidents (the Chippindale Report) as the "error". This terminology has led to confusion among some subsequent commentators who have taken the inspector's reference to "error" to mean the alteration 14 months later of the destination waypoint 27 miles east at 1.40 in the morning of the fatal flight.

However, the change of flight path which according to the Commissioner caused the crash was referred to by Air New Zealand not as an "error", but as the "correction" of a previous error which had been made 14 months before the crash. The inspector adopted the terminology of the airline.

This verbal sleight of hand enabled the airline to maintain in public that there was no "error" in the computer flight path at the time of the crash. How the airline employed this terminology was illustrated on 18 February 1980 when it was reported in an Auckland newspaper that wrong navigational information was fed into the aircraft's computer before the fatal flight to Antarctica. It was reported, the next day, that on 18 February the then chief-executive of the airline said that the computer had been fed the correct information. The result of the chief-inspector of air accidents, Mr R. Chippindale, in his report, adopting the same terminology, seems to have been that some eminent readers of his report have in consequence misunderstood key passages in it.

In 1978, the navigation section of the airline had, according to their evidence, accidentally altered by 27 miles the computer flight paths of all subsequent Air New Zealand flights to Antarctica. Thus, for the next 14 months, their destination waypoint was near the Dailey Islands in McMurdo Sound about 27 miles west of Mt Erebus.

The first flight to Antarctica after the "error", (which moved the destination waypoint 27 miles west) was on 7 November 1978. The briefing was held on 2 November, but the briefing officer asserts he believed the computer flight path still ran across Mt Erebus and brifed the crew accordingly. Mistake by the briefing officer in not ascertaining the flight path was 27 miles different from what he alleged he explained at the briefing. (Mistake 4)

The briefing officer handed out a copy of a track and distance diagram (B.517) a copy of which was produced to the Royal Commission as Exhibit 164 (fig. 11 on page 48). It shows a track running down roughly the centre of McMurdo Sound. Error by briefing officer in handing out this diagram if he believed the track ran across Mt Erebus. It is a remarkable coincidence that the computer flight path should be altered by error to run down McMurdo Sound at the same time a diagram showing a track down McMurdo Sound was distributed to the first flight of the season. (Mistake 5)

The briefing officer said (B.355): "I do not know why [Exhibit 164] was included in the briefing material". Mistake by officer in handing out this diagram without finding out why he was handing it out, when he knew it showed the wrong computer flight path. (Mistake 6)

Exhibit 164 was given to the briefing officer by the flight manager, line operations DC10/DC8 on 1 November 1978 (B.355), who accepts that it formed part of the 1978 briefing material (B.393). Mistake by flight manager in not noticing that the flight path ran down McMurdo Sound when he knew the "correct" flight path ran across Mt Erebus and giving that exhibit to the briefing officer. (Mistake 7)

While the flight path ran across Mt Erebus, pilots were given a diagram in the form of Annex J to the Chippindale Report which, when related to a topographical map, showed that the path ran across the mountain (fig. 10 on page 44). When the navigation section made the error of shifting the path 27 miles west down McMurdo Sound, the flight manager decided the copy he had of Annex J was so poor he called for another one from the navigation section, but instead of getting another he received a copy of Exhibit 164 showing a McMurdo Sound path (T.1383-1384). Mistake by flight manager in not noticing the two were different (T.1293-1296 and 1384). Mistake by navigation section in supplying a diagram showing a McMurdo Sound track, when asked for a diagram showing a Mt Erebus track. (Mistakes 8-9).

Counsel for the airline (memo between counsel, Mr McGrane and Pilots' Association counsel, Mr Davison of 16/9/80) said he had shown Exhibit 164 to the chief navigator in respect of 1978 flight briefings; but when the latter subsequently gave evidence, he gave what was tantamount to a denial of this (T.977, 979, 982). Mistake by chief navigator in forgetting discussion with counsel and other members of navigation section (or mistake by counsel). (Mistake 10).

The handwriting on Exhibit 164 was that of the navigation services superintendent (B.311), but neither he nor any other member of the navigation section had any record of, or could remember, how this McMurdo Sound diagram was given by them by mistake to the flight manager instead of the Mt Erebus diagram which he asked for. (Say three mistakes by navigation section) (Mistakes 11-13)

On the other hand the flight manager said his request would have been made orally, he could remember asking for it, but not which of the three or four gentlemen in the navigation section he asked, but he then said he could not recollect what he did (T.1294). The significance of Exhibit 164 is that is shows a flight path running down McMurdo Sound, about 27 miles away from the Mt Erebus track. (Mistake 14).

The briefing officer handed out a computer flight plan which showed the aircraft would be programmed on nav. track to fly down McMurdo Sound (B.517). Error by briefing officer in not detecting this conflicted with his belief that the track ran across Mt Erebus. (Mistake 15)

The briefing officer had received this McMurdo Sound flight path from the flight manager line operations DC10/DC8 on 1 November 1978 (B.353) as part of the 1978 briefing materials (Exhibit 196). The flight manager added handwritten comments to the flight plans (B.392). One mistake each by flight manager and by briefing officer in not noticing McMurdo waypoint had been altered by two degrees, equivalent to about 27 miles. (Mistakes 16-17)

In turn, the flight plans had been produced by the chief navigator (B.392). Mistake by chief navigator in producing printouts of computer flight plans, which contained what he said was a mistake in the McMurdo waypoint of two degrees made earlier by himself and handing them to the flight manager still without detecting the mistake. Mistake by flight manager in not checking the flight plans at that point. (Mistakes 18-19)

On the briefing for the fatal flight the McMurdo waypoint was either underlined or highlighted in some way (B.530). Mistake by briefing officer in underlining or highlighting a waypoint 27 miles in error; alternatively permitting the waypoint to be so marked, when in principle he handed out "sample" flight plans which had no relevance to the flights on which crews being briefed were to fly (T.1248). (Mistake 20)

The briefing officer, although intending to make it clear the track ran across Mt Erebus, left the pilot with the "general understanding that the route south of Cape Hallett would proceed to a position west of McMurdo Base" (B.517). (Mistake 21)

Subsequently, the airline flew to Antarctica on six occasions before the fatal flight, namely on 14, 21 and 28 November 1978 and 7, 14 and 21 November 1979. Not all the same errors can be specifically proved from the evidence in respect of every flight. Nevertheless, in broad terms, by the briefing of 9 November 1979, the briefing officer was still handing out flight plans from previous flights showing the track ran down McMurdo Sound, and still giving to those who attended the briefing the general understanding that the track ran down McMurdo Sound, while all along believing (according to his evidence) that he was making it plain that the track ran across Mt Erebus. Moreover, he was still handing out to aircrews maps of the flights which showed the track was down McMurdo Sound. Granted, these maps had been prepared for use of the passengers, but they were used to distribute to the air crews.

Exhibit 164 was not distributed at RCU briefing to the pilots of the 1979 flights. The replacement for it at briefings in 1979 was the RNC (radio navigation chart) (fig. 9 on page 43), which showed a track down McMurdo Sound, but its waypoints were different from those of Air New Zealand. Also handed out in lieu of Exhibit 164 was an extract from an Antarctic strip chart McMurdo Sound to Christchurch reproduced as Annex I to the Chippindale Report which showed a track to the same Byrd Reporting Point as did Exhibit 164, but the other waypoints were different (fig. 8 on page 42).

The number of errors on the lines stated above made by the briefing officer over this period of 14 months cannot be accurately stated owing to lack of evidence, but using the above as a basis of calculation is probably not less than 15. (Mistakes 22-36)

On 7 November 1979 the briefing officer himself went on an Antarctic flight (T.1238), but he did not discover the flight path ran down McMurdo Sound. Incidentally, this was the first flight to Antarctica the officer had been on and it was diverted to the South Magnetic Pole. (Mistake 37)

Flight plans for Antarctic flights were sent by teleprinter from Air New Zealand to McMurdo Air Traffic Control apparently before each flight. Thus, from 7 November 1978 to 21 November 1979, seven flight plans would have been transmitted, which showed the path ran down McMurdo Sound while the airline believed the flight path was 27 miles east. Seven mistakes in failing to detect the "error" being transmitted. (Mistakes 38-44)

On 9 November 1979 the RCU briefing supervisor briefed the air crew of the fatal flight of 28 November and the captain of the 14 November flight. By inspection of a map and the computer flight plan which were displayed the captain formed the opinion that the flight path ran down roughly the centre of McMurdo Sound to a waypoint near the Dailey Islands which he estimated must be about 10 miles west of the navigational aid near Williams Field called the TACAN. When the captain flew down to McMurdo, he noted that the distance between the computer flight path and the TACAN was about 27 miles, which he found surprisingly different from his previous estimate of 10 miles. As he thought other captains might make a similar mistaken estimate, he telephoned the flight manager, line operations DC10/DC8 on 15 or 16 November 1979 and told him the distance and suggested the information ought to be passed on to pilots of subsequent flights (B.249,430). The flight manager misinterpreted this telephone call. The flight manager first said (Exhibit 16; 10 December 1979) that the captain rang him and said "the McMurdo position was in error and should be 166°58'." In his brief of evidence he said the captain told him "the McMurdo waypoint ... would be better positioned at the TACAN." (B.406). Evidently the flight manager made an intermediate statement containing another mistake which had to be amended, but which was lost and never produced during the proceedings. (Mistake 45)

The flight manager passed on his understanding (B.406 and B.307) to the navigation services superintendent on 15 November. The superintendent referred to the then current NV90 computer printout (which did not hold the flight plans) and found the longitude of the McMurdo waypoint was 166°48.0'E (B.307). He said he failed to detect that the longitude used in the AIRS computer program which did hold the flight plans was 164°48'E. (Mistake 46)

The superintendent contacted the chief navigator who said that he consulted his ALPHA worksheet (but this did not store the actual flight plans) and found that it agreed with the McMurdo position contained on the NV90 printout and so he said he believed the change was approximately 2.1 nautical miles. He said he also failed to check the longitude in the AIRS computer program which stored the flight plans and which showed it was 164°48'E and so he said he failed to discover that the distance between the two waypoints was about 27 miles (B.239). (Mistake 47)

The superintendent explained to the flight services manager that the distance was 2.1 nautical miles (B.308 and B.407). The flight services manager said that he failed to ascertain that the waypoint had been in McMurdo Sound, about 27 miles to the west, for the past 14 months. The manager said he therefore instructed that the McMurdo position be amended in the belief he was instructing it to be shifted by only 2.1 nautical miles (B.407). (Mistake 48)

The chief navigator said (B.241) that he telephoned the flight services controller (flight despatch) on 20 November and instructed him to advise the crew of 21 November that the McMurdo waypoint was to be changed. This evidence appeared to corroborate the statement below of the flight services controller as to the date, but after the controller said his earlier statement was a mistake (T.1035-6), the airline counsel offered to recall the chief navigator (T.1036) and the Commissioner evidently interpreted this offer (M.90) as an offer to change his evidence and agree with the controller as to the date. (Mistake 49)

On 24 December 1979 the flight services controller told the chief pilot he was unable to offer any explanation why the change in the McMurdo position was not passed on to the flight planner before the flight left on 21 November (M.90). But much later he said he had not been told to make the change until 21 November (B.297-8 and M.90). (Mistake 50)

The flight services controller kept a personal log in his office and he entered under Wed.21 11 (meaning Wednesday 21 November) an item referring to the proposed change of the McMurdo waypoint. He later said (B.297-8) this was a mistake and the entries referred to Tuesday 27 November. (Mistake 51)

On 15 or 16 November 1979 a navigation instructor was told to make the change in the McMurdo waypoint, but he entered a figure in the wrong column of the ALPHA form (B.644) with the consequence that when the flight plan was teleprinted to McMurdo ATC (Air Traffic Control) the change in the flight plan co-ordinates was not revealed at ATC. (Mistake 52)

Teleprinter advice of flight plans contained a "flash ops" facility at the top of the message, the purpose of which was to draw the attention of the recipient to anything of note in the message. If Air New Zealand had used this facility on the flight plan of the fatal flight, then the accident would have been avoided. By not using this facility on the flight plan given to the aircrew immediately before departure, the airline was, in effect, telling the aircrew that there was nothing about the flight plan which was of particular importance. The aircrew thus had no reason to suspect that any waypoint had been altered.

If Air New Zealand had intended to reveal to the aircrew of 28 November 1979 that the flight plan differed from what they asserted had been explained on briefing then they should have done so by using the flash ops (T.1331). (Mistakes 53-54)

4. The 177 mistakes concerning height

The Chippindale Report, in conclusion 3.18, states that the Captain initiated a descent in contravention of his briefing by Air New Zealand.

According to paragraph 289 (k) of the Mahon Report, this conclusion is contrary to the evidence of the briefing officer, Captain Wilson, who said that, when briefing air crews for Antarctic flights in 1978 and 1979, he had told them that the practice on Antarctic flights was to descend to whatever level was authorised by McMurdo Air Traffic Control, and he said in his brief of evidence that he did not indicate any criticism of this course (B.387).

The Mahon Report explains that, when the pilot decided to descend to 1,500 ft in visual meteorological conditions (VMC), he was in fact acting in accordance with the authority given to him at his briefing.

The Cockpit Voice Recorder (C.63-83) shows the approval to descend given by McMurdo Air Traffic Control. McMurdo offered a radar let-down to 1,500 feet. The crew later asked for permission for further descent, or orbit descent VMC. McMurdo approved VMC descent and asked to be kept advised of altitude.

At 13,000 ft the crew asked approval to descent to 10,000 ft VMC. McMurdo approved and understood the crew was asking for a radar let-down through the cloud.

Then at 10,000 ft the crew asked for approval to let down and to proceed visually to McMurdo. They were told to maintain VMC and keep McMurdo advised of altitude as they approached, and to report at 10 miles from McMurdo. The Inspector himself did agree in evidence (T.81) that:

"Air Traffic Control did give a clearance to descend to 2,000 ft VMC and proceed visually to McMurdo."

The case presented by the airline and CAD (Civil Aviation Division) was that in descending beneath the level of Mt Erebus (12,450 ft, taken for safety as 14,000 ft, plus a 2,000 ft terrain clearance, equalling 16,000 ft) the pilots had breached their duty and so were responsible for the accident. The Chippindale Report into the crash made by the chief inspector of air accidents, as interpreted by its author, accepts that case.

The executive pilots were part of the airline management. The first two flights were made by executive pilots who denied they descended beneath 16,000 ft in the McMurdo area. Independent evidence from those on the flights suggested the contrary (M.74).

The Mahon Report found (M.75) (disregarding an irrelevant flight to the South Magnetic Pole) that all flights from and including 18 October 1977 descended to less than 6,000 ft in the McMurdo area and that, in the main, flights down McMurdo Sound and across the Ross Ice Shelf to the south of Mt Erebus were at 1,500 ft to 3,000 ft. This went on for two years, so that if the airline or CAD knew this, they would have condoned the practice and so taken responsibility for the height at which the airline's flights were actually conducted. Both denied any knowledge of low altitude flying.

The Mahon Report's opinions were (M.80) that the airline's management and flight operations division knew that flights ranged from 1,500 ft to 3,000 ft and some flew true south down McMurdo Sound; and CAD knew, probably from the outset, but at least from September 1978, that flights in the McMurdo area were well under 6,000 ft. Thus the Mahon Report opinions were that the evidence given to the contrary was untrue. If the Report is wrong, then the persons concerned were responsible for 177 mistakes, misunderstandings and omissions in not knowing what level the airline's pilots were flying at.

It was the airline's case as presented to the Commission that references to height meant heights beneath which pilots were not permitted to fly ("minimum safe altitudes"). On the other hand, with few exceptions, the air crews commonly believed that they were being instructed as to cloud break procedure, so that when they descended to low altitudes they believed they were complying with airline requirements. When an aircraft is flying above continuous cloud cover and the pilots desire to descend to clear air beneath that cloud cover, during their descent through the cloud ("the cloud break") they will be unable to fly visually but will have to descend on instruments. When the cloud break is carried out above terrain in the relevant area of say 3,000 ft, if the aircraft descended to 2,000 ft still in cloud then the danger is evident. Thus a safety clearance is imposed between the cloud base and terrain of say 2,000 ft, so that if an aircraft had descended to 5,000 ft without breaking out of cloud, then further descent is prohibited, and the crew must climb out. For further details, see page 200. The crew of the fatal flight made no cloud break, instead they found an adequate clear air gap in the cloud cover and descended visually with terrain beneath in view. The first Antarctic flights had computer flight paths running across Mt Erebus which was about 12,450 ft. This height was rounded up to 14,000 ft for safety allowing for altimeter inaccuracy under Antarctic conditions, and 2,000 ft clearance added, giving 16,000 ft on a sector approaching Mt Erebus. When a flight is instead over flat ice down McMurdo Sound the logic of such a height limitation did not seem relevant to the pilots.

It is instructive to examine the evidence first heard by the Commissioner during the inquiry but there is space for no more than one extract and this relates to evidence as to knowledge of height given by a witness called by the airline, the ex RCU (Route Clearance Unit, i.e. briefing) supervisor (T.380). No other airline witness admitted reading the newsletter or seeing the television item which this witness refers to.

Q. Did you have any knowledge of Antarctic flights descending below 6,000 feet before the accident occurred on 28 November 1979?

A. I do recall media comment prior to that period, yes.

Q. Did you not think it a matter you should take up with the Company [seeing] that you were involved with [briefing] of pilots on those routes?

A. I believe at that time I had no responsibility in that direction. By that time I was a line Captain DC10.

Q. So you didn't think it within your province to mention it to a person in authority within the Company?

A. I would expect that with the information although it was promulgated ample evidence would have been available to those in such a position to have taken what action they desired.

Q. Particularly what are you referring to when you say there would be ample evidence to those in authority?

A. I can recall some article in the staff news for example and I believe the other means that came to my attention was in a television news item.

Air New Zealand News of 30 November 1978 was produced and identified. The witness said that it was available to everybody in the airline and generally placed in crew rooms, staff cafeterias and the like.

The next day the witness was further examined and he asked if he could read the article again. The witness read:

A. "And as the DC10 cruised at 2,000 ft past the Antarctic's Mt Erebus and over the great ice plateau ... ". That would convey to me that the aircraft passed over Mt Erebus at 2,000 feet which would give me no great cause for alarm given the situation of visual conditions. I think I made reference in the transcript yesterday of the other means by which I heard about so called low flying, and if my memory serves me correctly it was in the form of a vague supposition by the television media.

Q. ... Are you telling us now that you did not know that these flights descended below 6,000 ft?

A. Apart from the manner I described yesterday that would be true.

Q. And the manner described was, to quote you, "Some article of the staff news and a television news item". Those were the two factors, were they not, that signalled to you that flights were descending below the stipulated 6,000 ft?

A. Yes that is correct.

Q. Are you now telling the tribunal that the article ... simply talks about overflying Erebus at 2,000 ft?

A. This article certainly could be read that way.

Q. I'm not asking how it can be read but how you read it at the time?

A. I must have interpreted it that way because it drew no undue concern from myself.

Q. Take the single phrase "as the DC10 cruised at 2,000 ft". What would that signify to your mind?

A. I would have to read it in conjunction with the rest of the sentence.

Q. I'm asking you to read it not in conjunction. What would that single phrase signify to you?

A. The aircraft had 2,000 ft clearance from whatever it was following. [sic]

Q. Now, what about the phrase "past Mt Erebus"? What would that signify to you as an individual phrase?

A. I would assume from my viewpoint ... that it would have reference to a vertical separation.

Q. So, in your view then, the phrase "as the DC10 cruised at 2,000 ft past the Antarctic Mt Erebus" properly translated, means "as the DC10 cruised at 16,000 ft over Mt Erebus"?

A. I would have to assume that *(witness explains why he so interprets phrase)* ...

Q. And you would agree, I take it, that to an uninstructed reader this paragraph would mean that the aircraft flew 2,000 ft above sea level, passed Ross Island, and over or towards the Ross Ice Shelf?

A. With due respect, the topography of the mountain is such that I can't quite envisage how you could fly at 2,000 ft past Mt Erebus and still be 2,000 ft above mean sea level.

Q. If this is how you read the article at the time, why did you see fit to draw attention to it yesterday when cross-examined by counsel on the topic of flights descending below 6,000 ft?

A. Because, if I understood counsel's question correctly, he asked me what reports I had heard of or seen and this report obviously came to mind.

Q. ... Was not [counsel] ... asking you about Antarctic flights descending below 6,000 ft before the accident of 28 November 1979? Was that not his topic?

A. Yes, I believe it was.

Q. And in immediate response to his question did you not refer to media comment upon that topic prior to then?

A. I did.

Q. And was not the media material to which you referred ... some article in the staff news and in television news item?

A. I recall reading that at the time, yes.

Q. And ... did you confirm that the news item in question was the article in the

Exhibit 148?

A. Yes, that would be so.

Q. Was not this whole discussion in the context of flights below 6,000 ft above sea level?

A. Yes, it was.

Q. Now that you've refreshed your mind as to what happened yesterday, as distinct from how someone may come to read Exhibit 148 [the staff newsletter], was it not your own view when you first read exhibit 148 that it referred to flights below 6,000 ft above sea level?

A. As I think I tried to explain my initial reading of this article, I conveyed the impression earlier that I did not consider it of great consequence at that time. It is probably only in the latter periods of time that more significance could be placed on the article.

Q. The reason you gave yesterday for the lack of concern was: "I believe that at that time I had no responsibility in that direction, I was a line Captain DC10". That's the reason you gave yesterday?

A. Yes.

Q. Wasn't the short point that you realised there was low flying below 6,000 ft above sea level, but you were no longer concerned with route qualification, were just a line Captain, and those in authority would have known full well what was going on? Isn't that what you told us?

A. Yes, I would agree with that.

The mistakes regarding height were:

The same supervisor who gave the above evidence was co-pilot on 18 October 1977 and he denied that the flight descended below 6,000 feet, even though a journalist on the flight had written a story in the *Auckland Star's 8 O'clock* on 22 October 1977 indicating a low level flight in McMurdo Sound well below Mt Erebus, which was 40 km away. The journalist, who was known to the co-pilot says he was told the height by either the pilot or co-pilot (M.75).

Regardless of whether the co-pilot or the journalist was mistaken, the article was published, and no action by the airline resulted. The Mahon Report (M.75) accepts that the co-pilot was mistaken and the flight did descend beneath 6,000 ft. Mistakes made by the chief executive and six executive pilots in not learning the flight descended to a low altitude, in particular, not learning it was reported in the press that it did so. Mistakes made by four CAD officers in failing to learn of those press reports. (Mistakes 1-11)

Not being told by the McDonnell Douglas President that his flight was at 3,000 ft. (Mistake by chief executive only, since it could be understood that executive pilots may not have been told.) One mistake. (Mistake 12)

Not reading the article sent by the President by letter to the chief executive. (Mistake by chief executive at least and, since the airline printed one million copies, arguably mistakes by others in failing to check it or read it while in company channels preparatory to printing.) One certain mistake, six probable mistakes. (Mistake 13)

Not reading or becoming aware of the contents of the President's article after it was printed and distributed (eleven mistakes). (Mistakes 14-24)

Not becoming aware that the airline had printed one million copies of an article by the President on low level Antarctic flights. Seven mistakes. (Mistakes 25-31)

Not being told by any of the airline executives on the Antarctic flights from 18 October 1977 to 21 November 1979 that they had been on flights under 16,000 ft. There were 11 flights over two years, but there was no evidence on how many executives and their wives were on each flight, only Mr Thomson's evidence that it was the usual practice. If one assumes, on average, one executive per flight (and I would welcome precise figures) that is mistakes by seven persons on 11 occasions (77 mistakes). The chief executive said (B.714) that from time to time he received informal reports which indicated that the flights were popular and had been conducted successfully, but he did not say from whom these reports came. He also said (B.714) that he had a long standing personal foible about the importance of proper terrain clearance. (Mistakes 32-108)

On 2 November 1979 an RCU briefing was given which was attended by the flight manager, line operations DC10/DC8, and two captains gave evidence (B.574 and B.547) that the flight manager led a discussion on the presentation of scenery to the passengers which involved previous crews in flying beneath 6,000 ft.

Mistake by flight manager in not instructing the crews being briefed that such flights were prohibited, when he was aware 6,000 ft was the minimum permissible altitude (B.389). (Mistake 109)

Second mistake by flight manager in subsequently forgetting he had been made aware at this briefing of low level flights beneath 6,000 ft (T.1353). (Mistake 110)

Failing to read the airline's newsletter which described Captain Keesing's flight at 2,000 ft. Seven mistakes. (Mistakes 111-117)

Failing to be aware that the briefing officer was briefing crews to approach McMurdo at whatever height was authorised by McMurdo Air Traffic Control.

This briefing officer took up his position on 5 December 1977 (B.350) and Antarctic briefings were subsequently held on 2, 13 and 16 November 1978, and on 2 and 9 November 1979, that is five briefings on each of which six persons failed to discover what height instructions the officer was giving — 30 mistakes. (Mistakes 118-147)

The evidence of the previous briefing officer as to learning of low level flights from a television news item and the staff newsletter dated 30 November 1978 has been quoted above. In October 1978 (B.353) and on 7 November 1978 that former briefing officer went over the Antarctic briefing material with the then current briefing officer, at the latter's request.

Mistake by former briefing officer after reading newsletter of 30 November in not telling his successor that flights were descending beneath 6,000 ft, contrary to the airline's rules. The mistake is compounded since it was at his successor's request for further assistance with the Antarctic programme that he had seen his successor on 7 November. (Mistake 148)

The briefing officer himself, failing to find out the airline's height limitations from his predecessor when consulting him twice regarding Antarctic flight briefing. Two mistakes each by present and former briefing officers (four mistakes). (Mistakes 149-152)

The briefing materials handed out to pilots by the briefing officer contained material under the heading "Descent to 6,000 feet for cloud penetration at McMurdo". See Ch. 9 Epilogue p. 200. Pilots who received it tended to believe it referred to cloud break or cloud penetration procedures. The airline subsequently asserted that this belief was a mistake. The briefing officer made the mistake of believing it permitted descent to any altitude approved by the McMurdo Air Traffic Control, and he failed to check its meaning with his predecessor who in turn failed to ensure his successor knew what it meant. (Mistakes 153-154)

There are two alternatives regarding aircraft reported 1,000 ft above glaciers. If one accepts the version of the CAD officer, then the airline officer failed to action the CAD report of breach of height restrictions. If one accepts the airline officer's version, then the CAD failed to bring height restrictions to the attention of the airline. Either way, one mistake. (Mistake 155)

Failing to hear on radio, or see the television news item, or read the story in two Auckland suburban newspapers, or the story in the *Auckland Star's 8 O'clock* all of which told of low level flights. From at least one printed source I personally knew at the time of the airline's low level flights, and I am puzzled that none of the airline's responsible employees or civil aviation responsible employees became aware of any of these stories. Eleven persons involved, seven airline, four CAD. It is not easy to place a figure on this number of mistakes. I would suggest a conservative figure would be not less than 22. (Mistakes 156 to 177).

5. Did the on-board aircraft radar show Mt Erebus?

The Chippindale Report conclusion, immediately before the summation of the "probable cause", states (C.53):

"The aircraft's radar would have depicted the mountainous terrain ahead."

This was taken by many to be an assertion of pilot error. Not only did its wording signify this, but also its location in the Report. It was placed immediately before the "probable cause" which was also widely understood as indicating pilot error.

This conclusion was based on paragraph 1.8.9 in which the Inspector explains that the aircraft was equipped with Bendix radar which has both "weather" and "mapping" modes.The Chippindale report says:

> "...Some pilots of previous Antarctic flights reported that the radar indications of high ground correlated well with the contours which they observed visually in VMC (visual meterological conditions). Expert opinion from the aircraft manufacturers was that the high ground on Ross Island would have been clearly indicated by the "shadow effect" had either pilot studied the radar presentation during the aircraft's descent to the north of (Ross) Island."

The inference of pilot error transpired to be misleading.

The Inspector relied on his notes of interviews with two pilots for his statement, Captain Simpson and First Officer Eaton (T.189). The latter never gave evidence, and when the Inspector had referred to his notes he said only that he had noted that First Officer Eaton "discovered that the definition of coastlines was poor owing to the sea ice being adjacent to those coastlines in his opinion." (T.189)

This statement does not support the proposition stated in the Chippindale Report, on the contrary it tends to support Captain Vette's text.

Captain Simpson gave no evidence on the topic.

Evidence was given by Captain Lawson that the weather radar did not show Mt Erebus or other high terrain (T.858, T.859, T.860).

The Commissioner took up the question of whether the radar would show Mt Erebus with the Avionics Division of the Bendix Corporation which manufactures the radar on DC10 aircraft. Bendix explained that the radar was designed to detect rain precipitation in cloud and, since there is no humidity in Antarctica, both ice and snow will normally be totally dry (M.122-3). The Commissioner's inquiries are reprinted above in Appendix F on pages 276 to 280.

The Bendix opinion was that, because the slopes of Mt Erebus were covered in snow and ice which were totally dry, the return from the mountain would be nil (M.125).

The Bendix experts said there had been no specific experiments and were prepared to concede, though they did not really believe in it, some kind of shadow effect, but they did not believe this hypothetical return would represent any warning of high ground to the aircrew (M.125).

A further point was made by the Bendix experts. Probably, the radar was in weather mode or stand-by — then, when the captain descended, he would not be interested in switching to the mapping mode if he could see the pack ice and water during his descent. Even if he had switched to the mapping mode, at about 3,000 ft, he would be too low for it to be of any assistance, and all he would get would be an insignificant return at the very bottom of his radar screen (M.125).

The Mahon Report says (M.125):

> "I discovered at Bendix that this special feature of the DC10 radar in ice-covered terrain had been notified to McDonnell-Douglas when they made an inquiry of Bendix some time after the disaster. It was also asertained at a later stage that the chief inspector had also been appraised of this information. I also found, again at a later stage, that the airline had been made aware by McDonnell-Douglas of the same information.
>
> While I did not expect the airline to produce evidence from Bendix which tended to absolve the air crew from any degree of fault, in that radar echoes returned by this special type of radar from dry snow and dry ice are nil, nevertheless it was unfortunate, in my opinion, that the chief inspector did not disclose these special features of the DC10 radar in his report. He should not have said, as previously quoted. 'The aircraft radar would have depicted the mountainous terrain ahead'. In the opinion of the Bendix avionic specialists — and they are world experts — that statement was not correct."

Direct and what appears irreconcileable conflict between the Mahon Report and the Inspector exists here.

The Inspector said:

> "I deny any knowledge of any amendment being passed to me on this opinion from McDonnell-Douglas or any other source. Certainly I did not hear this in evidence and the Judge did not tackle me about it." (Chippindale PR 3 Feb).

Mr Justice Mahon told me that the Inspector's denial was incorrect since the Commissioner had received from the Inspector a Memorandum which referred to the information in question, thus proving that the Inspector was aware of it.

The Inspector writes (PR 3 Feb):

> "Our expert's view was that in the case of the aircraft approaching the solid 12,500 ft high mountain at an altitude of 1,500 ft AMSL, the crew would see a shadow on the radar that could be nothing other than high ground ahead."

Can this expert be correct and the manufacturers of the radar wrong? Who is this expert from McDonnell Douglas? The Inspector was asked this question on three occasions during his evidence before the Royal Commission, but he was unable to remember the name (T.76, T.307, T.349).

The Inspector says (PR 3 Feb):

"While I did not know the name of the radar expert, the Air New Zealand counsel, who was cross examining me, did."

Counsel referred to was Lloyd Brown Q.C. and he denies the Inspector's statement.

The Inspector continues:

"Immediately after this incident, the McDonnell-Douglas lawyer who was in the court room gave me the expert's name."

The incident occurred on Thursday, 17 July 1980. On Friday the inspector was re-examined by David Baragwanath. Even though he had been impliedly criticised the previous day for having no record of the expert's name, the inspector still does not use it, and on Friday instead refers to "the person that I considered an expert who advised us at Long Beach" (T.349). Indeed, the Inspector had previously been asked the name by Mr Baragwanath on Tuesday 8 July, and had been unable to remember it.

In order to check this I spoke to John Katz, one of the two N.Z. counsel for McDonnell-Douglas, who told me that, though he did not know the expert's name, it was known to the American counsel for McDonnell-Douglas, named M.J. Foley.

Over a period of nine months I wrote to Mr Foley three times, requesting the information, but he failed to reply.

I also wrote to the Inspector himself, asking the name, but his reply was that:

"It is not our policy to release the names of our witnesses".

However, he added that I may be able to obtain the gentleman's name and address from Mr S. Lund, of Douglas Aircraft Company, who headed the Douglas team. Accordingly, I wrote to Mr Lund three times over nine months, but received no reply.

The Inspector concludes:

"Thus (the expert's name) was readily available to the Commissioner as I advised his counsel assisting at the time."

I therefore asked David Baragwanath for the name, but he told me that he did not know it and could not remember whether he had ever been told. The other counsel assisting, Gary Harrison, said he had never heard the name. The Commissioner told me that no name had ever been passed to him.

Before the Royal Commission it was revealed in cross examination that

the Inspector had no personal knowledge of the radar equipment or its capability (T.186), and had not seen the manufacturer's manual (T.308), that whether it was unreliable for terrain in the Antarctic was outside his competence (T.309), and the purpose of the radar was simply weather detection (T.186).

He further agreed that from his limited knowledge of the radar, it was prudent for the airline to prohibit its use as a navigational aid (T.306). The Inspector explained that he did not imply any criticism of the crew for possessing radar and not using it (T.308).

It transpired (T.308) that not only did the Inspector not know the person's name or his position, whether the person was a staff member or a manufacturer's representative, but the person did not claim authority on behalf of the company to offer an opinion. This happened on Thursday, but on Friday by when the Inspector now says he knew the position, he still did not clear up these matters during his continuing evidence.

It was after giving all the above evidence, that the Inspector said in his press release (PR 3 Feb):

> "Despite the Bendix expert's theory [i.e. the experts relied on by the Commissioner], he admits he had not tested it and again despite the theory that aircraft radio altimeters are unreliable in the same conditions that fitted to ZK-NZP worked perfectly prior to the accident both over the sea ice and the dry snow. Most of this dissertation on the use of radar is directed at using it for navigation in the mapping mode while flying at high levels. Our expert's view was that in the case of the aircraft approaching the solid 12,500-foot-high mountain at an altitude of 1500 feet AMSL the crew would see a shadow on the radar that could be nothing other than high ground ahead."

In referring to mapping mode at high levels, the Inspector first ignores the point made by Bendix about its use at low levels quoted by me from the Mahon Report, page 125.

He next puts forward his still unnamed expert's view (that mapping mode used at 1500 feet would show the mountain) to challenge the Bendix opinion that even 3000 feet would be too low for the radar return to be of any assistance.

6. Could "Wilson" mean Wilson Piedmont Glacier?

It was a major part of the final conclusion of the Chippindale Report that "the crew was not certain of their position". If that is taken to mean they were uncertain in their own minds of their position, then any evidence tending to show they believed they knew where they were would tend to disprove the Inspector's conclusion.

The Inspector included in his transcript of the cockpit voice recorder snatches of conversation which he asserted showed uncertainty of position and mounting alarm on the flight deck at the action of the captain in continuing to fly on.

The Commissioner agreed (M.44):

"that the Chief Inspector adopted as being accurate certain remarks... when it was by no means certain whether those exact observations were ever made and that he was persuaded to adopt that course because of his firmly held opinion that the crew had been uncertain of its position."

On page 15 of the Inspector's unpublished draft report, at time 0034:21 of the transcript appears:

? We're going down below cloud.
? Where's Wilson.
? Over to the right.

Now, the Wilson Piedmont Glacier is a major geographical feature of McMurdo Sound and stretches from Cape Roberts in the north to Cape Bernacchi in the south. Such a reference to it would indicate the crew believed they knew where they were, even if they were mistaken in their belief.

However, no reference to Wilson appears in the Inspector's published Report. Why?

The Inspector seems to agree in cross-examination that as a result of Farnborough advice he would come up with more content to the final CVR transcript compared with the interim transcript rather than less. The cross-examination continued (T.261):

Counsel: Why is it, then, that your transcript, as it appears in the final report, does not contain a number of passages which appeared in the interim report?

Chippindale: The purpose of the interim report, as you call it, and as I explained earlier in this inquiry, was to place in perspective the comments and possible areas of blameworthiness that I made to the parties that I considered [a]ffected by regulation 15 of the Accident Investigation Regulations.

Counsel: I am not asking you about that at all. What I am asking you is why is it you've omitted some passages in your report which appeared in your interim report, when one would have expected there to have been, if anything, more information in the final report of the CVR?

Chippindale: That was the question I was endeavouring to answer.

Counsel: Please do.

Chippindale: Give me the opportunity. As I was saying, the purpose of that interim report was simply to put the parties in the picture as far as the information available at that stage was concerned. The purpose of the transcript in the final report is to, in my opinion, support the conclusions of the Report itself, and in recognition of the sensitivity, particularly of the N.Z. Airline Pilots Association, I endeavoured to limit the extent of the transcipt to the minimum which I consider necessary for that purpose. Therefore, I continued editing the transcript almost continuously until the time I presented it to the Government Printer for publishing.

Comment:

Editing out the reference to Wilson is not beneficial to the crew, it is detrimental; hence pilot sensitivity cannot be the reason. This is one example of several passages beneficial to the pilots which were deleted from the transcript by the Inspector.

Counsel: So, when editing the transcript for the final report, did you endeavour to edit out passages which you considered were not relevant to the flight?

Chippindale: That was the sole aim of my revision of the transcript.

Here *Counsel* refers the Inspector to p.74, points out the omission of "Wilson" and asks:

"Would that be a reference to a dry valley area to the west of the aircraft, perhaps?"

Chippindale: We were not able to identify any place with the single name of Wilson.

Comment:

Was the "sole aim" of the Inspector relevance, or was it pilot sensitivity?

"Wilson Piedmont Glacier" is prominently marked on Appendix 3 to annex D of the Chippindale Report.

Is the Inspector seriously suggesting that a reference to "Wilson" could not be a reference to a major geographical feature of McMurdo sound, the Wilson Piedmont Glacier extending, according to the evidence, for some 30 miles, and that he actually examined maps to ascertain if any feature carried the single name "Wilson"?

7. Location of Taylor Valley

Paragraph 117 of the Mahon Report (M.43) reads:

"117. Counsel for ALPA were highly critical of the approach of the chief inspector to this whole question of the transcript of the CVR. Their submission was that the chief inspector had formed a preliminary view that the crew was "uncertain" of its position and was expressing "mounting alarm" and so forth, and that he construed the transcript, wherever possible, so as to give effect to that point of view. One example which was given to me had reference to a remark by Mr Mulgrew transcribed at Washington as follows:

"Taylor on the right now".

"This meant that Mr Mulgrew was pointing to the location of the Taylor Valley. In other words he was pointing to an area just south of that section of the Cape which he could see on the right. In the revised transcript published by the chief inspector, this remark by Mr Mulgrew is altered to a question by him, addressed to the captain, as to whether he will go to "the Taylor or Wright now". Then there is recorded a supposed answer by the captain "No I prefer here first." The captain's remark does not appear in the Washington transcript. In the view of counsel for ALPA the intention of the chief inspector in this respect was to avoid any suggestion that Mr Mulgrew had made a positive identification, because this ran counter to his controlling thesis that neither Mr Mulgrew nor the crew were quite sure as to the location of the aircraft. Mr Baragwanath, who as I have said heard the tapes in New Zealand, at Washington and at Farnborough, had this to say in the course of his final submissions:

"The point is that there is no evidence that this flight crew was in doubt as to its position.

"With that comment I entirely agree."

8. When did aircrew first express anxiety?

The CVR was transcribed in Washington by a team of three Air New Zealand officers of whom Captain Cooper was one. He was a DC10 Captain of some experience. He knew well Collins, Cassin, Brooks and Maloney, but did not know Mr Mulgrew. He explained that the fundamental requirements for transcribing CVR tapes laid down by the supervisor of the Washington laboratory were familiarity with the persons whose voices had been recorded, familiarity with the aircraft type, and familiarity with flight deck procedure and terminology.

Much publicity was given at the time to the Chippindale Report conclusion 3.24 — "The flight engineers endeavoured to monitor the progress of the flight and expressed their dissatisfaction with the descent toward a cloud covered area."

From the transcript analysis by the Commissioner and by Captain Cooper, the Inspector's conclusion seems unsupported by the evidence. I believe much weight should be given to the opinion of Captain Cooper where he states (B.449):

> "From my personal knowledge of the parties and my involvement in the transcript I believe that the first positive indication of anxiety or concern appears at the time of 0048:55 when Captain Collins asks "Have we got them on the tower?" This is before the time to comply with the instruction to report 10 DME and is just 35 seconds before the decision to climb was made. Therefore, I believe that isolated remarks identified by various witnesses at this Enquiry as indicating concern do not in my opinion when viewed against the above remarks indicate that any real concern was shown by any parties on the flight deck until 55 seconds before impact."

Captain Cooper's evidence is reprinted in *Impact Erebus* as Appendix B from pages 244 to 257.

9. Significance of Beaufort Island

The post accident Chairman of Air New Zealand, Mr Bob Owens, said (New Zealand Herald, 23 January 1982) he believed the Chippindale Report on the causes of the crash (it cites pilot error) was more correct than the findings of Mr Justice Mahon. (Mr Owens said three-fifths of the blame for the crash rested with the airline and two-fifths with the pilots).

The reason, said Mr Owens, was a photograph of Beaufort Island, taken by a Japanese passenger from the starboard side shortly before the crash. (Under the flight plan the Island should have been on the port side of the aircraft).

The co-pilot should have seen the Island and realised the aircraft was off track. This was made clear in Mr Chippindale's Report, Mr Owens said. (Auckland Star, 26 January 1982).

Has Mr Owens correctly interpreted the Chippindale Report on this issue? The Inspector told the *Star:*

"We analysed the photographs (of Beaufort and other landmarks) as a whole and used them mainly to show weather conditions at the time.

"I would not like to comment on Beaufort Island in isolation, therefore."

So one cannot find out from the Inspector whether Mr Owens has correctly understood the Chippindale Report, and upon this one issue rests Mr Owens' preference of the Chippindale Report analysis, over the Mahon Report. Mr Owens admitted in the press that he had not read the Mahon Report.

There are several photographs of Beaufort Island and a selection are set out and described in Appendix 4 of the brief of M.R. Sinclair (Further Record Material Vol. 2) (see text for photo 3, page 69 and photo 4, page 70).

Mr Owens, a qualified navigator, is not reported to have claimed any knowledge of Beaufort Island, McMurdo Sound, or polar flying.

One of the experts consulted by the Commissioner was G.W. Shannon who carried out a flying contract for a drilling site and thereby became reasonably familiar with the McMurdo region (M.69). In his view the crew probably thought Beaufort Island was Dunlop Island (M.118).

R.B. Thomson, Superintendent of Antarctic Division of D.S.I.R., a trained polar navigator with over 20 years experience in Antarctic operations, who visited Antarctica over 50 times, and spent over 5 years there, and who was able, working from photographs taken on the flight, to reconstruct the position of the aircraft within Lewis Bay, offered two reasons why the crew failed to recognise Beaufort Island. Either they failed to pay attention to surface detail or they were, through their inexperience in operating in the area, unable to recognise even quite prominent features (B.116).

There are two possibilities. The crew may have seen the land form but "perceived" it as another break in the sea ice. This, it will be noted from the text, was suggested by Captain Vette, and finds support in the views of Professor Day and Captain Ginsberg.

The crew may have seen the land form and "perceived" it as an island, but believed it was Dunlop Island, which is close to Victoria Land, and consequently believed its sighting provided further confirmation they were flying down McMurdo Sound. This was Mr Shannon's suggestion.

In the Mahon Report (M.115-116) the Commissioner (by comparing the actual track the aircraft took in Lewis Bay with the track the crew believed they were on in McMurdo Sound in reliance on the computer flight path) analysed how Beaufort Island would have been interpreted by the flight crew.

All on the flight deck, including Mr Mulgrew, would have known Beaufort Island must be well to the east, so that an island well to the west must have been a different island probably, as Mr Shannon thought, Dunlop Island.

Later in his Report (M.153) the Commissioner says that if Captain Collins had flown to McMurdo previously under supervision he would almost certainly have noted the distinctive feature of Beaufort Island. Had Captain Collins seen the Island previously, and had he identified it on the fatal flight, he would certainly have realised that his nav track had been changed. Here the Commissioner follows the view of Mr Thomson.

In effect, the Commissioner is adopting the second of Mr Thomson's reasons, that is the crew were unable to recognise Beaufort Island through their inexperience in operating in the McMurdo area, and adopting Mr Shannon's view that they mistook it for Dunlop Island.

Conclusion

Captain Vette wanted his book to be read by an independent observer who knew none of the parties, had not been involved in the proceedings, and was from outside the aviation industry, but who was prepared to read the voluminous evidence and study the documentary exhibits presented to the Royal Commission. I was fortunate that I met his specifications.

He gave me a free hand to comment on *Impact Erebus* in such manner as I saw fit; and this I have done in these notes.

He has explained to me the pressures exerted upon him to leave the airline. He became so disenchanted with the Air New Zealand response to the Commission findings that he resigned as a senior flight instructor on the Boeing 747s and gave up the opportunity of occupying the No. 1 position on the seniority list for at least the next five years.

Captain Vette has dared to stand by his conscience to tell the truth throughout the Erebus affair. He has shown me a letter written on behalf of participants accusing him of disloyalty. He was obliged to choose between honesty and loyalty. Besides, would not such "loyalty" have involved disloyalty to others? Of course, those others were considered expendible.

Counsel for the Collins family said that Air New Zealand and CAD advanced the banner of pilot error at every turn and that the airline simply ditched a captain and crew that had served them so well over so many years.

Reading the evidence supports counsel's submission. Towards the end of the proceedings it became clear that far from the pilot error case succeeding, the findings were more likely to hold airline error, coupled with some measure of CAD error. Skilled and experienced senior counsel for the airline in final submissions consequently argued that no attempt should be made to state the major cause of the accident in terms of importance or blameworthiness.

Pilots' families must realise that should their husband or father be killed in a crash, lifetime friends in the airline will turn against them, if they dare to disagree with the offical airline version.

It is clear that families of pilots must recognise that pilots are a disposable commodity. Once they have been killed in a crash it is in no one's interests to defend their actions. Consider the quotation from Peter Garrison which began this book:

> *Again and again, pilots are found to show little sympathy for their colleagues who are hurt or die; some simplifying explanation is immediately hit upon to reassure the others that the same fate will not be theirs.*
>
> Pilot Error

ROSS ICE SHELF
Mt EREBUS
Mt. TERROR
Scott Base
McMurdo Str
Cape
Mt. BIRD
Cape Tennyson
Cape Bird
LEWIS BAY
Beaufort Is.
Actual Track.
Expected Track.